AF697750
BOWLING GREEN STATE UNIVERSITY
DISCARDED
LIBRARY

Also in the Variorum Collected Studies Series:

GERARD L'E. TURNER
Scientific Instruments and Experimental Philosophy, 1550–1850

WILLIAM A. WALLACE
Galileo, the Jesuits and the Medieval Aristotle

GUY BEAUJOUAN
Par raison de nombres:
L'art du calcul et les savoirs scientifiques médiévaux

FRANZ ROSENTHAL
Science and Medicine in Islam

WALTER PAGEL
From Paracelsus to Van Helmont:
Studies in Renaissance Medicine and Science

ALLEN G. DEBUS
Chemistry, Alchemy, and the New Philosophy, 1550–1700:
Studies in the History of Science and Medicine

VIVIAN NUTTON
From Democedes to Harvey:
Studies in the History of Medicine

PAUL KUNITZSCH
The Arabs and the Stars

CURTIS WILSON
Astronomy from Kepler to Newton:
Historical Studies

DAVID A. KING
Islamic Astronomical Instruments

BRUCE S. EASTWOOD
Astronomy and Optics from Pliny to Descartes:
Texts, Diagrams and Conceptual Structures

Electricity and Experimental Physics in Eighteenth-Century Europe

Professor R.W. Home

R.W. Home

Electricity and Experimental Physics in Eighteenth-Century Europe

VARIORUM

BOWLING GREEN STATE UNIVERSITY LIBRARIES

BOWLING GREEN STATE UNIVERSITY
DISCARDED
LIBRARY

This edition copyright © 1992 by R.W. Home
All rights reserved.

Published by VARIORUM

Ashgate Publishing Limited
Gower House, Croft Road,
Hampshire GU11 3HR
Great Britain

Ashgate Publishing Company
Old Post Road
Brookfield, Vermont 05036
U S A

ISBN 0–86078–318–9

A CIP catalogue record for this book is available
from the British Library and the
US Library of Congress.

Printed by Bookcraft Ltd
Midsomer Norton, Nr. Bath,
Great Britain

COLLECTED STUDIES SERIES CS366

CONTENTS

This volume contains xii + 306 pages

PUBLISHER'S NOTE

The articles in this volume, as in all others in the Collected Studies Series, have not been given a new, continuous pagination. In order to avoid confusion, and to facilitate their use where these same studies have been referred to elsewhere, the original pagination has been maintained wherever possible.

Each article has been given a Roman number in order of appearance, as listed in the Contents. This number is repeated on each page and quoted in the index entries.

PREFACE

Historians of science tended at one time to be somewhat dismissive of the physics of the eighteenth century, seeing the period as lack-lustre in its achievements in comparison with the great ages of physical discovery that preceded and followed. Today, however, it is generally agreed that such a judgement is more a reflection of the limited outlook of whiggishly inclined historians than of anything else. Many, indeed, would now argue the converse view, namely that the period in question was actually of critical importance in the evolution of the subject because it was the period in which the basis of the modern physics discipline was laid. Yet how best to characterize what happened during these years, not to mention its longer-term significance, remains a matter of debate. The present volume will, I hope, contribute usefully to the discussion by bringing together a number of papers that are at present widely scattered that address various aspects of this very large question.

Electricity has always been recognized as one field to which eighteenth-century physicists did make substantial contributions. Benjamin Franklin, in particular, has long been famous for demonstrating the electrical nature of the lightning discharge and devising a means for protecting buildings against this most terrifying of Nature's weapons. Following the work of I.B. Cohen in his well known study, *Franklin and Newton* (Philadelphia, 1956), the importance of Franklin's work for electrical theory as well as practice came to be generally acknowledged. Indeed, in T.S. Kuhn's influential *Structure of Scientific Revolutions* (Chicago, 1962), Franklin's theoretical innovations were presented as constituting a veritable revolution in scientific thought, in which the study of electricity was transformed from an intellectual chaos into a field of investigation in which there was broad agreement as to fundamentals. These works formed the starting point of my own inquiry. In time, however, I became dissatisfied with their overwhelmingly Franklin-centred way of viewing things, and in several of the chapters that follow I take issue with this. Franklin remains an important figure in the story told here, but his work is not represented as bringing about a once-and-for-all revolution in electrical theory. On the one hand, the ideas of his predecessors, stretching back to Newton and Hauksbee in the earliest years of the century, are seen as constituting a more coherent approach to understanding the limited range of phenomena known at the time than was previously allowed; on the other, it is shown that the ideas he set out were not simply immediately adopted unaltered by his contemporaries and successors. Eighteenth-century electrical theory is thus depicted in this study as advancing not through a revolutionary upheaval so much as by a process of evolution within a complex web of interacting concepts.

The first few chapters have, as a major theme, the work of Isaac Newton and its influence on eighteenth-century physics. Newton in his later years became fascinated by electricity and allowed it a very considerable role in the workings of the world. I attempt to reconstruct the evolution of his thinking on the subject, and then to use what he wrote, most of which he left unpublished, and also his occasional comments on the subject of magnetism, to clarify his ideas more generally on the nature of matter and the sources of its activity. The writings of his eighteenth-century successors on the same subjects offer a convenient starting point for investigations of the merits of the common practice of portraying the science of this period as an epic contest between Cartesian, Newtonian and occasionally Leibnizian modes of thought, from which Newtonianism in due course emerged triumphant. The analysis suggests that labels such as these, which work by classifying the leading scientific thinkers of several generations into "schools" defined by the philosophical concerns of a previous age, are not only frequently difficult to apply in practice but carry little or no explanatory weight.

A major concern in this work, which I believe reflects a more appropriate way of analysing the science of the period, is to chart the changing views that were held at the time as to what actually constituted "physics", and more especially as to what constituted its proper method of inquiry. Of particular interest is a philosophically naive variety of experimental natural philosophy that here receives its most systematic discussion in Chapter VII. As I argue there, this approach, according to which fundamental physical principles could be directly discovered or exposed by well contrived experiments, enjoyed considerable support among eighteenth-century experimental investigators. Moreover, it had important implications so far as theory construction was concerned. Not only did it confine the range of theoretical entities employed to those that were in some sense "observable", it also emphasized the idiosyncrasies of individual experiments and thus tended to work against the development of physical theories expressed in a generalized mathematical form.

Generally speaking, therefore, physics was not a mathematical science at this period. And yet it became one in the early decades of the following century. The question of how this came about is the central theme of several chapters of the present work. The answer must, of course, be partly intellectual, since it is concerned with the invention of appropriate mathematical formalisms and the devising of physical theories suitable for mathematization. It must also, however, be partly social because, as is brought out in Chapter XIX in particular, whether the work of a would-be mathematizer is accepted as a valid approach in a given case is a matter for negotiation among those able to gain standing as authorities in the field in question. Both the intellectual and the social strands of the story are pursued below.

Much of what follows is predicated on the belief I have held since almost my first contact with the history of science, that the subject must concern itself with both the invention of theories and their subsequent reception and mediation by the wider scientific community to which they are directed. (This is not, of course, an exhaustive list of the kinds of question the historian may ask.) Theories frequently become embedded in the scientific literature, as I argue Franklin's did, in a form somewhat different from that envisaged by their first inventors. Moreover, the reception of a theory will depend on the structure of the relevant scientific community at the time and on the facility with which ideas can circulate within it. In the present case, with one of the central figures in the story, Aepinus, based in St Petersburg, far from the other major scientific centres of the age, the nature of his scientific base there and of the channels by which his ideas might disseminate to other parts of the learned world become questions of considerable importance to the history of electricity and magnetism. The answers provided to them below will also, I hope, be found of more general interest for the light they throw on both Russian intellectual life at the time and the structure of the wider eighteenth-century scientific world.

R.W. HOME

Melbourne, 1991

ACKNOWLEDGEMENTS

My interest in the history of eighteenth-century physics was aroused when I was a graduate student at Indiana University in the mid-1960s. In due course I completed a doctoral thesis on early eighteenth-century ideas about the nature of electricity that was published some years afterwards as *The Effluvial Theory of Electricity* (New York: Arno Press, 1981). The earliest articles reprinted here derive from work done during the preparation of this and owe much to my teachers there, especially my supervisor, Victor E. Thoren, and Richard S. Westfall who likewise read my thesis in its entirety. The remainder have been written during the many years I have subsequently spent at the University of Melbourne and have benefited from discussions with various colleagues among whom should be mentioned, in particular, Elizabeth Gasking, Diana Dyason, Hazel Maxian and Keith Hutchison. Much of the work has been supported by grants from the Australian Research Grants Committee (now the Australian Research Council).

Permission to reprint individual articles has been generously granted by the following: Kluwer Academic Publishers (I, VII, XI); Cambridge University Press (II); Springer Verlag (III); Science History Publications (IV); Australian National University Press (V); Taylor & Francis Ltd (VI, VIII); the British Society for the History of Science (IX, XIX); the Royal Society of London (X); the Science Council of Japan (XII); Franz Steiner Verlag Wiesbaden GmbH (XIII); the History of Science Society (XIV, XVI); *The Slavonic and East European Review* (XV); Oriental Research Partners (XVII) and the École Polytechnique, Paris (XVIII).

I

NEWTON ON ELECTRICITY AND THE AETHER*

It is well known that Newton became convinced towards the end of his life that electricity played a vital role in the operations of nature. In the famous final paragraph of the *Scholium Generale* that he added to the second edition of the *Principia*, published in 1713, he wrote of "a certain most subtle spirit which pervades and lies hid in all gross bodies." It was this active spirit that gave rise, he supposed, to the electrical attractions and repulsions that manifested themselves at sensible distances from most bodies after they had been rubbed, as well as to the cohesion of particles when contiguous. In addition, he surmised, it was the agency responsible for the emission, reflection, refraction, inflection and heating effects of light; and by its vibrations in "the solid filaments of the nerves," it carried sensations to the brain, and commands of the will from the brain to the muscles in order to bring about bodily motion.[1]

In Andrew Motte's 1729 English translation of the *Principia*, the electrical character of this strange substance was emphasized by the substitution of the phrase "this electric and elastic spirit" for Newton's unadorned "this spirit" (*hic spiritus*) in the final sentence. Yet Motte was by no means misrepresenting Newton's views here: on the contrary, the phrase *spiritus electricus* appears in one of the draft versions of the *Scholium Generale,*[2] and Newton later wrote in the margin of his own annotated copy of the second edition of the *Principia*, after the word *Spiritus*, the words *electrici & elastici.*[3] Moreover, the electrical character of the subtle spirit was already clear in the original published version of the Scholium, from the role Newton attributed to it of bringing about the motions traditionally subsumed under the rubric "electricity".[4] Being electrical, it is clearly not to be equated with the spirit of God, as Koyré once suggested it might be. Instead, it must be located unambiguously among the ranks of second causes.[5]

Soon after the publication of the *Scholium Generale*, Newton's views concerning the subtle spirit underwent some changes. These are reflected in the eight additional Queries, numbered 17–24, that he included in the new edition of his *Opticks*, published in late 1717.[6] In certain respects, the "Æthereal Medium" that he described in these Queries was like the "subtle spirit" of the earlier *Scholium Generale*. It shared with the subtle spirit such properties as rarity, subtlety, activity and elasticity. Like the subtle spirit, it

Z. Bechler (ed.), Contemporary Newtonian Research, 191–213
Copyright © 1982 by D. Reidel Publishing Company.
Reprinted by permission of Kluwer Academic Publishers.

was held responsible for the reflection, refraction, inflection and heating effects of light, and for the functioning of the nervous system in transmitting sensations from the sensory organs to the brain, and commands of the will from the brain to the muscles. No longer, however, was it said to be chiefly confined, as the "subtle spirit" was, to be pores of "gross bodies." On the contrary, Newton suggested that this new medium was by its elastic force "expanded through all the Heavens" and was densest, he thought, in spaces altogether devoid of ordinary matter. It was therefore a true aether, something the earlier "subtle spirit", being more localised, was not. The difference is important, and tends to be reflected in Newton's terminology.[7]

Henry Guerlac has argued that Newton's reintroduction of a genuine, universally disseminated aether at this point was inspired by an experiment performed before the Royal Society of London in 1716 by Jean Théophile Desaguliers, at Newton's instigation.[8] This experiment, on the comparative rates of cooling of thermometers in air and in vacuo, was described by Newton in one of the new Queries (Q. 18) added to the *Opticks* a year later; and Guerlac has drawn attention to a letter written by Desaguliers to Sir Hans Sloane many years afterwards, in which it is said in so many words that the experiment was carried out at Newton's behest.

In the course of the "aether" Queries, Newton applied his new conception to the explanation of optical phenomena in particular. Most of those which he had previously accounted for in terms of the "subtle spirit" were now said to depend upon the aether, and so, too, were the mysterious "fits" of easy reflection and easy transmission of light that he had discovered long before, but had always tended previously to leave unexplained. These, he now suggested, had their origin in vibrations excited in the aether by rays of light falling on the surfaces of pellucid bodies.[9]

It is notorious that elsewhere in the new Queries, in Query 21, Newton offered an explanation of gravity in terms of density gradients within his aether.[10] Some scholars have taken a similar explanation, in terms of the "subtle spirit," to be implicit in the *Scholium Generale,*[11] but this is to count for nought Newton's unambiguous statement in the immediately preceding paragraph of the Scholium that he feigned no hypotheses concerning the cause of gravity.[12] Furthermore, earlier drafts of the Scholium reveal that Newton in fact maintained a clear distinction between gravity, on the one hand, and the various effects he listed among those which would find their explanation in terms of the "subtle spirit."[13] With the apparent exception of electricity (and we shall see that in Newton's mind even this was in fact no exception), the latter all depended upon short-range forces only; apart from electricity, they occurred in the immediate vicinity of bodies, precisely where Newton said the "subtle spirit" was to be found. Newton rightly regarded gravity as being in a quite different category. If it was to be explained mechanically, a mechanical agent was required that extended, as the "subtle spirit" did not, over the entire space between widely separated material bodies. In other words, only when he replaced the "subtle spirit" by a universally disseminated medium did such an explanation become a possibility so far as Newton was concerned. And even after this, so Desaguliers tells us, Newton remained hesitant about identifying the "Medium acting upon light" with whatever it was that caused bodies to gravitate towards one another.[14]

The replacement of the "subtle spirit" of the *Scholium Generale* by the universal aether of the late Queries in the *Opticks* was attended by other, less striking changes in the explanatory role attributed to it by Newton. In particular, several of the phenomena that were said in 1713 to be explicable in terms of the "subtle spirit", namely cohesion, electrical attraction and repulsion, and the emission of light from luminous bodies, were not mentioned in 1717 among those brought about by aetherial action.

All these omissions were surely deliberate. Cohesion and the emission of light were, in Newton's general scheme of things, precisely the phenomena that depended upon the most intense and most short-range forces of all, one attractive, the other repulsive. Among the universal or quasi-universal forces of matter, these differed most from gravity in their mode of operation, and were therefore the least likely to be brought about by the same causal principle.

The omission of electricity was certainly intentional, as we learn from a passage in one of the newly inserted Queries, Query 22. Here, Newton confronts the question of how the motions of the heavenly bodies can continue undisturbed in an all-pervading medium such as the aether he is here invoking. His solution is to suppose that his aether is both extremely rare and extremely elastic; and to give this suggestion at least a modicum of plausibility, he cites instances already "known" in which matter retains its activity when reduced to a very low density. What is significant for us here is that one of his examples is the subtle matter which, he assumes, gives rise to electrical motions. "If any one would ask how a Medium can be so rare," he says, "let him . . . tell me, how an electrick Body can by Friction emit an Exhalation so rare and subtile, and yet so potent, as by its Emission to cause no sensible Diminution of the weight of the electrick Body, and to be expanded through a Sphere, whose Diameter is above two Feet, and yet to be able to agitate and carry up Leaf Copper, or Leaf Gold, at the distance of above a Foot from the electrick Body?"[15] There is no suggestion here that the aether and this electrical "exhalation" are one and the same thing. On the contrary, the electrical exhalation is taken for granted as something different, whose nature is well known and understood, and whose properties provide a useful analogy to buttress Newton's account of the aether. Unlike the "subtle spirit" of the *Scholium Generale*, the aether of the *Opticks* Queries is not an electrical spirit. By 1717, Newton's flirtation with the notion that electricity was a universal causative agency in the micro-realm was apparently at an end.

Or was it? It is plain from the passage just quoted that Newton's belief that electrical motions were brought about by *a* subtle spirit emitted by electric bodies was in no whit diminished. Furthermore, it was well known in his day that many if not all bodies could be rendered electrical by friction. All these bodies, at least, must therefore contain some kind of subtle and active matter that could it appropriate circumstances be extruded from them. In addition, the question of what caused cohesion and the emission of light remained unanswered. It seems likely that Newton continued to hold that these effects were brought about by an electric spirit that pervaded the pores of all gross bodies. Even if he did not, however, he certainly continued to believe in the existence of the electric spirit itself, as will be made abundantly clear in what follows. Hence, the universally disseminated aether of the 1717 *Opticks* Queries ought not to be seen as a transmogrification of the more localised electric spirit of 1713. Rather, it was a *second* active medium introduced into the discussion at this point, to which certain of the functions previously ascribed to the electric spirit had been transferred, and to which various others had been attributed in addition.[16]

This interpretation, if correct, prompts an obvious question. If, in 1717, Newton really saw the electric spirit and the aether as two different kinds of active medium, why did he not make this clearer to readers of the new edition of *Opticks*? Certain unpublished drafts of proposed additions to the Queries suggest that he initially intended to do so. In the end, however, his habitual caution won out, and the relevant passages were suppressed.

At one point in the preparation of the new edition, Newton brought together on a single sheet a series of insertions he intended to introduce into the original set of Queries published in 1704. Included amongst these was the well known passage, destined for Query 8, describing some of Francis Hauksbee's striking experiments with electrified glass globes. Almost all the proposed insertions were in due course included without further amendment in the new printing. The following, however, which Newton at first intended to insert as a new Query at "Pag. 134 lin 1" (that is, immediately after Query 8), was omitted:

Do not electric bodies by friction emit a subtile exhalation or spirit by which they perform their attractions? And is not this spirit of a very active nature & capable of emitting light by its agitations? And may not all bodies abound with such a spirit & shine by the agitations of this spirit within them when sufficiently heated? ffor if a long cylindrical piece of Ambar be rubbed nimbly it will shine in the dark & if when it is well rubbed the finger of a man be held neare it so as almost to touch it, the electric spirit will rush out of the Ambar with a soft crackling noise like that of green leaves of trees thrown into a fire, & in rushing out it will also push against the finger so as to be felt like the ends of hairs of a fine brush touching the finger. And the like happens in glass. If a long hollow tube of flint glass about an inch be rubbed nimbly with a paper held in the hand till the glass grows warm, it will in rubbing emit light & the face or any other tender part of the skin being held neare that part of the glass where it has been most rubbed, the electric spirit which is excited by the friction will rush out of the glass with a cracking noise & push against the skin so as to be felt, & in pushing emit light so as to make the skin shine like rotten wood or a glow worm. And if the glass was held neare pieces of leaf brass scattered upon a table the electric spirit wch issued out of the glass would stir them at the distance of 6, 8 or 10 inches or a foot, & put them into various brisk motions, making them sometimes leap towards the glass & stick to it, sometimes leap from it with great force, sometimes move towards it & from it several times with reciprocal motion, sometimes move in lines parall (*sic*!) to the tube, sometimes remain suspended in the air, & sometimes move in various curve lines. Which motions shew that this spirit is agitated in various manners like a wind. And if a broad plate of glass be placed between the tube & the pieces of brass, yet the Tube will attract them them (*sic*!), tho not so strongly as when the plate of glass is taken away. Which shews that the electrick spirit is so subtile as readily to pass through glass tho not so readily as through the Air. And whilst it pervades dense bodies so easily, why may it not be latent in them all in some measure or other, tho those only emitt it by friction in which it abounds most copiously? And since it easily emits light by agitation, why may it not emit light in all dense bodies heated red hot & thereby cause them to shine?[17]

No clearer statement could be desired of the rationale underlying Newton's linking of the emission of light with the activity of the electric spirit, nor of the grounds on which he concluded that this spirit was present in greater or lesser amounts in all gross matter. The experiments upon which Newton bases his arguments were due, like those with the glass globe, to Hauksbee, and date from the period 1706–1709. So far as its contents are concerned, therefore, the new Query could have been drafted at any time between about 1710 and late 1717. Yet the fact that the other passages on this page had reached their final form by the time the page was written out suggests that the whole was of relatively late composition, and perhaps even post-dated Newton's revived interest in a universal aether as an explanatory mechanism. More conclusive is the fact that Newton subsequently re-numbered his draft Query as "Qu. 18B." Since a Query of this kind would have been quite out of place between Queries 18 and 19 of the 1706 Latin *Optice* (that is, between Queries 26 and 27 in the 1717 English version), it must be that Newton intended it to go between the new Queries 18 and 19 of 1717, that is, between two of the early "aether" Queries in that edition. In other words, the re-numbering must have been done *after* the aether Queries themselves were drafted.

The extra Query would, in fact, have fitted quite logically at this point. It is precisely in Query 18 that Newton described Desaguliers' thermometer experiment and argued on that basis for the existence of a subtle medium, or aether, "expanded through all the Heavens."[18] There is not the slightest hint in the proposed Query 18B that the electric spirit discussed there was connected in any way with this universally disseminated agent. Rather, the new Query would have served immediately to introduce Newton's second subtle medium, an electric spirit that was also responsible for the emission of light from bodies under a variety of circumstances. In other words, it would have made plain his belief that there were in fact two different subtle media at work in the world.

For some reason, however, Newton was still not satisfied, and on the second of the pages into which the sheet was divided, he tried again, this time numbering it "Qu 24" and placing it after the last of the "aether" Queries and before the first of the re-numbered Queries from the Latin *Optice* (Query 22 of the eventual printed version was not included at this stage, so that the final "aether" Query was numbered 23, not 24):

Qu 24. Do not electric bodies by friction emit a subtile exhalation or spirit by wch they perform their attractions? And is not this Spirit – – – & thereby cause them to shine. Qu: 25 Are there not other original properties of light . . .[19]

Another attempt, which Newton numbered "Qu. 23", is recorded after this on the same page. It concentrates upon the detailed mechanism underlying electrical attraction and repulsion, and was presumably intended by Newton to preface his discussion in the proposed "Qu 24" of the involvement of electricity in the production of light. It includes an explicit suggestion that the electric spirit may be but one of several "subtle invisible Mediums which may have considerable effects in the Phaenomena of nature":

Qu. 23. Is not electrical attraction and repuls; performed by an exhalation wch is raised out of the electrick body by friction & expanded to great distances & variously agitated like a turbulent wind, & wch carrys light bodies along with it & agitates them in various manners according to its own motions, making them go sometimes towards the electric body, sometimes from it & sometimes move with various other motions? And when this spirit looses its turbulent motions & begins to be recondensed & by condensation to return into the electric body doth it not carry light bodies along with it towards the Electrick body & cause them to stick to it without further motion till they drop off? And is not this exhalation much more subtile than common Air or Vapour? For electric bodies attract straws & such light substances through a plate of glass interposed, tho not so vigorously. And may there not be other Exhalations & subtile invisible Mediums which may have considerable effects in the Phaenomena of nature?

And the same things have been found by rubbing a long & large cylinder of glass or Ambar with a paper held in ones hand, & continuing the friction till the glass grew warm.[20]

That electricity continued to be linked, so far as Newton was concerned, with the general phenomenon of cohesion is suggested by his choosing to append at this point a somewhat modified opening sentence for Query 31:

Qu. 31. May not the mutual attraction of the small parts of bodies be sufficiently strong to make them cohere & compose very hard bodies.

In the end, he did not use this, but he did add to the final sentence of the first paragraph of the same Query a remark that has since puzzled many an historian: "and perhaps electrical Attraction may reach to such small distances, even without being excited by Friction."[21] At this stage, he seems finally to have abandoned the idea of including an extended Query on the electrical spirit in the new edition of his book. In the end, all that remained was the brief allusion to an electric "exhalation" quoted earlier.

The significance of the document just described has been obscured by the presence in the same box of Newton's papers of a second set of draft Queries, also dealing with electrical matters, of which the first is again numbered 24. Quite understandably, there has been a tendency to assume that the two sets were connected at the drafting stage.[22] I believe, however, that they were not, and that the second set was in fact drafted somewhat earlier, probably at about the same time as the *Scholium Generale*. It reads as follows:

Quaest. 24. May not the forces by wch the small particles of bodies cohere & act upon one another at small distances for producing the above mentioned phænomena of nature, be electric? ffor altho Electric bodies do not act at a sensible distance unless their virtue be excited by friction, yet that virtue may not be generated by friction but only expanded, for the particles of all bodies may abound with an electric spirit wch reaches not to any sensible distance from the particles unless agitated by friction or by some other cause & rarefied by the agitation. And the friction may rarefy the spirit not of all the particles in the electric body but of those only wch are on the outside of it: so that the action of the particles of the body upon one another for cohering & producing the above-mentioned phænomena may be vastly greater then that of the whole electric body to attract at a sensible distance by friction. And if there be such an universal electric spirit in bodies, certainly it must very much influence the motions & actions of the particles of the bodies amongst one another, so that without considering it, philosophers will never be able to give an account of the Phænomena arising from those motions & actions. And so far as these phænomena may be performed by the spirit wch causes electric attraction it is unphilosophical to look for any other cause.

Quaest 25 Do not all bodies therefore abound wth a very subtile active potent elastic spirit by wch light is emitted refracted & reflected, electric attractions & fugations are performed, & the small particles of bodies cohære when contiguous, agitate one another at small distances & regulate almost all their motions amongst them selves, ffor electric – – – uniting the thinking soul & unthinking body. This spirit may be also of great use in vegetation, wherein these things are to be considered, generation, nutrition & præparation of nourishment. . . .

The Query then goes on to consider these aspects of living matter at some length.

In the passage quoted, Newton's dashes again indicate that he intended to take over *in toto* at this point a passage he had already written out elsewhere. The missing passage has been preserved as an un-numbered draft Query on the second half ot the sheet, and turns out to be quite extensive:

Quaest: Do not all bodies abound with a very subtil active vibrating spirit by wch light is emitted reflected & refracted, electric & magnetic attractions & fugations are performed, the small particles of bodies cohære when continguous, agitate one another at small distances & regulate almost all their motions amongst themselves as the great bodies of the Universe regulate theirs by the power of gravity? ffor electric bodies could not act at a distance without a spirit reaching to that distance. And by several experiments shewn by Mr Hawksby before ye R. Society it appears that a cylindrical rod of glass or hard wax strongly rubbed emitts an electric spirit or vapour wch pushes against the hand or face so as to be felt, & upon application of the finger to ye electric body crackles & flashes, & that the electric spirit reaches to ye distances of half a foot or a foot from the glass or above & passes readily through the solid body of a plate or vessel of glass, the electric body attracting things beyond the glass; & that if a globe of glass be nimbly turned round upon an axis & in turning rub upon a man's hand to excite its electric virtue, the the (*sic*) hand if the glass be empty of air shines through the glass wth a purple light, if some air be let into the glass, the whole cavity of ye glass appears illuminated wth flashes of a whiter light; if the Air be let in freely the glass emitts an electric vapour or spirit wch may be felt by the hand & wch in dashing upon the hand or upon white paper or a handkerchief at the distance of a quarter of an inch or half an inch from the glass or above, illuminates the hand or paper or handkerchief with a white light while the glass continues in motion, the spirit by striking upon those bodies being agitated so as to emit the light. & that if some threds of cotton or worsted yarn hanging by one end at a little distance from one another be attracted at the other end towards the glass, & a mans finger be advanced towards the attracted ends of the threds, the threds will recede from the finger, & this they will do as well when they are within the glass as when they are without it. There is therefore an electric spirit by wch bodies are in some cases attracted in others repelled & this spirit is so subtile as to pervade & pass through the solid body of glass very freely in both cases, & is capable of contraction & dilatation expanding it self to great distances from the electric body by friction. & thefore (*sic*) is elastic & susceptible of a vibrating motion like that of air whereby sounds are propagated. & this motion is exceeding quick so that the electric spirit can thereby emit light. And that wch emits light in the experiments above mentioned, may emitt it in all shining bodies whenever sufficiently agitated either by heat or by putrefaction. And the Medium wch emitts light may also be able to refract & reflect it as was noted above. This spirit may be also the Medium by whose vibrating agitations stirred up within dense bodies, the bodies receive heat & communicate it to contiguous bodies; the vibrations being propagated from one body into another where the bodies are contiguous, but reflected at the surface where they are not contiguous & by reflections kept within the hot body. The like vibrations may be excited in the bottom of the eye by light & propagated thence through the solid capillamenta of the optick nerves into the sensorium for causing vision & the like of other senses. The like vibrations may be also propagated from the brain through the solid fibres of the spinal marrow & its branches into ye muscles for agitating & expanding the liquors therein & thereby contracting the muscles to cause the motions of animals. For as (*sic*) liquors are expanded by heat & by consequence by the vibrating agitations of this spirit. If the agitations be of short continuance they expand the liquors without heating them for want of time to do it. If lasting (as in running a race, or in supporting a burden without external motion of the body) they heat the body by degrees & at length excite sweat. This spirit therefore may be the medium of sense of animal motion & by consequence of uniting the thinking soul & unthinking body.[23]

As in the later, numbered version of the Query, Newton in this remarkable document then goes on to discuss the activity of living matter. His first paragraph is relevant here, because in it he explicitly invokes the electric spirit as the source of this activity:

The vegetable life may also consist in the power of this spirit supposing that this power in substances wch have a vegetable life is stronger then in others & reaches to a greater distance from the particles. ffor as the electric vertue is invigorated by friction so it may be by some other causes. And by being stronger in the particles of living substances then in others it may preserve them from corruption & act upon the nourishment to make it of like form & vertue wth the living particles as a magnet turns iron to a magnet & fire turns its nourishment to fire & leaven turns past to leaven. ffor the living particles may propagate the vibrating motions of this spirit into into (*sic*) the contiguous particles of the nourishment & cause ye spirit in those particles to vibrate & act after ye same manner & by that action to modify the nourishment after the same manner with the living particles.[24]

There are two reasons for thinking that these passages are unconnected with those quoted previously, and were written somewhat earlier. First, their implications are rather different, because instead of being concerned with the electric spirit merely as the cause of electrical motions and the emission of light, as the others were, they assign to the spirit a very general explanatory function, much more typical of Newton's opinions at the time he composed the *Scholium Generale* than of his attitude a few years later when he wrote the "aether" Queries. More concretely, it emerges that these alternative Queries 24 and 25 were intended to go not after the final aether Query of 1717, Q. 23, but at a point indicated by Newton's note, "conjungi queant ut cohærescant. p. 340. lin. 27", which turns out to be in the middle of what is now Q. 31, but which in the Latin *Optice* was Q. 23. They were thus drafted before ever Newton drew up the aether Queries and was led thereby to change his numbering scheme. The point where the new Queries were to be inserted comes at the end of Newton's long discussion of the various short-range forces at work in the world. He apparently intended to break off the existing Query at this point and argue for the electrical origin of these forces. Only then, presumably, would he have turned as he did in the original version to wider issues such as the tendency of the world gradually to run down, were not certain forces or active principles constantly at work in it.[25]

Professor Guerlac has described another bundle of papers from this fascinating collection, in which Newton wrote out a set of "Observations" to be included in a new edition of the *Opticks*. Several of these Observations deal with electricity. As in the passage quoted above, most of the experiments are Hauksbee's, including the production of light from a rubbed globe, the prickling sensation, crackling sound and light obtained when a finger is brought near a well rubbed piece of glass, and the motion of light bodies to and from an electrified tube; and on this occasion Newton also meant to include an account of an experiment he himself had reported to the Royal Society many years earlier.[26] He at first intended his "Observations" to come after the eleven Observations already set out in Book III of his *Opticks*, and he numbered them, accordingly, from XII to XV. Later, when Desaguliers' experiments convinced him anew of the existence of an aether, he decided to add emphasis to his discussion by setting it apart as a separate Part II of Book III. He included descriptions of Desaguliers' two leading experiments (the two-thermometer experiment and a variant of the old experiment in which a guinea and a feather were let fall together in a vacuum), called these Observations I and II, and renumbered the rest as Observations III to VI. Guerlac has on very plausible grounds dated the original drafting to late 1715 or 1716, while the amendments must have been made to it after Desaguliers devised his improved "guinea and feather" experiment in 1717.

A paper of Newton's headed "De vi electrica" was published a few years ago. A reference without further qualification to "Prop. XCIV, XCV & XCV (*sic*) Lib. I" indicates that, though now located among the manuscripts relating to the *Opticks*, this was destined initially for the *Principia*. It, too, discusses the "spirit hid in all bodies, by means of which light and bodies act upon each other mutually", that same spirit which is "the cause of electrical attraction and not only reflects, refracts and inflects light but also emits it," and whose vibrations in the nerves transmit sensations and the commands of the will. "Thanks to the same spirit the particles of bodies attract each other mutually at short distances, even without friction." Some further details are also provided, and these indicate – as do the links with the *Principia* and the wide range of phenomena ascribed to the subtle matter – that the document was drafted before Newton's reacceptance of a universal aether. "This spirit is not terminated in the external parts of bodies at a mathematical surface," we are told, "but becomes rarer gradually, and the more rare part of it spreads out from bodies on all sides to short distances and gradually comes to an end." Newton also suggests that we sense different colours as a result of vibrations of different "magnitude and number" being excited in the spirit "hidden within the capillaments of the optic nerve," and he spells out again the connections he envisages between the subtle spirit and the reflections, refractions, and alternate fits of easy reflection and easy transmission of light.[27]

To summarise: we may discern in the many papers dealing with electricity from this late period of Newton's life – papers in which the influence of Hauksbee's striking experiments is everywhere apparent – two different patterns of explanation. Newton found in Hauksbee's experiments proof that ordinary "gross" matter is everywhere pervaded by an active subtle spirit, responsible for the attraction of cohesion and, when excited by friction, for the attraction called electrical. When set vibrating, it brought about the emission of light. At first Newton believed this same spirit to be responsible also for the reflection, refraction, inflection, heating effects, and alternate fits of easy transmission and easy reflection of light, not to mention the operation of the nervous system in our bodies, and perhaps other functions of living matter as well. The *Scholium Generale* dates from this period, as do the draft "De vi electrica" and the extensive new Queries 24 and 25 meant to be inserted into Q. 23 of the Latin *Optice*. Soon afterwards, Newton drew up for inclusion in Book III of the *Opticks* the set of four additional Observations incorporating Hauksbee's leading results.

Following Desaguliers' success in November 1716 with the two-thermometer experiment, Newton took up again the idea of a universally disseminated aetherial medium. It was to this that he now attributed both the workings of the nervous system and the various optical phenomena previously ascribed to the electrical spirit. He also inclined to the view that this aether was the cause of gravity. Yet he did not altogether abandon the notion of a subtle electrical spirit pervading the pores of ordinary matter. On the contrary, he continued to attribute to the action of such a spirit not only specifically electrical effects but also the emission of light from luminous bodies and, probably, cohesion. A Query concerning the cause of gravity was drafted by July 1717,[28] and it must have been soon after this that the earlier set of Observations was recast to include an account of Desaguliers' work.[29] But then Newton had second thoughts, and decided to throw his new material into the form of Queries. The two-thermometer experiment went without difficulty into a new Query, Q. 18. The discussion of electricity was at first appended to Q. 8, but in the end only the section dealing with the production of light remained there. The passages in which Newton argued again for the existence of a subtle electric spirit were first relocated so as to come immediately after the account of the two-thermometer experiment, as Q. 18B, then moved again to the end of the "aether" Queries, and eventually omitted altogether. Only a hint of Newton's real views on this subject lasted into print, namely the passage in Q. 22 quoted above.[30]

Newton's intense interest throughout this period in the phenomena of electricity and the causes underlying them is manifest from the documents quoted above. Moreover, it seems he did not hesitate to make known to his colleagues at the Royal Society his hopes in this direction. In a French work from later in the century, he is quoted on the authority of Martin Folkes (F.R.S. 1714, and later President of the Society) as having urged the Fellows more than once to continue to investigate the subject with the utmost diligence: "Mes yeux s'éteignent, mon esprit est las de travailler, c'est à vous à faire les plus grands efforts pour ne pas laisser échapper un fil qui peut vous conduire."[31]

Yet the roots of Newton's fascination with electricity during this late period of his career have remained unclear. Zev Bechler has suggested that Newton seized on it as "the fundamental force of the micro-world" in very much a *post hoc* fashion, as the only one of the three forces known to act in the macro-world – gravity, magnetism and electricity – that fitted the technical requirements of his optical theory.[32] The documents we have quoted, however, tell a different story.

To begin with, they fully confirm Guerlac's suggestion that Hauksbee's experiments inspired Newton's renewed interest in the subject.[33] They show, in particular, that Newton found in Hauksbee's work direct experimental proof that a subtle electric spirit filled the pores of ordinary matter. In fact, most of the evidence he cites derives immediately from Hauksbee.[34] The electric spirit, Newton argues, can when excited by friction be detected directly by various of our senses. It can be felt pushing against our skin when we hold an electrified rod near our face or hand, and it can be both heard and seen (as a spark, we would say) when the point of a finger approaches the rod. If we use an electrified globe instead, the spirit can once again be felt "dashing upon the hand", and the hand and other bodies are illuminated by a white light, "the spirit by striking upon those bodies being agitated so as to emit the light." Besides these new and striking effects, however, Newton takes the age-old attraction of light objects towards an electrified body as proof enough in itself of the existence of the electric spirit. In a passage which squares extremely ill with a popularly held view of his general scientific approach, he baldly asserts that "electric bodies could not act at a distance without a spirit reaching to that distance."[35]

The remainder of Newton's argument follows very naturally, and not at all after the manner suggested by Bechler. Having already decided that agitating the electric spirit led to the emission of light, Newton found it a relatively small step to suppose that a similar cause lay behind the emission of light in other circumstances, that is, that "that wch emits light in the experiments above mentioned, may emitt it in all shining bodies whenever sufficiently agitated either by heat or by putrefaction." Since the spirit can be excited in most bodies by friction, and all bodies become luminous if heated sufficiently, the spirit must be universally present in matter. And from this apparently empirically well-founded position, the very general role that Newton ascribed to the electric spirit in bringing about the phenomena of the micro-world followed by a simple application of that much-abused philosophical principle, Ockham's razor. The relevant passage is worth quoting again in its entirely:

> And if there be such an universal electric spirit in bodies, certainly it must very much influence the motions & actions of the particles of the bodies amongst one another, so that without considering it, philosophers will never be able to give an account of the Phænomena arising from those motions & actions. And so far as these phænomena may be performed by the spirit wch causes electric attraction it is unphilosophical to look for any other cause.

Even Newton's subsequent change of heart on this last point is fully consistent with the position set out here. Desaguliers' experiment, without at all undermining the earlier parts of Newton's argument, nevertheless convinced Newton of the existence of a second subtle active medium. This made the application of Ockham's razor less straightforward, and meant that Newton had to choose between the two active spirits in assigning a cause to the various optical phenomena with which he had so long been concerned. But that is all.

In Q. 31, Newton cited electricity, along with gravity and magnetism, as examples of attractive forces already known to act in the world. "These Instances," he said,

> shew the Tenor and Course of Nature, and make it not improbable but that there may be more attractive Powers than these. For Nature is very consonant and conformable to herself .... The Attractions of Gravity, Magnetism, and Electricity, reach to very sensible distances, and so have been observed by vulgar Eyes, and there may be others which reach to so small distances as hitherto escape Observation; and perhaps electrical Attraction may reach to such small distances, even without being excited by Friction.[36]

Historians have generally assumed from this that for Newton the causes of magnetism and electricity were as mysterious and unknown as he generally took that of gravity to be. I have argued elsewhere, however, that in the case of magnetism this was not so, and that we ought to take literally Newton's caveat, "What I call Attraction may be perform'd by impulse, or by some other means unknown to me. I use that Word here to signify only in general any Force by which Bodies tend towards one another, whatsoever be the Cause."[37] Though sparse, all the available evidence suggests that, far from regarding magnetism as another unexplained force acting at a distance, Newton thought he knew the cause. The evidence indicates, in fact, that he attributed the power of the magnet to circulating streams of a peculiar subtle matter that passed axially through magnetized bodies and then returned through the external air before resuming its circulation once more. This theory was not original to Newton. It derived from the similar but more complicated scheme set out in Descartes' *Principia philosophiae* (1644), and was widely accepted by leading scientists in Newton's day. Similar conclusions ought to be reached, I believe, in the case of electricity. Here, too, Newton accepted a mechanistic explanation of the phenomena at hand. Here, too, the opinions he expounded were similar to those of his leading contemporaries, and resembled those set out in Descartes' *Principia philosophiae*. Electrical attraction and repulsion were not, for Newton, examples of unexplained forces acting at a distance.

In Descartes' day, the only generally recognized electrical phenomenon was the attraction that bodies such as amber, jet, sealing wax and glass exerted, when briskly rubbed, on nearby light objects. Descartes sought to reduce this, like all other natural phenomena, to motions and impacts among particles of matter. First, he suggested that the attraction might be brought about by a process analogous to what happens when a drop of viscous liquid hangs on the end of a rod, and the rod is shaken; part of the drop is seen to stretch out, and then to return promptly to the drop. Just so, Descartes envisaged the rubbing of a body as causing little branches composed of the smallest parts of the substances to spread out into the surrounding air: as these would, he said, remain joined together, they had to return immediately to the body, and in doing so could carry along any small objects that got entangled with them.[38]

Unfortunately, this account was incompatible with what Descartes had said earlier in his book concerning the nature of glass – yet glass, too, could be readily electrified by friction. To resolve the difficulty, Descartes invoked the finest of the three kinds of matter out of which he supposed the world to be constructed, his so-called "first element." When glass was rubbed, he said, thin strips composed of this fine and very mobile matter were forced out of the glass into the surrounding air. There, however, they found no pores fit to receive them, so they returned immediately to the glass, carrying with them any small bodies with which they became entangled. Nor, Descartes asserted, was this mode of attraction peculiar to glass: on the contrary, most cases of electrical attraction were brought about by this means, not by the mechanism he had first described.[39]

To modern eyes, Descartes' two mechanisms do not seem very different. In both, the electricity of bodies (that is, their ability to attract nearby light objects) is explained in terms of the "sweeping" action of some fine matter, initially ejected from the bodies in question as a result of their being rubbed, then returning to these bodies again. To seventeenth-century authors, however, the differences were important, and highlighted what remained for almost a century one of the chief unresolved questions in electrical theory. In the mechanism that Descartes preferred, the stuff involved was a universally disseminated subtle matter, or aether, whereas in the alternative model it was the matter of the rubbed body itself which brought about the effect.

Among Descartes' successors, opinions differed as to the relative merits of these two proposals. Orthodox Cartesians such as Rohault naturally tended to be guided by their master's opinions.[40] Boyle, on the other hand, confessed himself "very inclinable" to the view that ascribed the attractions to "the Emission of the finer parts of the attrahent," and in doing so he cited Digby, Browne and Gassendi as other proponents of it (in fact crediting

Digby with being its originator).[41] A generation later, Hauksbee, too, settled for this alternative. "The Effluvia which are provok'd from the Glass," he said, "seem to be, and are nothing else but part of the same Body exerted from it by rubbing." When, soon afterwards, he discovered that glass apparently allowed free passage to the effluvia emitted by rubbed sealing wax, he was most surprised, and could only suppose that "the Figure of the Parts of Glass and Sealing-Wax, are much alike, otherwise I cannot conceive how the Effluvia of one can penetrate and pass with such ease the Body of the other, and there to act as if it was one and the same with it."

In his early papers, Hauksbee tended, like Descartes, to regard the effluvia as solid threads of fine matter. At some time prior to the publication in 1709 of his *Physico-Mechanical Experiments on Various Subjects*, however, he changed his mind and came to see them instead as making up a subtle fluid. Boyle had also seen them in this light; but whereas Boyle had regarded them as "streams" which brought about the electrical attractions by condensing and returning to the glass, Hauksbee now ascribed the attractions to the inward motion of *air*, consequent upon the "Emission and Discharge of the Electrical Matter" from the electrified body. In Hauksbee's mature theory, in other words, there were two distinct streams of fluid matter in the vicinity of an electrified body, subtle matter from the rubbed body moving outwards and air moving inwards to take its place.[42]

Newton, in a brief discussion of electricity in his famous 1675 letter to the Royal Society setting out his "Hypothesis explaining the properties of light," made clear his allegiance at that stage of his career to a pattern of explanation similar to that accepted by his contemporaries. He first described an experiment in which small pieces of paper trapped in an air space beneath a piece of glass would "be attracted and move nimbly to & fro" with all manner of motions when the upper surface of the glass was rubbed with coarse cloth. To explain this, he, like Descartes but unlike English contemporaries such as Boyle, had recourse to a universal subtle matter, and concluded from the experiment that "there is something of an æthereall Nature condens'd in bodies." This, Newton supposed, was excited by the rubbing of the glass, and spread into the surrounding air. There it activated the little pieces of paper before recondensing into the glass again:

> Now whence all these irregular motions should spring I cannot imagine, unless from some kind of subtill matter lyeing condens'd in the glass, & rarefied by rubbing as water is rarified into Vapour by heat, & in that rarefaction diffused through the Space round the glasse to a great distance, & made to move & circulate variously & accordingly to actuate the papers, till it returne into the glasse againe & be recondensed there. And as

this condensed matter by rarefaction into an aethereall wind (for by its easy penetrating & circulating through Glass I esteeme it aethereall) may cause these odd motions, & by condensing againe may cause electricall attraction with its returning to the glass to succeed in the place of what is there continually recondensed; so may the gravitating attraction of the Earth be caused by the continuall condensation of some other such like aethereall Spirit . . .[43]

Newton's extensive discussions, quoted above, from a much later period of his life reveal that so far as the cause of electrical attraction was concerned, his view had changed hardly at all during the intervening period. Contrary to what has been suggested by the Halls,[44] Newton never envisaged the "electric spirit" as a Franklin-style fluid *collected* by friction on electrifiable bodies. On the contrary, he maintained with his contemporaries that electrification consisted in the *excitation* of matter already present in bodies, and he took it for granted that the various electrical motions were brought about in a wholly mechanical way by this agitated matter. As Newton put it, the exhalation raised out of bodies by friction "carrys light bodies along with it and agitates them in various manners according to its own motions."[45] Or again,

. . . the Agent or Spirit wch glass emits by friction, is agitated with various motions like a wind, carrying along with it little light bodies by means of wch you may know wch way it moves. And since for the most part it carries those bodies towards the glass & makes them stick to it, we may thence conclude that it is the spirit by which Electric bodies perform their attraction. The friction rarefies it & makes it expand it self into much more room then while it was in the glass. But it quickly begins to condense again & in condensing & shrinking into less room it returns into the glass & in returning carries along with it little light bodies & makes them stick to the glass untill it be all condensed & returned into the glass where it was before the friction.[46]

Newton frequently referred to the cause of the electrical motion as a "spirit". This does not mean, however, that he regarded it as some kind of immaterial agency, and electricity, therefore, as an irreducibly non-mechanical force.[47] In his 1675 letter to the Royal Society, Newton used the phrases "subtill matter" and "æthereall Spirit" interchangeably. In his later writings he sometimes calls the curious substance he is discussing an "exhalation", at other times a "vapour". Both terms imply a material nature, as Newton himself makes clear elsewhere, in a draft intended for the *Principia*:

Vapours and exhalations on account of their rarity lose almost all perceptible resistance, and in the common acceptance often lose even the name of bodies and are called spirits. And yet they can be called bodies in so far as they are the effluvia of bodies and have a resistance proportional to density.[48]

Likewise, he says the electric spirit "expands", and subsequently "condenses" and "shrinks" back into the body from whence it came. It can be "agitated", whereupon it "will rush out . . . with a soft crackling noise" and "push against the finger". On occasion Newton describes its "turbulent motions", which he compares to "a wind". He also says it is "susceptible of a vibrating motion like that of air". These are all terms appropriate to a material agency, but of very dubious application to an immaterial one. The most natural reading in every case – and the number of cases adds conviction here – is the one that sees Newton's ideas about electricity as firmly set in the same "Cartesian" mould as those of most of his contemporaries. To be sure, the electric spirit is "much more subtile than common Air or Vapour"; but the difference is one merely of degree, not kind.

There remains the difficult question of what Newton had in mind when in the 1717 edition of the *Opticks* he added to the sentence, "The Attractions of Gravity, Magnetism, and Electricity, reach to very sensible distances, and so have been observed by vulgar Eyes, and there may be others which reach to so small distances as hitherto escape Observation", the remark, "and perhaps electrical Attraction may reach to such small distances, even without being excited by Friction."[49] Hawes has argued that "Newton envisaged two distinct types of electric force", and maintains that in addition to the usual friction-induced attraction over appreciable distances that is effected by the electric spirit, for Newton "it is the small particles of matter which themselves possess an electric force, and by means of their inherent electric force they are able to act upon one another at small distances."[50] This is, in my opinion, seriously to mistake his meaning. Granted, Newton on several occasions mentions a number of experiments involving what we now call cohesion and surface tension, to show that a short-range attractive force that he describes as electrical acts between the particles of matter. This does not for a moment imply, however, that he conceived of this attraction as uncaused and inherent in the particles of matter themselves. On the contrary, in the final paragraph of the 1713 *Scholium Generale*, in the document "De vi electrica", and again in the draft "Quaest 25" quoted above, he expressly attributes the attraction, for reasons discussed already, to the action of the electric spirit.

What, then, is the distinction between "electrical attraction unexcited" and the ordinary friction-induced electrical force? Both, it appears, are brought about by the immediate action of the electric spirit, but whereas in the case of ordinary electrical attraction the spirit must first be made (by the friction) to spread across the intervening space until it reaches the bodies to be attracted, in the other the particles concerned are already within the range of activity of their associated electric spirit. In other words, in the case of the short-range force, the friction is simply unnecessary!

We have seen that in the case of ordinary electrical attraction, the electric spirit acts in a wholly mechanical way to push little light bodies towards an electrified object. Newton unfortunately nowhere tells us enough about the spirit's role in bringing about short-range forces for us to reach any definite conclusions about his view of its mode of action in this case. It is possible that, at this level, he envisaged the electric spirit acting by means of certain (unexplained) short-range forces between its particles and those of ordinary matter.[51] If what I have suggested above is correct, however, and short-range attractions are really no more than ordinary electrical attractions acting in a situation where no initial friction is necessary, then every analogy would suggest that Newton probably saw these forces, too, as caused mechanically by some kind of activity, perhaps vibrations,[52] in the electric spirit.

Such a conclusion cuts sharply across the tendency of recent Newtonian scholarship to stress the centrality of unexplained forces in Newton's mature philosophy of nature. It now appears that in several important cases – magnetism, electricity, and perhaps even short-range cohesive and chemical forces – Newton remained closer to the mechanistic modes of thought of his predecessors than has been appreciated. This is so especially in respect to the actual mechanisms he envisaged as causing ordinary magnetic and electrical attractions, but also in the general similarity between the function he alloted, in the passages quoted above, to his "very subtil active vibrating spirit," and that ascribed by Descartes to his "subtle matter" or "first element". In both cases, it seems, the substance in question provided the "go" for a wide range of hidden mechanisms on whose action the workings of nature depended.

There were differences, of course, between the Cartesian "first element" and Newton's "subtle spirit". Since Descartes' universe was a plenum, his subtle matter was necessarily dense. Also, its particles could not act either on each other or on particles of other matter except through contact. By contrast, Newton described the electric exhalation as "rare", which suggests that he conceived its particles to be widely separated. He also said that it was elastic, and if his discussion in the *Principia* of the elasticity of air is any guide, he would have taken this to mean that it was composed of particles mutually repelling each other at a distance. This, certainly, is how he proposed in 1717 to explain the elasticity of his newly re-introduced aether.[53]

These differences are important. So, too, is Newton's methodological emphasis on determining the laws of action of the various forces at work in the world. Unlike Descartes, he did not concern himself primarily with discovering the underlying causes of these forces. There is no inconsistency, however, in his nevertheless taking an interest in those underlying causes, or in his adherence in a number of cases to the general pattern of explanation provided by Descartes. Here, as elsewhere, we ought not to allow our recognition of Newton's extraordinary achievements in certain branches of science to blind us to the fact that in many respects he remained a child of his time.

NOTES

* Research for this paper was assisted by a grant from the Australian Research Grants Committee.

1 Isaac Newton, *Mathematical Principles of Natural Philosophy*, trans. Andrew Motte, rev. Florian Cajori (Berkeley/Los Angeles, 1934), p. 547.

2 A. R. and M. B. Hall (eds.), *Unpublished Scientific Papers of Isaac Newton* (Cambridge, 1962), pp. 357, 359. See also *idem*, 'Newton's Electric Spirit – Four Oddities', *Isis* **50** (1959), 473–476.

3 A. Koyré and I. B. Cohen, 'Newton's "Electric and Elastic Spirit" ', *Isis* **51** (1960), 337.

4 On the seventeenth-century meaning of the word "electricity", see N. H. de V. Heathcote, 'The Early Meaning of *Electricity*: Some *Pseudodoxia Epidemica*', *Annals of Science* **23** (1967), 261–275.

5 Koyré, *From the Closed World to the Infinite Universe* (Baltimore, 1957), p. 234. Cf. A. R. and M. B. Hall, 'Newton's Electric Spirit', pp. 475–476.

6 Isaac Newton, *Opticks, or a Treatise of the Reflections, Refractions, Inflections & Colours of Light* (New York, 1952: based on 4th ed., London, 1730), pp. 347–354. That the new edition was in print before the end of the year, and that it had already created quite a stir, is evident from the remark in a London newsletter dated 19 December 1717 that "Sir Isaac Newton has advanced something new in the last edition of his Optics, which has surprised his physical and theological disciples" (Historical Manuscripts Commission, *Report on the Manuscripts of His Grace the Duke of Portland*, V [Norwich, 1899], 550).

7 Sometimes, however, Newton was not so careful with his choice of words. For example, in an unpublished draft, discussed in more detail below, he referred to a "subtle Aether or Aetherial elastic spirit," an "aetherial spirit or spiritual effluvium" contained within gross bodies (University Library Cambridge, Add. MS. 3970, f. 626; quoted by Henry Guerlac, 'Newton's Optical Aether,' *Notes and Records of the Royal Society of London* **22** (1967), 45–57, p. 48.

8 Guerlac, *op. cit.*, pp. 48–52.

9 Newton, *Opticks*, p. 348. Both Guerlac, 'Newton's Optical Aether', *passim*, and R. S. Westfall, *Force in Newton's Physics* (London/New York, 1971), p. 395, have remarked upon Newton's emphasis on explaining optical phenomena in the "aether" Queries.

[10] Newton, *Opticks*, pp. 350–352.
[11] Cf. Joan L. Hawes, 'Newton and the "Electrical Attraction Unexcited"', *Annals of Science* **24** (1968), 121–130; p. 123.
[12] Newton, *Mathematical Principles* . . . , p. 547. Cf. Koyré, *Newtonian Studies* (London, 1965), pp. 35–36, for the rendering of "fingo" as "feign" rather than "frame". J. L. Heilbron, *Electricity in the 17th and 18th Centuries* (Berkeley, 1979), p. 241, argues that Newton's "optical-gravitational aether" is actually incompatible with the earlier subtle spirit: "As the agent of cohesion, the electrical spirit should be present in bodies in proportion to their densities, while the optical-gravitational aether must stand rarer in denser bodies".
[13] University Library Cambridge, Add. MS. 3965.12, ff. 350–365, quoted by Westfall, *Force in Newton's Physics*, pp. 392–393.
[14] Desaguliers to Sir Hans Sloane, 4 March 1730/31, quoted by Guerlac, 'Newton's Optical Aether', p. 51.
[15] Newton, *Opticks*, p. 353.
[16] Newton's distinguishing between aether and electric spirits at this period has been noted by J. E. McGuire, 'Force, Active Principles, and Newton's Invisible Realm', *Ambix* **15** (1968), 154–208; pp. 179–81.
[17] University Library Cambridge, Add. MS. 3970, f. 295. This and subsequent passages from Newton's papers are published by permission of the Syndics, Cambridge University Library.
[18] Newton, *Opticks*, pp. 348–349.
[19] University Library Cambridge, Add. MS. 3970, f. 293^v. Although bearing a folio number ahead of that given to the other half of the sheet, this was evidently drafted later, because the set of dashes in "Qu 24" as written out here clearly refers back to the longer version quoted previously.
[20] *Ibid.*
[21] Newton, *Opticks*, p. 376.
[22] University Library Cambridge, Add. MS. 3970, f. 235. Cf. Westfall, *Force in Newton's Physics*, p. 416.
[23] University Library Cambridge, Add. MS. 3970, ff. 241^v – 241^r. Joan L. Hawes has quoted extracts from this document in her article 'Newton's Two Electricities', *Annals of Science* 27 (1971), 95–103; pp. 96–97. (Folios 235 and 241 are the two halves of a single folded sheet.)
[24] *Ibid.*, f. 241^r.
[25] Cf. Newton, *Opticks* (1952), pp. 397ff.
[26] University Library Cambridge, Add. MS. 3970, ff. 623–29, discussed in Guerlac, 'Newton's Optical Aether'. Cf. Newton to Henry Oldenburg, 7 December 1675; in H. W. Turnbull *et al.* (eds.), *The Correspondence of Isaac Newton* (Cambridge, 1959–77), I, 362–86; pp. 364–65.
[27] University Library Cambridge, Add. MS. 3970, ff. 427–428^v; published in Newton, *Correspondence*, V, 362–65. In the accompanying English translation (pp. 365–68), Newton seems to attribute the "fits" to vibrations in the *air*. This, however, is not the sense of the Latin original.
[28] Cf. Newton, *Opticks*, 'Advertisement II', p. [cxv].
[29] Guerlac, noting that Desaguliers first reported his improved "guinea-and-feather" experiment to the Royal Society on 24 October 1717, has argued that "unless Newton

was privy to certain preliminary experiments", he could not have recast the set of Observations, and included an account of this experiment, until then. In fact, Desaguliers did perfect the experiment somewhat earlier, and doubtless promptly communicated the result to Newton. By September, Desaguliers had his apparatus working so well that he was able to demonstrate the experiment at Hampton Court before the King and the Princess of Wales (*Philosophical Transactions* **30** [1717], 717–20). Not until October, however, did the Royal Society reassemble after its customary summer recess, and so not until then could Desaguliers formally report his results to that body.

30 An unresolved puzzle, in the light of this analysis, is Newton's decision to leave the final paragraph of the *Scholium Generale* unaltered in the third (1726) edition of the *Principia*. It was presumably soon after the appearance of the previous (1713) edition that he thought about describing the subtle spirit more explicitly as "electric and elastic". Later, it seems he envisaged deleting the entire paragraph (Koyré and Cohen (eds.), *Isaac Newton's Philosophiae Naturalis Principia Mathematica: The Third Edition (1726) with Variant Readings* [Cambridge, 1972], II, 764). Does his eventual decision not to do so indicate that towards the end of his life he changed his mind yet again about the aether, and reverted to his previous ideas about a subtle spirit associated with bodies? But if so, why did he not then include the additional adjectives describing this spirit, as Motte did shortly afterwards in his English translation?

31 Comte de Tressan, *Essai sur le fluide électrique, considéré comme agent universel* (Paris, 1786), I, xliv–xlv. Though published very late in the century, this work was in fact written much earlier, in the 1740s (cf. Bibliothèque Nationale Paris, MS fr. 12280).

32 Bechler, 'Newton's Law of Forces which are Inversely as the Mass: a Suggested Interpretation of His Later Efforts to Normalise a Mechanistic Model of Optical Dispersion', *Centaurus* 18 (1973–74), 184–222; pp. 215–16.

33 Guerlac, 'Francis Hauksbee: expérimentateur au profit de Newton', *Archives internationales d'histoire des sciences* **16** (1963), 113–28; *idem*, 'Sir Isaac and the Ingenious Mr Hauksbee', pp. 228–53 in I. Bernard Cohen and René Taton (eds.), *Mélanges Alexandre Koyré: Tome I, L'aventure de la science* (Paris, 1964); *idem*, 'Newton's Optical Aether'.

34 Hauksbee's results were first published in numerous articles in the *Philosophical Transactions* in the years 1705–11, and were subsequently brought together in his *Physico-Mechanical Experiments on Various Subjects* (London, 1709; 2nd enlarged ed., 1719). I have discussed his work in my paper, 'Francis Hauksbee's Theory of Electricity', *Archive for History of Exact Sciences* **4** (1967), 203–17. Gad Freudenthal points out ('Early Electricity between Chemistry and Physics: The Simultaneous Itineraries of Francis Hauksbee, Samuel Wall, and Pierre Polinière', *Historical Studies in the Physical Sciences* **11** (1981), 203–229; p. 215) that Newton also drew upon the experiments of Dr Samuel Wall.

35 See above, p. 198.

36 Newton, *Opticks*, p. 376.

37 *Ibid.* Cf. R. W. Home, '"Newtonianism" and the Theory of the Magnet', *History of Science* **15** (1977), 252–66.

38 Descartes, *Principia philosophiae* (1644), in C. Adam and P. Tannery (eds.), *Oeuvres de Descartes* (Paris, 1897–1913), VIII, 311.

39 *Ibid.*, pp. 312–314.

40 Jacques Rohault, *A System of Natural Philosophy* (London, 1723; reprinted New York, 1969), II, 186–187.

41 Boyle, *Experiments and Notes about the Mechanical Origine or Production of Electricity* (London, 1675: reprinted Oxford, 1927), pp. 3–6.
42 R. W. Home, 'Francis Hauksbee's Theory of Electricity', *passim.*
43 Newton, *Correspondence*, I, 365.
44 A. R. and M. B. Hall (eds.), *Unpublished Scientific Papers of Isaac Newton*, p. 209.
45 University Library Cambridge, Add. MS. 3970, f. 293^{v}.
46 *Ibid.*, f. 628. Newton's failure to concern himself here with explaining electrostatic repulsion suggests that, like those of his predecessors who had noticed it, he regarded the effect as a mere mechanical rebounding and not a real phenomenon. Hauksbee thought otherwise, for reasons of which Newton must have been well aware.
47 Cf. McGuire, 'Force, Active Principles, and Newton's Invisible Realm', p. 176, where Newton's assertion, "And so far as these phænomena may be performed by the spirit wch causes electric attraction it is unphilosophical to look for any other cause" (cf. p. 198 above) is taken to mean that " 'it is unphilosophical to look for any other cause' of electrical attraction." As indicated earlier, I, on the contrary, see the passage as an invocation of Ockham's razor with respect to the cause of "the motions & actions of the particles of the bodies amongst one another" more generally. I do not see it as an abandonment of the search for a mechanical explanation of electricity.
48 University Library Cambridge, Add. MS. 3965.13, f. 437^{v}; quoted by McGuire, 'Body and Void in Newton's De Mundi Systemate: Some New Sources', *Archive for History of Exact Sciences* 3 (1966), 206–48, p. 219 (Latin original p. 245).
49 Newton, *Opticks*, p. 376.
50 Hawes, 'Newton and the "Electrical Attraction Unexcited" ', pp. 125–26; and *idem*, 'Newton's Two Electricities', *passim.*
51 Heilbron, Note 12 above, p. 240.
52 See above, p. 198. Another document tells us that the "agitation" of the spirit *softens* hard bodies and converts them into liquids (University Library Cambridge, Add. MS. 3970, f. 604). The same document adds chemical forces to the range of phenomena explained in terms of the electric spirit. Relevant passages are quoted, and an English translation provided, in McGuire, 'Force, Active Principles, and Newton's Invisible Realm', pp. 177–78. Since elsewhere in this document Newton attributes reflection, refraction and so on to the electric spirit, the whole presumably pre-dates his return to the idea of a genuine aether in late 1716.
53 Cf. Newton, *Mathematical Principles . . .*, pp. 295–302, and *Opticks*, p. 352.

II

Force, electricity, and the powers of living matter in Newton's mature philosophy of nature

One of Newton's most widely quoted methodological pronouncements appears in the preface he prepared for the first (1687) edition of his *Principia*. "The whole burden of philosophy seems to consist in this," Newton there wrote: "from the phenomena of motions to investigate the forces of nature, and then from these forces to demonstrate the other phenomena."[1] He made the same point in his other great work, his *Opticks,* towards the end of the long, final Query that he added to the Latin edition published in 1706:

> To tell us that every Species of Things is endow'd with an occult specifick Quality by which it acts and produces manifest Effects, is to tell us nothing: But to derive two or three general Principles of Motion from Phaenomena, and afterwards to tell us how the Properties and Actions of all corporeal Things follow from those manifest Principles, would be a very great step in Philosophy, though the Causes of those Principles were not yet discover'd.[2]

The procedure thus outlined has been seen as encapsulating the very essence of the "Newtonian revolution" in science. Though Newton's world was composed, like that of mechanical philosophers such as Descartes and Gassendi, of material corpuscles in motion, no longer was Newton's science constrained, as the earlier mechanical philosophy had been, to reducing all natural phenomena to empirically inaccessible, and hence merely imagined,

[1] Isaac Newton, *Mathematical Principles of Natural Philosophy,* trans. Andrew Motte, rev. Florian Cajori (Berkeley: University of California Press, 1934), pp. vii–viii.

[2] Newton, *Opticks, or a Treatise of the Reflections, Refractions, Inflections and Colours of Light* (New York: Dover, 1952), pp. 401–2.

motions and impacts of particles. On the contrary, as the passages just quoted reveal, Newton admitted a further explanatory principle into his science, namely force. For Newton, armed as he was with a clearly understood principle of inertia, force became that which changed (or tended to change) the motions of bodies; individual forces came to be measured in terms of the changes of motion they produced; and the chief object of science became the discovery of the various forces acting in the world.[3] As Newton himself demonstrated so convincingly in *Principia* with respect to gravity, such forces were at least in some cases empirically determinable in a way that corpuscular mechanisms were not. Elsewhere in *Principia,* and even more so in *Opticks,* Newton held out the hope that many other kinds of force might eventually become equally as well characterized as gravity now was.

At least as important, in the eyes of many Newtonian scholars, is the fact that, in Newton's hands, "force" became a quantifiable concept. His approach thus held out the prospect of a truly mathematical physics in which various natural effects would be shown to follow in a rigorously demonstrative and quantitatively exact manner from mathematically expressed force laws. This is, of course, what Newton himself achieved in *Principia* in relation to gravity. Though success on a similar scale in other areas of physical inquiry proved elusive, Newton, by focusing on the concept of force, had dramatically enlarged man's expectations regarding the degree of precision possible in science.

On this much, historians of science are generally agreed. In other respects, however, and especially on the question of what Newton's real attitude was to the various forces he invoked, opinion is divided. Were passages such as those quoted above merely methodological in import, or did they carry with them, as well, an ontological commitment on Newton's part? In other words, when Newton so ostentatiously refrained from offering mechanical explanations of the kind demanded by his critics for the forces of which he wrote, did he in fact do so not only because he believed that such explanations were unnecessary within the context of his inquiry, but also because he believed that these forces had a real

[3] The evolution of the concept of force during the seventeenth century is described in R. S. Westfall, *Force in Newton's Physics: The Science of Dynamics in the Seventeenth Century* (London: MacDonald, 1971).

existence of their own as true actions at a distance, independent of all mechanical explanation?

Westfall is one of those who has argued most strongly for the latter view. More than this, he has offered a reconstruction of the development of Newton's thought from which such an attitude on Newton's part toward his forces emerges as a natural consequence.[4]

Newton, Westfall has shown, was in his early years profoundly influenced in his thinking about the physical world by the ideas of the seventeenth-century mechanical philosophers – of men such as Descartes, Gassendi, Gassendi's English mouthpiece Walter Charleton, and Robert Boyle.[5] In line with this, we find Newton as an undergraduate embracing (though with a hint of occasional reservations perhaps inspired by Henry More) the fundamental credo of the mechanical philosophy, the reducibility of natural events to the motions and impacts of particles of matter. Similarly, a decade and more later, we find him trying to account for a wide range of natural phenomena in typically mechanistic style in terms of interactions between particles of matter and an all-encompassing material aether (or aethers).[6]

It is now, however, generally recognized that by the early 1680s, Newton had abandoned his former belief in a universal dense aether and had begun to speak instead in what we now see as characteristically "Newtonian" style of forces acting between particles at a distance.[7] Westfall claims to have found an explanation for this change of outlook on Newton's part in the alchemical investiga-

[4] Richard S. Westfall, *Never at Rest: A Biography of Isaac Newton* (Cambridge: Cambridge University Press, 1980), chaps. 7–9.

[5] Richard S. Westfall, "The Foundations of Newton's Philosophy of Nature," *British Journal for the History of Science, 1* (1962), 171–82; idem, *Never at Rest*, pp. 83–93. Cf. also A. R. Hall, "Sir Isaac Newton's Notebook, 1661–65," *Cambridge Historical Journal, 9* (1948), 239–50; and J. E. McGuire and Martin Tamny, eds., *Certain Philosophical Questions: Newton's Trinity Notebook* (Cambridge: Cambridge University Press, 1983).

[6] Newton, "An Hypothesis explaining the Properties of Light," in H. W. Turnbull et al., eds., *The Correspondence of Isaac Newton* (Cambridge: Cambridge University Press, 1959–77), *I*, 362–86; "De Aere et Aethere," in A. R. and M. B. Hall, eds., *Unpublished Scientific Papers of Isaac Newton* (Cambridge: Cambridge University Press, 1962), pp. 214–20 (English trans., pp. 221–8); Newton to Robert Boyle, 28 February 1679, in *Correspondence, II*, 288–95.

[7] Henry Guerlac, "Newton's Optical Aether: His Draft of a Proposed Addition to His *Opticks*," *Notes and Records of the Royal Society of London, 22* (1967), 45–57; J. E. McGuire, "Force, Active Principles, and Newton's Invisible Realm," *Ambix, 15* (1968), 154–208.

tions that engaged so much of Newton's attention at this period. In alchemy, Westfall has argued, Newton found an idea "that refused to be reconciled with the mechanical philosophy. Where that philosophy insisted on the inertness of matter, such that mechanical necessity alone determines its motion, alchemy asserted the existence of active principles in matter as the primary agents of natural phenomena."[8] Westfall finds hints of the idea even in the generally mechanistic "Hypothesis explaining the Properties of Light" of 1675, for example in Newton's willingness to ascribe the immiscibility of oil and water to "some secret principle of unsociablenes" between them. However, he dates the full-blown transformation in Newton's thinking to a few years later when, armed with this insight drawn from alchemy, Newton turned anew to the analysis of the planetary motions:

> As it appears to me, Newton's philosophy of nature underwent a profound conversion in 1679–80 under the combined influence of alchemy and the cosmic problem of orbital mechanics, two unlikely partners which made common cause on the issue of action at a distance. . . . Henceforth, the ultimate agent of nature would be for him a force acting between particles rather than a moving particle itself.[9]

Just as widely recognized among Newtonian scholars as the change in Newton's approach in the early 1680s is the fact that, beginning in about 1707 (that is, shortly after the publication of the Latin *Optice*) Newton once more began actively to consider possible causes for many of the forces he had invoked. In particular, Newton, fascinated, it seems, by various electrical experiments devised by Francis Hauksbee, began to explore the notion that many natural powers were manifestations of the activity of a subtle electric spirit that pervaded gross bodies. Hints of these speculations appear in the final paragraph of the "General Scholium" that Newton appended to the second (1713) edition of *Principia,* and in certain passages in the Queries published at the end of the

[8] Westfall, *Never at Rest,* p. 299.

[9] Ibid., p. 390. Cf. also Westfall, *Force in Newton's Physics,* chap. 7. McGuire has likewise seen a definite ontological commitment in Newton's attitude towards his forces, arguing, indeed, that these become the chief components of the universe as Newton conceived it in the years immediately after the first publication of *Principia:* "Not only did Newton enrich his ontology by including forces, but in the nineties they became essential to his natural philosophy. By 1706, he seemed to consider them, rather than matter, to be the primordials of nature" (McGuire, "Force, Active Principles, and Newton's Invisible Realm," p. 161).

1717 edition of *Opticks*. They survive in more ample form in extensive unpublished drafts intended for these editions of Newton's two great books. By 1717, however, Newton's thinking had evolved still further, and he now attributed many of the phenomena previously accounted for in terms of the electric spirit to a new form of universal aether, the activity of which might even, he suggested, be sufficient to explain the force of gravity.[10]

Historians such as Westfall and McGuire, who have argued strongly for an ontological rather than a merely methodological commitment to the doctrine of forces on Newton's part in the years after 1680, have found these later vacillations something of an embarrassment. McGuire ignores the many indications that Newton provides that he sees the electric spirit as a material agency, and treats it instead as a generalized force, "an electrical *arche* connecting mind with matter"; and he dismisses the 1717 aether on the ground that it does not conform to his general view of Newton's philosophy of nature:

> it is difficult to suppose that Newton took it seriously. It was subject to obvious conceptual inconsistencies; it was a flagrant example of the sort of intermediate entity which Newton had always tended to reject; and more significantly it repudiated his basic metaphysics of God in an empty universe.[11]

Westfall, too, treats these later views of Newton's as aberrations, arising, he suggests, from a "growing philosophical caution" that "reflected at once the impact of Hauksbee's electrical experiments on Newton and perhaps also the effect of unrelenting criticism by mechanical philosophers on an aging man no longer able to sustain a revolutionary position in the face of general opposition."[12] He sees Newton's invocation of an electric spirit as a retreat toward a

[10] Newton, *Opticks,* Queries 17–24 (pp. 347–54). The "classic" papers on Newton's flirtation with the electric spirit are by Henry Guerlac: "Francis Hauksbee: expérimentateur au profit de Newton," *Archives internationales d'histoire des sciences, 16* (1963), 113–28; "Sir Isaac and the Ingenious Mr. Hauksbee," in I. B. Cohen and R. Taton, eds., *Mélanges Alexandre Koyré* (Paris, 1964), *I,* 228–253; Guerlac, "Newton's Optical Aether" (note 7). See also Joan L. Hawes, "Newton and the 'Electrical Attraction Unexcited,' " *Annals of Science, 24* (1968), 121–130 and "Newton's Two Electricities," *Annals of Science, 27* (1971), 95–103; and, more recently, R. W. Home, "Newton on Electricity and the Aether," in Z. Bechler, ed., *Contemporary Newtonian Research* (Dordrecht: D. Reidel, 1982), pp. 191–213.

[11] McGuire, "Force, Active Principles, and Newton's Invisible Realm" [see note 7], pp. 176, 187.

[12] Westfall, *Never at Rest,* pp. 644, 793.

mode of explanation "which at least appeared acceptable to conventional mechanical philosophers," and concludes that "there is no satisfactory explanation of his return to such fluids. . . . It is hard to imagine Newton acquiescing in such a retreat at an earlier, more vigorous age."[13] Consistent with this view, the preparation of the new edition of *Opticks,* with its famous Queries about the aether, is allotted a scant two pages in Westfall's 874-page biography, and these are consigned to a chapter entitled "Years of Decline."[14]

To be sure, there are aspects of the "aether" Queries of the 1717 *Opticks* that give the historian pause. Years before, in composing the concluding sections of Book II of *Principia,* Newton had marshaled overwhelming arguments against the possibility of any dense Cartesian-style aether filling all space. The motions of the planets in accordance with Kepler's laws were, he showed, inconsistent with the existence of such a medium. Hence the new aether of 1717 was, of necessity, of a very different character. It was both exceedingly rare, in order that it might offer no perceptible resistance to the motions of the planets through it, and exceedingly elastic, in order that it might transmit vibrations with the speed required if it were to fulfill the role in the theory of light that Newton wished to ascribe to it. Newton also supposed, however, that it was composed of particles. It followed inevitably, on Newtonian principles, that those particles exerted powerful forces of repulsion upon each other.[15] Hence, even if the 1717 aether offered a satisfactory basis for explaining gravity (and there is some evidence that Newton remained less than fully convinced on this score),[16] it by no means met the fundamental objection leveled by mechanical philosophers against Newton's physics: its reliance on unexplained forces acting at a distance. All the evidence we have suggests that Newton's "aether" Queries do indeed represent a defensive response of an old man to sustained criticism.

There are even indications that Newton subsequently repented their publication. Though historians have often failed to notice the fact, the new Queries were strictly inconsistent with the final paragraph of the concluding General Scholium of the 1713 *Principia,*

[13] Ibid., pp. 793, 747. [14] Ibid., pp. 792–4. [15] Newton, *Opticks,* p. 352.

[16] J. T. Desaguliers to Sir Hans Sloane, 4 March 1730/31, quoted by Guerlac, "Newton's Optical Aether" [see note 7], p. 51.

because various phenomena – the reflection, refraction, inflection, and heating effects of light, and the transmission of sensations through the nerves to the brain and commands of the will from the brain to the muscles – which were there attributed to the action of "a certain most subtle spirit which pervades and lies hid in all gross bodies," were in the 1717 Queries ascribed to a universally disseminated aether. This latter was no mere transmogrification of the earlier subtle spirit, but was a second and quite distinct subtle matter. Even after reintroducing the aether, Newton continued to believe, as well, in the existence of the subtle spirit associated with matter, and to ascribe certain natural effects such as electricity to it;[17] but now he transferred some of the other actions previously ascribed to the subtle matter to the aether. At the very least, therefore, some modification of the General Scholium was called for. In fact, Newton envisaged deleting the entire final paragraph dealing with the subtle spirit from the next edition of *Principia*.[18] When that edition finally saw the light of day, however, in 1726, he did not do so; instead, the subtle matter continued to be invoked to explain the same effects as in 1713. This suggests that, at the very end of his life, Newton may have changed his mind yet again about the aether, and reverted to his previous mode of explanation of the various optical and physiological effects in terms of a subtle spirit associated with bodies.

If we may thus accept Westfall's judgment as to the relative insignificance of the 1717 aether for our general understanding of Newton's thinking about forces, the situation is, I believe, very different with respect to his writings on the subtle spirit. In the various published hints and unpublished drafts on this subject, we surely have before us no last-minute concessions by a declining 74-year-old, as the aether Queries appear to be, but the fruit of at least ten years of speculation and inquiry by a Newton still capable of prodigious feats of intellectual activity. Although much of what he had to say on the matter was consistent with views he had held at an earlier period of his career, these were now given a new lease of life by Hauksbee's remarkable experiments. There is ample evi-

[17] Home, "Newton on Electricity and the Aether" [see note 10], passim.

[18] A. Koyré and I. B. Cohen, eds., *Isaac Newton's Philosophiae Naturalis Principia Mathematica: The Third Edition (1726) with Variant Readings* (Cambridge: Cambridge University Press, 1972), *II*, 764.

dence of the renewed interest and enthusiasm with which Newton pursued the work.[19] Furthermore, this interest pre-dates the main Continental onslaught on his doctrine of forces, and so cannot be dismissed as a mere defensive response to that criticism. Rather, it should be seen as yet another constructive phase – perhaps, indeed, the last major constructive phase – in his thinking about matter and its powers. If this be accepted, it then becomes of considerable interest to discover precisely what views Newton arrived at concerning the subtle spirit and, more particularly, what implications these views might have for our broader understanding of his philosophy of nature.

I have argued elsewhere that, even at this late period of his life, Newton's ideas regarding the various modes of action of the subtle spirit depended much less than has generally been allowed on his invoking unexplained forces acting at a distance. I have suggested, indeed, that in many respects his ideas concerning the subtle spirit even then more closely resembled those of that arch-mechanist, Descartes, than they did those of a programmatic "Newtonian," systematically reducing each class of natural phenomena, in turn, to the action of some force or other.

Consider, first of all, the case of electricity, which lies at the heart of Newton's discussions of the subtle spirit. Newton sets out his ideas on this subject at length in a number of surviving documents. One of these is a long, unnumbered draft Query that predates the introduction of the aether hypothesis of 1717. "Do not all bodies abound," Newton here asks rhetorically, "with a very subtil active vibrating spirit by w^{ch} . . . the small particles of bodies cohære when continguous, agitate one another at small distances & regulate almost all their motions amongst themselves as the great bodies of the Universe regulate theirs by the power of gravity?" In particular, Newton believes that diffusion of this subtle matter into the space surrounding a rubbed body such as glass is what brings about electrical attraction, "for electric bodies could not act at a distance without a spirit reaching to that distance." Hauksbee's experiments provide, according to Newton, ample evidence that this diffusion occurs:

[19] Guerlac, "Francis Hauksbee: expérimentateur au profit de Newton" [see note 10], passim.

[B]y several experiments shewn by Mr Hawsksby before ye R. Society it appears that a cylindrical rod of glass or hard wax strongly rubbed emitts an electric spirit or vapour wch pushes against the hand or face so as to be felt, & upon application of the finger to ye electric body crackles & flashes, & that the electric spirit reaches to ye distances of half a foot or a foot from the glass or above . . . ; & that if a globe of glass be nimbly turned round upon an axis & in turning rub upon a man's hand to excite its electric virtue, . . . the glass emitts an electric vapour or spirit wch may be felt by the hand & wch in dashing upon the hand or upon white paper or a handkerchief at the distance of a quarter of an inch or half an inch from the glass or above, illuminates the hand or paper or handkerchief with a white light while the glass continues in motion, the spirit by striking upon those bodies being agitated so as to emit the light. . . . There is therefore an electric spirit by wch bodies are in some cases attracted in others repelled & this spirit is so subtile as to pervade & pass through the solid body of glass very freely . . . , & is capable of contraction & dilatation expanding itself to great distances from the electric body by friction. & thefore *[sic]* is elastic & susceptible of a vibrating motion like that of air whereby sounds are propagated. & this motion is exceeding quick so that the electric spirit can thereby emit light.[20]

The introduction of the aether hypothesis did not in any way affect Newton's thinking about what lay behind the various phenomena of electricity. This we discover from a passage which Newton at one point numbered "Qu. 18B," intending it to be inserted between two of the early "aether" Queries of the new edition of *Opticks:*

Do not electric bodies by friction emit a subtile exhalation or spirit by which they perform their attractions? And is not this spirit of a very active nature & capable of emitting light by its agitations? And may not all bodies abound with such a spirit & shine by the agitations of this spirit within them when sufficiently heated? ffor if a long cylindrical piece of Ambar be rubbed nimbly it will shine in the dark & if when it is well rubbed the finger of a man be held neare it so as almost to touch it, the electric spirit will rush out of the Ambar with a soft crackling noise like that of green leaves of trees thrown into a fire, & in rushing out it will also push against the finger so as to be felt like the ends of hairs of a fine brush touching the finger. And the like happens in glass. If a long hollow tube of flint glass about a inch be rubbed nimbly with a paper held in the hand . . . , the electric spirit which is excited by the friction will rush out

[20] University Library, Cambridge, Add. MS. 3970, fols. 241^{v}–241^{r}.

of the glass with a cracking noise & push against the skin so as to be felt, & in pushing emit light so as to make the skin shine like rotten wood or a glow worm. And if the glass was held neare pieces of leaf brass scattered upon a table the electric spirit wch issued out of the glass would stir them at the distance of 6, 8 or 10 inches or a foot, & put them into various brisk motions, making them sometimes leap towards the glass & stick to it, sometimes leap from it with great force, sometimes move towards it & from it several times with reciprocal motion, sometimes move in lines parall *[sic]* to the tube, sometimes remain suspended in the air, & sometimes move in various curve lines. Which motions shew that this spirit is agitated in various manners like a wind.[21]

The thoroughly mechanistic flavor of these passages is reinforced by yet another draft Query, numbered "Qu.23" and apparently also intended to accompany the "aether" Queries of the published edition. Whereas in the passages quoted above, Newton focuses chiefly on the new electrical phenomena discovered by Hauksbee, he here discusses, in detail, the cause of the electrical forces of attraction and repulsion:

Qu.23. Is not electrical attraction and repuls; performed by an exhalation which is raised out of the electrick body by friction expanded to great distances & variously agitated like a turbulent wind, & wch carries light bodies along with it & agitates them in various manners according to its own motions, making them go sometimes towards the electric body, sometimes from it & sometimes move with various other motions? And when this spirit looses its turbulent motions & begins to be recondensed & by condensation to return into the electrick body doth it not carry light bodies along with it towards the Electrick body & cause them to stick to it without further motion till they drop off?[22]

The import of passages such as these is, in my view, unmistakable. Newton does not regard electricity as a true action at a distance. On the contrary, the various electrical effects are, in his view, brought about in a straightforwardly mechanical way following the agitation, by friction, of subtle spirit residing in the pores of electrifiable bodies such as amber and glass. (As the argument proceeds, Newton in fact concludes that the spirit resides in the pores of all bodies whatsoever.) Set in motion by the friction "like a turbulent wind," the subtle spirit simply sweeps along any light objects it encounters until, as it slows, it recondenses and

[21] Ibid., fol. 295. [22] Ibid., fol. 293^v.

carries such objects with it towards the electrified body. If it strikes against larger bodies, it may be set vibrating, and it then emits light in just the same way as a vibrating column of air generates sound. This spirit is no generalized immaterial force, but a material agency. To be sure, Newton usually referred to it as a "spirit," but on occasion he called it a "vapour" or "exhalation," terms which in Newton's vocabulary certainly implied materiality. Likewise, the word "spirit" could be used in this way, as he himself made clear in a draft intended for *Principia:*

> Vapours and exhalations on account of their rarity lose almost all perceptible resistance, and in the common acceptance often lose even the name of bodies and are called spirits. And yet they can be called bodies in so far as they are the effluvia of bodies and have a resistance proportional to density.[23]

As I have argued elsewhere, Newton's ideas about electricity, as set out in passages such as those quoted above, were in almost every respect typical of his day. The notion that electrical effects were caused by subtle matter, originally contained in the rubbed body, being excited by the friction, spreading out as effluvia into the surrounding air and sweeping along any light objects lying in the way, had been widely held at least since the time of William Gilbert. It continued to be generally espoused until the 1750s, when it began to be displaced by Benjamin Franklin's notion that the phenomena in question were brought about by static accumulations or "charges" of subtle electric fluid transferred by friction from one body to another.[24] In Newton's day, the main point of contention was not the correctness of the effluvial picture in general terms but the nature of the subtle matter or matters involved. In particular, whereas some maintained that the effluvia were composed of the matter of the rubbed body itself ("the finer parts of the attrahent," as Boyle put it),[25] others, including both Descartes and Newton, held that the stuff involved was a distinct sub-

[23] University Library Cambridge, Add. MS. 3965.13, fol. 437^{v}; quoted by J. E. McGuire, "Body and Void in Newton's De Mundi Systemate: Some New Sources," *Archive for History of Exact Sciences, 3* (1966), 206–48; p. 219 (Latin original, p. 245).

[24] Home, "Newton on Electricity and the Aether" [see note 10], pp. 204–7. Cf. idem, *The Effluvial Theory of Electricity* (New York: Arno Press, 1981), and J. L. Heilbron, *Electricity in the 17th and 18th Centuries: A Study of Early Modern Physics* (Berkeley: University of California Press, 1979).

[25] Robert Boyle, *Experiments and Notes about the Mechanical Origine or Production of Electricity* (London, 1675; reprinted Oxford University Press, 1927), p. 6.

tle matter – whether the Cartesian "first element" or Newton's "electric and elastic spirit" – common to all bodies. Others again, most notably Niccolò Cabeo and his Jesuit followers, and also Hauksbee, thought that the air, too, was involved in bringing about the attraction.[26]

Magnetism was a power that Newton often cited, together with electricity, as an example of a force known to act in the world. He saw it, however, as a power that was limited in its operation, being confined, so he thought, to iron and some of its ores. Accordingly, only rarely did he mention it in connection with the electric spirit; in particular, magnetism is strikingly absent from the list of effects ascribed in the 1713 General Scholium to the action of this spirit, or in the 1717 "aether" Queries to the action of the aether.[27] For Newton, magnetism seems to have been a separate power, subject to its own laws. Here too, however, such evidence as we have suggests that Newton did not see the magnetic force as a true action at a distance. On the contrary, it appears that, as in the case of electricity, he assumed the existence of an underlying mechanism of unmistakably Cartesian provenance.

Descartes's invention of a mechanism that could account for the archetypically occult power of the magnet had been one of the major early triumphs of the mechanical philosophy. According to his scheme, associated with any magnet were circulating streams of subtle material effluvia of a distinctive kind. These streams passed axially through the magnet, emerging from one pole and returning to the other through the surrounding air before resuming their course. The various known magnetic effects were then explained in terms of interactions between such streams of matter, or between a single stream and pieces of iron that came in its way.

This general picture quickly won wide acceptance and, indeed, remained predominant for a century and more, even though various details propounded by Descartes were challenged and in some cases largely rejected.[28] Newton's early adherence to at least the

[26] On Hauksbee's ideas, see R. W. Home, "Francis Hauksbee's Theory of Electricity," *Archive for History of Exact Sciences*, 4 (1967), 203–17.

[27] On at least one occasion, however, Newton does inexplicably include magnetism in such a list (University Library Cambridge, Add. MS. 3970, fol. 241; quoted by Joan L. Hawes, "Newton's Two Electricities" [see note 10], p. 97). I find it difficult to understand this except as a slip of the pen.

[28] R. W. Home, introduction to *Aepinus's Essay on the Theory of Electricity and Magnetism* (Princeton, N.J.: Princeton University Press, 1979), chap. 4, "Magnetism."

broad outlines of the theory is clearly displayed in an unpublished manuscript dating from 1666 or 1667, in which he actually sketches the likely patterns of flow of the subtle matter in various circumstances.[29] In another paper, which Westfall dates to the same period, Newton discusses magnetism at greater length than anywhere else. Throughout, he assumes the existence of peculiar "streams" in the vicinity of a magnet; indeed, he assumes (though in terms somewhat different from those used by Descartes) two separate and "unsociable" streams entering a magnet at its two opposite poles and passing through it in opposite directions.[30]

It is tempting to see these papers as characteristic of Newton's youthful flirtation with the mechanical philosophy, and to suppose that he would later have abandoned such unrestrained hypothesizing, as he did in relation to the cause of gravity. The evidence, however, suggests otherwise, for although it is nowhere specific enough to show that Newton remained faithful to the details of his early explanatory scheme, it does unmistakably reveal that even after he had abandoned the search for a mechanical explanation for gravity, he continued to believe that magnetic effects were brought about by the action of subtle effluvia of some kind. Even in the 1690s, at the height of Newton's disillusionment with mechanism, David Gregory recorded, while visiting him, his host's view that the magnetic virtue "seems to be produced by mechanical means."[31] A casual allusion to "magnetick Effluvia" in the first edition of *Opticks*, published ten years later, tells the same story.[32] Finally, from the period with which we are chiefly concerned in this chapter, in one of the "aether" Queries of 1717, Newton uses the existence and activity of the magnetic effluvia, which he again takes as beyond dispute, to justify by analogy assumptions he is making about the nature of the aether: "If any one would ask how a Medium can be so rare," he suggests, "let him tell me . . . how the Effluvia of a Magnet can be so rare and subtile, as to pass through a Plate of Glass without any Resistance or

[29] University Library Cambridge, Add. MS. 3974, fols. 1–3.

[30] University Library Cambridge, Add. MS. 3970, fols. 473–4. Cf. Westfall, *Force in Newton's Physics* [note 3], p. 332.

[31] Newton, *Correspondence, III,* 335 (English trans., p. 338). The entry in Gregory's diary is dated 5–7 May 1694.

[32] Newton, *Opticks* (1st ed., 1704), Book II, p. 69. Cf. reprint ed. (New York, 1952), p. 267.

Diminution of their Force, and yet so potent as to turn a magnetick Needle beyond the Glass."[33]

Newton frequently cited electricity and magnetism as other instances, besides gravity, of forces known to act in the world, to support the suggestion that there were many other forces at work as well. Whenever he did so, however, he surrounded the argument with disclaimers, of which the one near the beginning of Query 31 is the most famous:

> How these Attractions may be perform'd, I do not here consider. What I call Attraction may be perform'd by impulse, or by some other means unknown to me. I use that Word here to signify only in general any Force by which Bodies attract one another, and what are the Laws and Properties of the Attraction, before we enquire the Cause by which the Attraction is perform'd.[34]

Historians have tended to place little weight on such remarks, and to dismiss them as little more than symptoms of Newton's habitual caution in expressing his views publicly. In the light of the evidence just presented, however, they take on a new significance. Whatever his attitude toward gravity might have been, Newton, we discover, did not at all suppose that either electricity or magnetism was an irredeemably nonmechanical power. On the contrary, he thought both that there was an underlying mechanism in each case, and that he knew what this was. The disclaimer in the passage just quoted must be taken seriously; Newton believes that some of the attractions of which he speaks *are* "perform'd by impulse." In other words, what is being proposed here is not a general philosophy of nature in which forces have ontological primacy, but rather a methodology only.

According to this methodology, in studying magnetism, for example, one ought above all to try, as Newton did,[35] to discover the mathematical law governing the magnetic force. Such a methodological priority is, however, in no way inconsistent with one's simultaneously having views about the existence and nature of a mechanism that might bring about the attraction in accordance with that law. Nor, for that matter, is having views about the

[33] Newton, *Opticks* (reprint ed.), p. 353. For further discussion of Newtonian views on magnetism, see R. W. Home, " 'Newtonianism' and the Theory of the Magnet," *History of Science*, 15 (1977), 252–66.

[34] *Opticks*, p. 376.

[35] Newton, *Mathematical Principles of Natural Philosophy*, p. 414.

existence of such a mechanism inconsistent with allowing that this may perhaps itself ultimately be found to rest upon the behavior of a fluid the particles of which exert unexplained forces on each other at a distance. Whether that be so, or whether they act on each other only by contact, remains undetermined, in the case of magnetism, at the point reached by Newton's inquiry. In the case of electricity, on the other hand, he seems rather closer to the view that the particles of the subtle matter that provides the mechanism do indeed exert forces on each other at a distance.[36]

Conclusions such as these are further reinforced when we turn to what Newton had to say on another subject, of which the views he expressed have yet to attract systematic study: namely, the powers of living matter. Here, too, Newton's slowness to ascribe an ontological significance to the forces he discovers at work in the world becomes apparent. Once again, he appears much more inclined to assume fairly orthodox mechanistic explanations than one might have expected of a committed "Newtonian" bent on reducing natural events to the actions of independently existing forces. This is not to say, however, that Newton is here bowing to the conceptual constraints imposed by the mechanical philosophy. On the contrary, after reducing a variety of life processes to mechanisms of one kind or another, he eventually concludes that there is associated with living matter a peculiar power for which he offers no hint of an explanation, a power which appears to lie behind the various mechanisms he has described, but which itself appears to be intrinsically nonmechanical.

Newton's best-known statements about living processes comprise, first, anatomical descriptions of the optic system that offer no additional insights into his philosophy of matter[37] and, second, various comments about the operation of the nerves in transmitting sensations from the "organs of sense" to the brain and commands of the will from the brain to the muscles. These latter comments appear in writings from all periods of his career, the best known, however, being his late-in-life remarks in the final paragraph of the 1713 General Scholium and in the last two of the

[36] Home, "Newton on Electricity and the Aether" [see note 10], p. 209.

[37] *Opticks*, pp. 15–17. Cf. also three letters of Newton's to William Briggs, 1682–5, in his *Correspondence, II*, 377–8, 381–5, 417–9; and his "Description of the Optic Nerves and their Juncture in the Brain," published in D. Brewster, *Memoirs of the Life, Writings and Discoveries of Sir Isaac Newton* (Edinburgh, 1855), *I*, 432–6.

"aether" Queries of 1717 (Queries 23 and 24). In 1713, sensation and the commands of the will were held to be transmitted by vibrations of the subtle electric and elastic spirit "mutually propagated along the solid filaments of the nerves, from the outward organs of sense to the brain, and from the brain into the muscles." By 1717, the vibrations were held to occur not in the subtle spirit but in the newly introduced "Æthereal Medium": In all other respects, however, the account remained as before.[38] In both cases, the explanations were conceived in wholly mechanistic terms, and bore obvious affinities to the opinion, commonly held at the time, that the nerves functioned as conduits for "animal spirits" flowing between the brain and the various sensory and motor parts of the body.[39] Insofar as they offered scope for characteristically "Newtonian" interpretations, they did so only by admitting the possibility, noted earlier, that the fluid in question might derive the elasticity requisite to sustain the vibrations to which Newton referred, from a repulsive force operating between its particles.[40]

Much more interesting from the point of view of this paper are, first, a manuscript entitled "De vita et morte vegetabile" and, second, sections of some associated draft Queries, prepared by Newton for a new edition of *Opticks* at about the time he composed the concluding General Scholium for the second edition of *Principia,* but never in fact published. In these papers, Newton confronts directly, albeit briefly, the nature of the distinction between living and nonliving matter. In doing so, he reveals yet again the extent to which his physical ideas were in tune with those held more generally in his day. At the same time, the discussion sheds new light on the place of unexplained forces in his mature philosophy of nature.

The passages in question are, in fact, linked with those discussed earlier in which Newton set out at length his ideas on electricity; indeed, the most important of them occur in the very same draft of the Queries from which most of the passages quoted above

[38] Newton, *Mathematical Principles of Natural Philosophy*, p. 547; *Opticks*, pp. 353–4.

[39] Cf. René Descartes, *Treatise of Man*, trans. T. S. Hall (Cambridge, Mass.: Harvard University Press, 1972), pp. 21 ff.

[40] In Newton's earliest surviving discussion of the operation of the nerves, in the "Hypothesis explaining the Properties of Light" of 1675, he invokes the additional nonmechanical concepts of sociableness and unsociableness to explain how impulses could be confined to the nerves and how, once arrived at a muscle, they could cause this to contract (Newton, *Correspondence*, *I*, 368–9).

concerning electricity were drawn. Here as nowhere else in Newton's papers we see revealed the full extent to which his powerful imagination had been captivated by Hauksbee's astonishing experiments. Both the traditional electrical attraction of light objects and the striking new effects studied by Hauksbee arose, Newton had concluded, from the agitation of a very active subtle spirit associated with the particles of bodies. The role of friction in bringing about the phenomena was not, he thought, to generate the "virtue," but only to expand it, the subtle spirit under normal circumstances "reach[ing] not to any sensible distance from the particles."[41] Now, however, he goes on to consider the possibility that there may be other causes, too, by which "the electric vertue is invigorated." One such cause, he says, may be the power of life:

> The vegetable life may also consist in the power of this spirit supposing that this power in substances wch have a vegetable life is stronger then in others & reaches to a greater distance from the particles. ffor as the electric vertue is invigorated by friction so it may be by some other causes. And by being stronger in the particles of living substances then in others it may preserve them from corruption & act upon the nourishment to make it of like form & vertue wth the living particles as a magnet turns iron to a magnet & fire turns its nourishment to fire & leaven turns past to leaven.[42]

As Newton himself makes plain, he is concerned here not with all properties of living things but only with those that had, since ancient times, been technically distinguished as constituting the vegetative aspects of life. These amounted to the ability of living

[41] University Library Cambridge, Add. MS. 3970, fol. 235; quoted by Westfall, *Force in Newton's Physics* [see note 3], p. 394, and Home, "Newton on Electricity and the Aether" [see note 10], p. 197.

[42] University Library Cambridge, Add. MS. 3970, fol. 241; quoted by Home, "Newton on Electricity and the Aether" [see note 10], p. 199. Folios 235 and 241 are the two halves of a single, folded sheet. A long unnumbered Query that begins on fol. 241^{v} was drafted first. This carries over to fol. 241^{r} and then, briefly, to fol. 235^{v}. At this stage, however, Newton evidently turned the sheet over and started again on fol. 235^{r}, writing out new Queries which he numbered 24 and 25, intending them to be inserted into the middle of what is now Q. 31 but in the Latin *Optice* of 1706 was Q. 23. A series of dashes in the proposed Q. 25 indicate that the long first paragraph of the unnumbered Query on fol. 241^{v} is to be taken in at this point. The manuscript headed "De vita et morte vegetabile" is now marked as fol. 237 in the same bundle, with drafts on fols. 238 and 240. It deals with the same subject matter but, as we shall see, in a rather different way. The first few paragraphs and the concluding sentences of this paper have been published by Westfall, *Force in Newton's Physics* (note 3), pp. 417–8.

matter, plant or animal, to maintain its organized form and to organize other matter likewise, whether in the processes of nutrition and growth, whereby existing tissues or organs are renewed or enlarged through the accretion of new matter, or in reproduction, where new living individuals are generated. In these passages, Newton does not discuss the additional powers that distinguish animate from mere vegetative life, or rational animals (i.e., human kind) from the beasts. For Newton, the subtle spirit, the agitation of which through friction gives rise to the phenomenon of electricity, also sustains the organic character of living matter by activity brought about in another way. When, at death, that activity ceases, putrefaction sets in; that is, the organized structures that characterized the matter in question during life now fall into decay.

In order to uphold this general position, Newton in these documents sets out at some length his general views on nutrition, growth, and generation, and contrasts these essentially constructive processes with the destruction and dissolution that constitute putrefaction. The ideas he displays are, in fact, very typical ones for his day.

So far as the reproduction of living forms is concerned, Newton reveals himself as an "ovist," that is, an upholder of that version of the preformation theory that held that the primordium or embryo of the young was present, already formed, in the female egg, and that the role of the male parent was merely to provide an indirect stimulus for its growth.[43] On this view, the process of generation reduces to one of growth of the embryo from suitable nutriment and, at an appropriate stage, a budding off of the embryo from the mother's body. Indeed, Newton draws the analogy with budding explicitly:

> Generation is nothing else then separating a branch from the tree and giving it better nourishment. If a separated branch takes root in the earth or a separated twigg or bud by grafting or inoculation is nourished from ye root of a young stock, it grows into a new tree as big as the tree from which it was separated being better nourished from a young root than from an old one. The seed of a tree has the nature of a branch or twig or

[43] For a lively account of the preformation theory and its intellectual milieu, see Elizabeth Gasking, *Investigations into Generation: 1651–1828* (London: Hutchinson, 1967), Chapters 2–5. Also see Jacques Roger, *Les sciences de la vie dans la pensée française du XVIIIe siècle* (Paris: Armand Colin, 1963).

bud. While it grows upon ye tree it is a part of the tree: but if separated and set in the earth to be better nourished, the embryo or young tree contained in it takes root & grows into a new tree. [In] like manner the egg of a female wth the embryo formed inside while it grows in the ovarium is a branch of ye mothers body & partakes of her life, yet the Embryo is as capable of being separated from the mother & growing great by due nourishment as a branch or twigg or bud or seed of a tree is of being separated from the tree & growing into a new tree.[44]

Newton's adherence to the preformation theory, and to the ovist version in particular, is made plain by his reference to "the egg of a female wth the embryo formed inside," growing in the ovarium prior to the intervention of the male parent. This theory was widely held during the last decades of the seventeenth century, though it was being challenged at the end of the period by the animalculist alternative according to which the preformed individual derived not from the female but from the male parent, being present in the spermatozoa newly discovered in the male semen by Leeuwenhoek. Newton does not mention the spermatozoa, which suggests that he adhered to the fairly common opinion that, far from being the essential part of the semen, they were in fact parasites. For him, the male semen played an entirely different role, namely the essentially chemical one of reacting with the female juices to produce the nourishment required by the embryo if it were to grow:

by the act of generation nothing more is done then to ferment the sperm of the female by ye sperm of ye male that it may thereby become fit nourishment for ye Embryo. ffor ye nourishment of all animals is prepared by ferment & the ferment is taken from animals of the same kind, & makes the nourishment subtile & spiritual. In adult animals the nourishm^{t} is fermented by the choler and pancreatic juice both wch come from the blood. The Embryo not being able to ferment its own nourishment wch comes from the mothers blood has it fermented by the sperm wch comes from ye fathers blood, & by this nourishment it swells, drops off from ye Ovarium & begins to grow with a life distinct from that of ye mother.[45]

According to this line of thinking, nutrition and growth become *the* fundamental biological phenomena. Nutrition, in turn, was for Newton a two-stage affair. First, a fermentation occurs,

[44] University Library Cambridge, Add. MS. 3970, fol. 235. [45] Ibid.

brought about by the digestive juices or, in the act of generation, by the male semen. This chemical process breaks down the nutriment into minute, extremely subtle parts:

Now in all fermentation wch generates spirits, the ferment abounds wth a suppress acid wch being more attracted by the other body forsakes its own to rush upon & dissolve ye other & by the violence of the action breaks both its own particles & the particles of ye other body into smaller particles & these by their great subtilty volatility & continual digestion resolve ye whole mass into as subtile parts as it can be resolved by putrefaction.

Then, these separated parts reassemble under the organizing influence of the body being nourished:

And when the nourishment is thus prepared by dissolution & subtillation, the particles of the body to be nourished draw to themselves out of the nourishment the particles of the same density & nature wth themselves. ffor particles of one & the same nature draw one another more strongly then particles of different natures do.[46]

Newton alludes at this point to the formation of mineral deposits in the earth as a known instance in which the "like attracts like" principle operates in nature. Yet, such a principle remains insufficient for his purposes because it does not explain that assembling of the new materials into appropriate structures, which manifestly occurs in the nourishment and growth of living creatures. Newton argues that the growth of crystals provides a model here:

And when many particles of the same kind are drawn together out of ye nourishment they will be apt to coalesce in such textures as the particles wch drew them did before because they are of the same nature as we see in the particles of salts wch if they be of the same kind always crystallize in the same figures.

The analogy is one that was commonly drawn upon by mechanical philosophers. Newton, however, is intellectually rigorous enough to recognize that it remains but an analogy, and does not in any sense constitute an explanation. It is precisely at this point, in order to provide an explanation for the remarkable power that living forms possess of organizing new matter into existing structures, that he invokes the force of electricity:

[46] Ibid., fols. 235–235^{v}.

And for faciliating *[sic]* this assimilation of ye nourishment & preserving the nourished bodies from corruption it may be presumed that as electric attraction is excited by friction so it may be invigorated also by some other causes & particularly by some agitation caused in the electric spirit by the vegetable life of the particles of living substances: & the ceasing of this vigour upon death may be the reason why ye death of animals is accompanied by putrefaction.[47]

Throughout the different versions of this long draft Query, whenever Newton mentions electricity, he presents it as an effect produced by the agitation of a subtle spirit that pervades the pores of ordinary matter. By contrast, in the document "De vita et morte vegetabile," the electric spirit is not mentioned. Instead, Newton refers repeatedly to the "electrical force" associated with vegetating matter, without any further hint as to its ontological status. It is therefore tempting to date this paper to the period before Hauksbee's spectacular experiments rekindled Newton's belief in the existence of the subtle spirit and his faith in its explanatory possibilities. Other features of the paper seem firmly to link it, however, with the "subtle matter" Queries. Not only are there close parallels between the arguments presented in the two documents, "De vita et morte vegetabile" displays the same striking conviction – which must surely postdate Hauksbee's work – that we find in the "subtle matter" Queries, that electricity is a power universally associated with matter that lies behind many well-known effects not normally thought of as electrical. In this case, the document would stand as a suggestive instance of Newton confining his discussion to forces and offering no explanation for these, even though he actually had an elaborate explanation worked out at the time.

Whether or not this be allowed, however, these discussions by Newton of the vegetative power of living matter are revealing in other ways. In particular, as indicated earlier, they show that here, too, as with (or so I have argued) electricity and magnetism, Newton preferred to adopt a mechanistic (or at least pseudomechanistic) theory current in his day, the preformation theory, rather than invoking yet another category of unexplained forces to explain the effects in each case.

47 Ibid., fol. 235^v.

Newton, I would suggest, is in fact much more unwilling than has often been supposed to invoke a specific kind of in-principle-inexplicable force to account for each different category of natural event. Unexplained forces remain full of mystery for him; indeed, his conviction that the most famous of them, gravity, brought him close to the activity of God himself is well known,[48] and there is no reason to suppose that others would have had less profound implications.

Prima facie, then, one might expect that Newton would have recourse to entities of this kind only for the most universal of natural powers; and this does, indeed, seem to be the pattern that emerges. Neither the "amber effect" – the drawing of little light objects to rubbed amber or glass – nor the attraction of iron to a magnet, is an effect of sufficient universality; but a force of repulsion acting between the particles of an all-pervading subtle matter and ultimately responsible not just for the macroscopic electrical attraction but for the wide range of natural phenomena listed in the 1713 General Scholium, might be. Similarly, the various specific forces associated with living forms are unlikely to be the end point of the analysis, whereas a very general power of activating the subtle electric spirit even without friction might be what universally distinguishes living from nonliving matter. Something presumably does; and whatever it be, it is likely to be closely linked to the activity of God as the Creator of life and to be irreducibly nonmechanistic in character.

Meanwhile, Newton's scientific method leads him to invoke forces of all kinds in studying natural processes, and encourages him to focus on discovering the mathematical laws according to which those forces act rather than on explaining their action. In these cases, I believe, as I have indicated above, that we ought to take more seriously than has usually been done Newton's oft-expressed caveat about the causes lying behind the forces of which he spoke. Some of these forces might, he suggested, be "perform'd by impulse."[49] If the interpretation I am suggesting is correct, Newton would in fact have thought that a mechanism was involved in most instances. I have presented evidence, indeed, that he did think so in certain leading cases. He would also not have been surprised, however, if, as these mechanisms were investi-

[48] Westfall, *Force in Newton's Physics* [see note 3], pp. 395ff. [49] *Opticks*, p. 376.

gated, they were found ultimately to depend upon a small number of irreducibly nonmechanical powers of quasi-universal application. At that point, he would have said – but only at that point – we are indeed approaching a knowledge of the first cause, which is God.

III

Francis Hauksbee's Theory of Electricity

Introduction

FRANCIS HAUKSBEE remains today a rather mysterious figure in the history of science. Neither the date nor the place of his birth is known. He appeared abruptly on the scientific stage as the demonstrator of experiments for the Royal Society on December 5, 1703 — the same day on which NEWTON presided at a meeting of the Society for the first time — in 1705 he was elected a Fellow of the Society, and in late May or early June, 1713, he died. Beyond this, nothing is known of his life.[1]

In his span of less than ten years at the Royal Society, HAUKSBEE was an exceedingly active experimenter. A long series of articles describing his experiments is to be found in the *Philosophical Transactions* of the Society for that period, and in 1709 he brought his early papers together in slightly modified form and published them under the title *Physico-Mechanical Experiments on Various Subjects* ...[2] Some of his later papers were included in a second and posthumous edition of this work which appeared in 1719. Internationally he was known not only to Continental subscribers to the *Phil. Trans.*, but also through an Italian version of his book which was published in 1716. No French edition appeared until 1754, but his electrical work at least had already received wide publicity in France by 1733 at the latest, for in that year CHARLES-FRANÇOIS DE CISTERNAI DUFAY included a long account of it in the first memoir on electricity which he read to the Académie Royale des Sciences in Paris.

In a recent article,[3] Professor GUERLAC has discussed the relationship evident between HAUKSBEE's work and some of NEWTON's later thoughts as revealed in the additions he made to the Queries in the second (1706) and third (1717) editions of the *Opticks*, and in his phrasing of the General Scholium which he added to the second (1713) edition of the *Principia*. In particular, Professor GUERLAC has traced the role played in NEWTON's thought by HAUKSBEE's work

[1] For even this small amount of information, we are indebted to Professor HENRY GUERLAC. See his articles, "Francis Hauksbee: expérimentateur au profit de Newton," *Archives internationales d'histoire des sciences*, **16**, (1963), 113—128; and "Sir Isaac and the Ingenious Mr Hauksbee," pp. 228—253 in I. BERNARD COHEN & RENÉ TATON (eds.), *Mélanges Alexandre Koyré: Tome I, L'aventure de la science*, (Paris: Hermann 1964).

[2] FRANCIS HAUKSBEE, *Physico-Mechanical Experiments on Various Subjects, containing an Account of several Surprizing Phenomena touching Light and Electricity, producible on the Attrition of Bodies. With many other Remarkable Appearances, not before observ'd.* (London: R. Brugis, for the Author, 1709).

[3] GUERLAC, *op. cit.*, "Francis Hauksbee: expérimentateur ..."

on capillarity and on electricity; he has suggested that it may well have been this work that led to NEWTON'S renewed interest in the aether as a mechanism for the forces he believed responsible for the operations of nature. In the present paper, I intend to approach HAUKSBEE'S work from a different direction. Restricting the discussion to his study of electricity, I hope to be able to show that this was important not only by virtue of its influence on NEWTON, but also in its own right. HAUKSBEE, I shall maintain, not only discovered a number of interesting facts concerning electricity, he also developed a workable theory to explain them. This theory was generally accepted within the scientific community for a period of more than twenty-five years; in particular, it provided the theoretical structure which guided DUFAY throughout his extremely important investigations during the 1730's, and even though it was DUFAY'S work which ultimately showed it to be inadequate, it was not until 1745 that an alternative theory was developed in sufficient detail to take its place.

Electricity prior to Hauksbee

At the time that HAUKSBEE began his electrical researches, the simple fact of attraction was the only recognized electrical phenomenon. In ancient times, this had been further limited to the attraction rubbed amber shows for straw, but in 1600, WILLIAM GILBERT had announced that many other substances besides amber acquire the same power when rubbed, and that they will then attract not just straw but any light objects at all.[4] GILBERT had explicitly denied the occurrence of any electrical repulsion, however, and although several people observed the repulsion later in the seventeenth century, it was only as a result of HAUKSBEE'S work early in the next century that it was recognized as a genuine phenomenon rather than a mere mechanical rebounding.[5]

GILBERT had found the attraction to be inhibited by the interposition of anything, even just a sheet of paper, between amber and straw, and this led him to conclude that it must be caused by some material agency, an effluvium of some sort, passing between attractor and attracted:

> It is probable therefore that amber does exhale something peculiar to itself, which allures bodies themselves, not the intermediate air.[6]

[4] WILLIAM GILBERT, *On the Magnet.* Edited by DEREK J. PRICE, (New York: Basic Books, Inc., 1958). A facsimile edition of the translation by SILVANUS P. THOMPSON (London: Chiswick Press 1900), which itself is a replica in all but language of the original Latin edition published in London under the title *De Magnete, magneticisque Corporibus, et de magno Magnete Tellure* ... in 1600.

[5] CABEO has often been credited with first observing electrical repulsion, but he regarded it as a mere mechanical rebounding of the attracted objects from the amber (DUANE & DUANE H. D. ROLLER, "The Development of the Concept of Electric Charge," in JAMES BRYANT CONANT (ed.), *Harvard Case Histories in Experimental Science* (Cambridge, Massachusetts: Harvard University Press, 1964), II, 556). PRIESTLEY gave the credit for the discovery to GUERICKE (JOSEPH PRIESTLEY, *The History and Present State of Electricity, with Original Experiments* (2nd edition, London, 1769), p. 9), but it has now been shown that GUERICKE had no conception of his experiments being specifically electrical; rather, he felt he was illustrating various features of an all-pervasive force, responsible for terrestrial gravity and the motion of the moon, among other things (N. H. DE V. HEATHCOTE, "Guericke's Sulphur Globe," *Annals of Science*, **6** (1950), 293—305).

[6] GILBERT, *op. cit.*, p. 55.

As is indicated by the following two passages in which he elaborated upon the subject, he seems to have envisaged the mode of action of this effluvium as being closely analogous to what we would call surface tension in a liquid:

> A breath therefore, proceeding from a body which has been compacted from humour or from a watery liquid, reaches the body to be attracted; the body that is reached is united with the attracting body, and the one body lying near the other within the peculiar radius of its effluvia makes one out of two; united, they come together into the closest accord, and this is commonly called attraction.[7]
> All bodies are united and, as it were, cemented together in some way by moisture; so that a wet body, when it touches another body, attracts it, if it is small. So wet bodies on the surface of water attract wet bodies. But the peculiar electrical effluvia ... entice corpuscles.[8]

The concept of an electrical effluvium was retained throughout the seventeenth century, and indeed, as we shall see, well into the eighteenth century. Perhaps unintentionally,[9] DESCARTES contributed to the popularity of the idea by providing a further analogy with surface tension phenomena which was often cited by his successors.[10] He envisaged the rubbing of a body as causing little branches composed of the smallest parts of the substance to spread out into the surrounding air; as these remained joined together, they had to return immediately to the body, and in doing so could carry along any small objects that got entangled with them. This process he compared with the behavior of a drop of viscous liquid on the end of a rod; when the rod is shaken, part of the drop stretches out, then returns to the main body of liquid.[11]

It was within the context of this effluvium tradition that HAUKSBEE carried out his electrical researches. He discovered a large number of new phenomena associated with electrification, and some of these, he thought, provided him with conclusive evidence for the reality of the effluvia. At the same time, each new phenomenon required explanation, and this a vague analogy with surface tension effects could not provide. Consequently HAUKSBEE was driven to expound upon the theory in a much more detailed manner than his predecessors had done. First of all, he had to consider in detail what motions the effluvia might have, and he had to show how these motions could account for his observations. Ultimately, as we shall see, he had also to concern himself with the nature of the matter of which the effluvia were composed. How successful was he with these theoretical speculations? The worth of any theory can only be appreciated in the light of the facts it had to explain, and so in this case we must consider the facts which HAUKSBEE had at his disposal before we can answer that question.

---

[7] *Ibid*, p. 56.

[8] *Ibid*, p. 57.

[9] I say "perhaps unintentionally" because DESCARTES in fact preferred a different account, one expressed in terms of the principles of his extraordinary universe. RENÉ DESCARTES, *Les principes de la philosophie*, part IV, arts. 184—186, in CHARLES ADAM & PAUL TANNERY (eds.), *Oeuvres de Descartes*, (Paris: Léopold-Cerf 1897), *Vol. IX*.

[10] BOYLE, for example, preferred the same picture of electrical action, and he cited DIGBY, BROWNE and GASSENDI as other proponents of it (in fact crediting DIGBY with being its originator). ROBERT BOYLE, *Experiments and Notes about the Mechanical Origine or Production of Electricity* (London, 1675; facsimile edition by Burndy Library, New York, 1945), pp. 3—4.

[11] DESCARTES, *loc. cit.*

Hauksbee's contribution

It has already been remarked that HAUKSBEE was the first to recognize repulsion as a genuine electrical phenomenon; a number of experiments led him to this conclusion. First, in observing the motions of small pieces of leaf brass in the presence of a rubbed tube more closely than had usually been done, he saw, as had CABEO,[12] that sometimes the pieces were thrown off the tube after they had touched it. Unlike CABEO, however, he saw this as a definite "throwing off", not as a mere rebounding:

Sometimes the Bodies Attracted would adhere to the Tube, and there remained quiet. Sometimes would be thrown violently from it to good Distances: Sometimes in their Motions towards, and sometimes even touching it, they would suddenly be Repell'd back to the distance of 4 or 5 Inches, repeating the same several times with great Velocity in a very surprising manner.[13]

His interpretation was supported by a related experiment. Some dry lamp-black was placed on a sheet of paper, and a glass tube, previously rubbed, was held above it: he found that the lamp-black was thrown into a state of violent agitation, with particles of it ascending *and descending* with great rapidity. Further, he observed that the speed of those descending was such that "their striking the paper was very audible."[14] In the original paper, he did not explicitly draw any conclusion from this, but in commenting on the experiment in his book, he remarked that, since no noise was audible when the particles simply fell under their own weight, they must in this case have acquired an additional velocity, which could only mean that an active force of repulsion was exerted on them by the tube.[15]

In yet another experiment, he found that he could hunt a piece of leaf brass around the room with his electrified tube, the brass always keeping its distance from the tube (after it had first come close to it, presumably). In this experiment, he made another important discovery — if the piece of brass touches some other object, it will no longer be repelled; rather, "it is attracted and returned again several times with great swiftness."[16] Something similar to this was recognized in a comment regarding the original experiment of leaf brass in the presence of a rubbed glass tube, a comment which does not appear in the original account in the *Philosophical Transactions* of the Royal Society, but only in his *Physico-Mechanical Experiments:*

They [the pieces of leaf brass] would often repeat this alternate rising and falling; the Attractive and Repulsive forces (whatever they are,) exerting themselves as it

[12] See footnote 5.

[13] FRANCIS HAUKSBEE, "An Account of an Experiment made before the Royal Society at Gresham College, touching the Extraordinary Elistricity (sic!) of Glass, producable on a smart Attrition of it; with a Continuation of Experiments on the same Subject, and other Phenomena," *Phil. Trans. Roy. Soc. London,* **25**, (1706), 2327. HAUKSBEE's other papers on electricity appear in the *Phil. Trans.* as follows: *Phil. Trans.*, **24**, (1705), 2129—2135; **24**, (1705), 2165—2175; **25**, (1706) 2277—2282; **25**, (1707), 2372—2377; **25**, (1707), 2413—2415; **26**, (1708), 82—86; **26**, (1708), 87—92; **26**, (1708), 219—221; **26**, (1708), 391—392; **26**, (1708), 439—443. All but the last two of these were included (in revised form) in the first edition of the *Physico-Mechanical Experiments.*

[14] *Ibid*, p. 2331.

[15] HAUKSBEE, *Physico-Mechanical Experiments,* p. 49.

[16] HAUKSBEE, *Phil. Trans.*, **26**, 85.

were in turns; the one drawing up, and the other beating down these light bodies; and that for several times, one after the other.[17]

HAUKSBEE devised a number of experiments illustrating what we would call today the direction of the electric field around an electrified glass globe, and also around a cylinder, and these led to yet another identification of a repulsion. In electrifying these objects, he used a machine he had devised with which he could rotate the glass very rapidly and electrify it by applying his hand to its surface as it rotated.[18] In the first series of experiments of this kind, a wire hoop to which several loose threads were attached was held over either a glass cylinder or a glass sphere in such a way that the threads hung down with their loose ends about an inch from the glass. When the glass was electrified, the threads arranged themselves so that they were all pointing radially inwards towards the glass. This result in itself was interesting enough. What is revelant for the present discussion is what happened when a finger was brought near: "if one's Finger ... be approach'd near the pointing ends of the Threads, ... they would flee and avoid a touch from it."[19]

GILBERT had found that the interposition of anything stopped the electrical influence, and HAUKSBEE's initial experiments with the threads confirmed this, for he found that "if any thing is held between the Glass and the directed Thread, then the Thread immediately looses hold of the Effluvia, and retires to its first position; yet upon withdrawing the Interpos'd Body, ... it will again return to its Tendency."[20] However, in an important modification of the experiment, the threads were attached at the centre of a hollow glass hemisphere, and then it was found that when the glass was rubbed, on either the inside or the outside, the threads again arranged themselves radially. Once again, it was found that when a finger was brought near, the threads were repelled, despite the fact that there was now a layer of glass between finger and thread. In other words, the electrical power can act through at least one kind of solid matter. Further, he found that simply by blowing towards the glass from several feet away, he could put the threads inside into motion.[21]

HAUKSBEE confessed to finding these observations "most surprising". Accordingly, he designed new experiments to test the phenomenon further. First, he laid a piece of leaf brass in a space between two pieces of wood; bringing an electrified tube near, he found that it was unable to influence the brass through the wood, though as soon as the wood was removed, the brass was attracted to the tube as usual. Next, he showed that "if a piece of Paper was held to touch the upper part of the Tube, it would not attract at all, altho' approach'd very near; but so soon as the Paper was remov'd, it recover'd its Sphere of Activity, ... giving a brisk motion to those Bodies, which just before were Quiescent." Taking

---

[17] HAUKSBEE, *Physico-Mechanical Experiments*, p. 43.

[18] For a complete description of the machine, and illustrations of it, see the *Physico-Mechanical Experiments*, particularly Plate VII. The machine served as a prototype for the German electricians HAUSEN, BOSE and WINKLER when they developed their well-known electrical machines in the 1740's.

[19] HAUKSBEE, *Phil. Trans.*, **25**, 2372.

[20] *Ibid.*

[21] HAUKSBEE, *Phil. Trans.*, **26**, 82.

the hollow globe with threads attached at its centre, he showed that, without rotating or rubbing the globe at all, the threads inside could be made to move by bringing an electrified object near the outside of the globe. Finally, he covered some leaf brass with a thick glass dish, and found that he could set the pieces of brass in motion by holding a piece of rubbed sealing-wax over the dish, or even by simply rubbing the upper surface of the dish.[22] Clearly, the last two results confirmed the original observation, that glass is transparent to the electrical action. The other experiments, however, seemed to indicate that glass is exceptional in this regard.

The last of the experiments just described was not, in fact, new, for in a letter to HENRY OLDENBURG which was read before the Royal Society in 1675, NEWTON had described an almost identical observation. He had laid a piece of glass set in a brass ring on a table so that a small volume of air was enclosed by the ring between glass and table; in this air space, he had put some tiny scraps of paper. He had found that if he then rubbed the upper surface of the glass with a coarse cloth, the pieces of paper "began to be attracted and move nimbly to & fro." Further, "upon Sliding my finger on the upper Side of the Glasse, though neither the glass, nor inclosed Air below, were moved thereby, yet would the papers, as they hung under the glasse, receive some new motion, inclining this way or that way accordingly as I moved my finger."[23]

NEWTON had regarded his experiment as evidence that "there is something of an aethereall nature condens'd in bodies," and he had gone on to argue that, just as electrical attraction was due to an electrical spirit condensing and returning to the glass as it did so, so also might gravitation be due to the condensation in the Earth of an aether. HAUKSBEE was never concerned with large-scale speculations of this kind: rather, he was concerned with explaining the electrical phenomena themselves, and thus he had to give a full account of the effluvia and the way in which they were supposed to act, something NEWTON had never done. Before he did so, however, it was essential for him to explain any appearance of arbitrariness in the phenomena. This he had occasion to do immediately after describing the motions of pieces of leaf brass near a rubbed tube, for, as he pointed out, not all the various motions he described always occurred. Variations of this sort he ascribed to changes in atmospheric conditions, and particularly to moisture:

> When the Air is clogg'd, either with humid and aqueous, or other more gross and solid parts, rais'd up from the vast Fund of Terrestrial matter here below, there's no doubt but the resistence these fine Effluvia then meet with in their way, must be much greater than when the Air is free, and no such Impediments lie to oppose them in their passage. For The Effluvia, how subtile soever they can be imagin'd to be, are yet Body and Matter, and must therefore be liable to the common Laws of Bodies, which is to be resisted in some proportion to the strength and density of the Medium. Neither is it improbable, but that in a moist Constitution of the Air, the watery parts may run together, and condense on the surface of the Tube; and so choak up and obstruct the passage of the Matter, which otherwise would be emitted very vigorously

[22] *Ibid*, pp. 82—92.

[23] ISAAC NEWTON, "An Hypothesis explaining the Properties of Light discoursed of in my severall Papers," pp. 362—386 in H. W. TURNBULL (ed.), *The Correspondence of Isaac Newton, Vol.* I, *1661—1675*, (Cambridge: at the University Press, 1959).

from thence. And indeed, I have always observ'd, that moisture is a great Enemy to all Experiments of this kind.[24]

HAUKSBEE was not the first to recognize the role of condensation on the glass. The other role he suggested for moisture was more novel, namely that the increase it brought about in the density of the medium should lead to an increased resistance to the motion of the effluvia because of "the resistance of the Effluvia by certain little Bodies interposed." To illustrate this idea further, he cited the ability of a thin muslin cloth to destroy the attraction between glass and leaf brass when interposed between them:

This Phenomenon, I think, will easily be allow'd to be the meer Effect of the Obstruction of the passage of the Effluvia by the fine Threads of the Linnen; which threads either wholly stop'd and detain'd, or else broke and obtunded the force of those active parts emitted from the Tube by Friction; so that they could not shake or agitate the light Bodies, that lay expos'd to them.[25]

The whole scheme of effluvia being resisted in their motions by the medium through which they move seems thus far to be both eminently reasonable and in good agreement with the observed facts; we can thus imagine how astounded HAUKSBEE must have been when he made the discovery to which reference has already been made, that glass is apparently transparent to the effluvia, for the density of glass is, of course, far greater than that of muslin.

Several passages which have been quoted already have indicated HAUKSBEE's commitment from the start to the effluvium theory. As was suggested earlier, however, he did not commit himself to that theory unthinkingly; rather, he gave several arguments purporting to establish the reality of the effluvia. One of these was based on a phenomenon he discovered which today is called the "electric wind":

They [the effluvia] ... were ... plainly to be felt upon the Face, or any other tender part, if the rubb'd Tube were held near it. And they seem'd to make very nearly such sort of stroaks upon the Skin, as a number of fine limber Hairs pushing against it might be supposed to do.[26]

The effluvia could be detected by senses other than touch, too; they could be both seen and heard when a glass tube was rubbed by hand in the dark, for

If another Hand was held near the Tube, a Light would evidently break forth from it, and That accompanied with a Noise, resembling that of the cracking of a green Leaf in the Fire; tho' not so loud ... Farther; if any other Object (besides the Hand) was brought near the Tube, tho' it did not touch it, yet a Light would fix upon it, and give much the same appearance as upon the Hand ...[27]

Again, when a cylinder was turned on the machine and rubbed,

I was surpriz'd with the appearance of a brisk and vigorous Light continued between the point of my Finger and the Glass. It was not only plainly visible on the Finger; but besides, seem'd as it were to strike with some force upon it, being easily to be felt by a kind of gentle pressure, tho' the moving Body was not touched with it by near half an inch. This Light seem'd to issue from the Glass with a considerable Noise, ... and 'twas easy enough to distinguish it from the noise made by the working of the Engine ...[28]

[24] HAUKSBEE, *Physico-Mechanical Experiments*, p. 44.
[25] *Ibid*, p. 45.
[26] *Ibid*, p. 46.
[27] *Ibid*, p. 48.
[28] *Ibid*, p. 51.

Observations such as these, then, convinced HAUKSBEE that the effluvia of which he spoke were not in any sense part of a mere speculative hypothesis, but rather, that they had a real and demonstrable existence. His further experiments were designed to shed more light on the properties of these effluvia, not to cast doubt on their existence.

The various experiments with directed threads in particular stimulated a great deal of discussion about the effluvia. For a start, one of them provoked yet another argument for a material agency in electrical phenomena:

> What we may observe from hence therefore, is, ... the certain Dependence of this Phenomenon (viz. the extension of the Threads) upon the action of some Matter, whose Direction is in strait lines towards the Glass. For upon the interposition of any Body between the Threads and the Glass, they lose their regular extension, and hang as their own weight causes them. And what can such Interposition of a Body possibly do in this case, but only interrupt the course of the Matter which is the cause of the Extension, and consequently make way for the gravity (which was overcome by a superior force before) to exert its self again? [29]

Both observation and argument are in fact essentially the same as those given by GILBERT a century earlier: because the electrical effect is interrupted by the interposition of any barrier, its cause must be a material one. HAUKSBEE's experimental arrangement led him to recognize a second feature of the phenomenon, though, one which GILBERT's methods could not even hint at; in all the experiments, the direction of motion of the matter is in straight lines, if it be granted that "the Threads were directed, according to the course of the matter."[30] Furthermore, the directions assumed by the threads in those experiments where they lay outside the glass indicated that the matter moves *towards* the glass, not away from it. This was a recurring problem — the attraction of leaf brass to a rubbed tube indicated the same direction of motion of the matter, yet in each case, the effluvia must originate in the glass. Earlier workers had bypassed the problem by using analogies with surface tension, but now, as HAUKSBEE tried to give a more detailed account, the insufficiency of that analogy became more and more clear, while the problem became more and more pressing.

The repulsion of the threads when a finger is brought near was originally interpreted by HAUKSBEE as indicating that

> It [the mass of effluvia] seems very much to resemble or emulate a Solid, since Motion may be given to a Body, by pushing the Effluvia at some distance from it.[31]

Shortly afterwards, he expressed the same idea thus:

> The Parts of the Effluvia are stiff, and continu'd, that when any part of them are pusht, all that are in the same Line suffer the same Disorder.[32]

The fact that the repulsion continued when a layer of glass intervened between finger and thread he took to indicate that there is a continuity of these "stiff" effluvia through the glass:

> Allowing a Continuum of Parts, ... the Effluvia within, and those without, are all of a Piece, (for they are both begot by the same Attrition), that when the Effluvia are

[29] *Ibid*, p. 58.
[30] *Ibid*, p. 61.
[31] HAUKSBEE, *Phil. Trans.*, **25**, 2374.
[32] HAUKSBEE, *Phil. Trans.*, **26**, 83.

pusht, or disturb'd without, the Effluvia within, in the same direction are so too, and consequently the Threads which are upheld and directed by 'em.[33]

It was at this stage in his investigations that HAUKSBEE related the experiments already described concerning the transparency of glass to the electrical action. The fact that things like wood and paper could apparently destroy the electrical power showed, he thought, that "the Effluvia of Glass act not but in a Continuum of their Parts"; when the wood or paper interrupts the continuity of the effluvia, the electrical action ceases. On the other hand, to account for the electrical transparency of glass, he suggested that

> The Effluvia which are provok'd from the Glass, seem to be, and are nothing else but part of the same Body exerted from it by rubbing; therefore (I think) can be no Impediment to the Motion of its own Effluvia, for otherwise I do not see how the Effluvia within can be produc'd by an Attrition without ...[34]

Now, though, he discovered that a piece of rubbed sealing-wax could influence threads or pieces of leaf brass through glass. This he found "very amazing", for if, as he had just argued, the effluvia of a body are "nothing else but part of the same body," how is it that the effluvia of one substance (wax) should so readily pass through another substance (glass) and yet retain that continuity necessary for their action? He could only suggest that

> the Figure of the Parts of Glass and Sealing-Wax, are much alike, otherwise I cannot conceive how the Effluvia of one can penetrate and pass with such ease the Body of the other, and there to act as if it was one and the same with it.[35]

To illustrate further the stiffness of the effluvia, HAUKSBEE devised another experiment. Previously, he had shown that if an evacuated glass globe was rubbed with the hand, a bright light would issue from within. Now, he arranged such a globe on one of his machines, and put another, unevacuated, globe close beside it on another machine. He then rotated both globes, and applied his hand to the unevacuated one; he found that a light was produced on that portion of the evacuated globe nearest the other. This he explained in terms of the effluvia driven out from the rubbed globe themselves rubbing the evacuated one: they must then be sufficiently stiff to be "capable to Act or Perform the Office of a Solid Body."[36]

The passages just quoted were published in the *Philosophical Transactions* for the first six months of 1707. Between that time and the composition of the *Physico-Mechanical Experiments*, a period of only two years, HAUKSBEE seems to have changed his mind completely about the nature of effluvia.[37] In the latter work, the last experiment described above was no longer explained in terms of the solidity of the effluvia; this solidity was in fact explicitly denied:

> In other Experiments, Lights were produc'd by the attrition of one solid Body against another: but here a Light is produc'd by the friction of a very subtile Fluid upon a Solid.[38]

---

33 *Ibid.*

34 *Ibid.*

35 *Ibid*, p. 220.

36 HAUKSBEE, *Phil. Trans.*, **25**, 2415.

37 This change of mind seems to have been overlooked by historians. ROLLER & ROLLER, for example, who have paid more attention to HAUKSBEE'S work than most, describe his ideas about the solidity of the effluvia in some detail in their *Harvard Case History* (*op. cit.*), but make no reference whatever to his later ideas.

38 HAUKSBEE, *Physico-Mechanical Experiments*, p. 64.

That is, the effluvia were now supposed to make up a subtle fluid. The experiment was now taken to demonstrate "the Force and Vigorous Action of the Effluvia." Thus, "this may come in, amongst many others, as an instance of the powerful Effects of small Bodies [*i.e.* the particles of the fluid], when put into brisk and vigorous motions."[39] Again, when he came to explain those experiments in which the threads were repelled from a finger, the account he gave in the *Physico-Mechanical Experiments* was very different from that he had given earlier, the later discussion once again being in terms of the motion of a fluid rather than a solid. On the other hand, the experiments were still taken to show that there is a continuity between the matter inside the glass and that without. HAUKSBEE'S full discussion was as follows:

'Tis manifest, there's a Communication between the Medium without, and That within the Glass ... Not only a Communication, but a Continuity of the Matter which occasions the Motion of the Threads. The Progress of it seems to be in a streight and direct track; in which the Matter is push'd by the shortest Course, from the Approach'd Body to the Threads that are shaken by it. And if the Threads are mov'd by influence of any Matter emitted from the Glass, it appears to be impossible to explain how they should be so, and at such distances, without a Continuity. So that the Case seems to be thus; That the Effluvia pass along, as it were in so many Physical Lines, or Rays; and all the Parts that compose them, adhere and joyn to one-another, in such manner, that when any of 'em are push'd, all in the same Line are affected by that Impulse given to others.[40]

Though he did not say so, the situation he now envisaged when the finger was brought near would seem to be analogous to the distortion of the lines of streaming in a fluid as it flows past a pointed obstruction.

In discussing the ability of rubbed sealing-wax to affect the threads or leaf brass through glass, HAUKSBEE remained as vague as before. After repeating his account of the experiments, he merely remarked that "this shews the Penetration, Subtilty, and very great Activity of the Effluvia (at least of these) Electrical Bodies."[41] He added, however, that the experiments would not succeed if the atmosphere were humid, and this, he supposed, was due to moisture condensing on the glass and obstructing the passage of the effluvia.

HAUKSBEE'S changed opinion here seems to have been a result of an increased concern on his part with the nature of the matter involved in electrical action. It has already been pointed out that until this time the surface tension analogy was often relied upon as a substitute for a detailed explanation. NEWTON had based his speculations about electricity upon a "subtill matter" which, he supposed, was emitted upon being rarefied by the rubbing; it would then return to the electric and recondense. BOYLE had described a very similar theory and had attributed it to DIGBY — he had regarded the oil-drop analogy as illustrating just such a process. His account was as follows:

According to this Hypothesis, the Amber, or other Electrick, being chaf'd or heated, is made to emit certain Rayes or Files of unctuous Steams, which, when they come to be a little cool'd by the external air, are somewhat condens'd, and having lost of their former agitation, shrink back to the body whence they sallied out ...[42]

[39] *Ibid.*
[40] *Ibid*, p. 111.
[41] *Ibid*, p. 125.
[42] BOYLE, *op. cit.*, p. 3.

There was one major difference between Boyle's concept and Newton's, however; to Newton, the effluvia were composed of a subtle fluid which was probably his universal aether, while to Digby and Boyle, they consisted of "the finer parts of the attrahent."[43] Hauksbee shared the latter view; we have already seen his concern as to how the effluvia from one substance (wax) could penetrate another (glass), a concern which would be quite meaningless if the effluvia were composed of a universal aether which could presumably penetrate anything. For Hauksbee, however, the air was involved too; a number of experiments he had performed concerning the relationship between the presence of air and the electrical action had led him to this conclusion.

It was his investigation of what was known as the "barometric light", the flashes of light seen in the space above the mercury in a barometer when the barometer was shaken, that first led Hauksbee to study the relationship between air and electricity. He soon realized that what generated the light was the friction between the mercury and the glass of the barometer tube, and this led him to study the effect of rubbing other substances *in vacuo*, including amber. Eventually he was led to compare the effect of rubbing a glass tube (or globe) when it was evacuated, and when it was filled with air. He found that when it was rubbed after having been evacuated, it could induce hardly any motion at all in leaf brass placed nearby, but "when the Air was let into the Tube again, the attractive power was immediately restor'd." At the same time, he noted very different optical effects in the two cases. When the tube had air in it, he could draw a spark by bringing his hand near; on the other hand, when the tube was exhausted, a vivid light appeared inside it when it was rubbed, but a spark could no longer be drawn.[44]

To account for these and the other results already described, Hauksbee framed a number of propositions which he incorporated into an appendix added to the *Physico-Mechanical Experiments*;[45] it is among these that he expressed his views about the role of the air in electrical phenomena. These propositions, indeed, mark the climax of Hauksbee's theoretical speculations.

In Proposition I, Hauksbee asserted that

> Within the Body of the Glass, are contain'd and lodg'd certain Parts of Matter, of considerable Force and Activity, which by their Motions and Percussions are the Causes of all these Effects.

That there is some matter emitted from the glass as a result of the friction is, of course, the crux of the effluvium theory. Hauksbee argued again that this emission was "too plain to be question'd," since the matter emitted could be detected by almost all the senses, as shown earlier. To account for the observed irregular motions imparted to pieces of leaf brass, in Proposition II he proclaimed that "the Motion of this Matter is not equable and regular, but disorderly, fluctuating and irregular."

In the next few propositions he made clear his interpretation of the experiments described above relating to the presence of air. First, in Proposition III, he argued that "the Air contiguous to the inner Surface of the hollow Tube,

[43] *Ibid*, p. 6.
[44] *Ibid*, p. 47—48.
[45] *Ibid*, pp. 185—194.

has an Influence on Operations of the Effluvia." Such a conclusion of course seems rather obvious from his results; to blunt any objections which might have been offered on the grounds that the air did not seem to be necessary in the case of a solid glass tube, he remarked that

> the Proposition does not assert, that the Effluvia can in no case exert themselves with vigour, without the concurrent assistance of the Air: but it asserts, that the contiguous Air had some advantageous Influence in the Case of the hollow Tube.

In Proposition IV, it was suggested that the observed effects with the hollow tube could be accounted for by supposing that the heat generated by the friction rarefies the air nearest to the outer surface of the glass. Since the air inside the tube is further from the source of heat, it will be less rarefied, so that "the Spring of the less-rarefied Air within, [will be] superior to the Pressure of the more-rarefied Air without": the equilibrium of the electrical matter will thus be disturbed, and it will move in the direction of least resistance, outwards. On the other hand

> when the Tube is exhausted of its Air, and consequently the Balance lost on the inside; all the Attrition that can be given will not be sufficient to bring the Effluvia out against an incumbent Pressure, as long as the inward Cavity is clear of Air, and there is no Counter-force to oppose their Conatus or Tendency that way.

This is not the only way in which the air was supposed to act, however. In Proposition V, he stated that "as the internal Air is necessary to the Action of the Effluvia, so is the external too," and in Proposition VI, we learn why he thought this should be the case:

> The external Air appears to be necessary to carry the little Bodies ... towards the Tube. For if by Heat and Rarefaction, consequent upon the Attrition, the Medium contiguous to the Tube be made specifically lighter; then of course, to keep up the balance, the remoter Air, which is denser, must press in towards the Tube, and so carry away (in the Torrent) the little Bodies lying in its way.

In other words, he supposed that there were *two* distinct types of matter involved in the electrical action, each having its own peculiar motion — the electrical matter moving out from the tube, and, as a result, air moving towards the tube. In Proposition VII, and in four propositions concerning the experiments with directed threads, he elaborated this concept of a double action. For example, the inwardly-directed positions assumed by the threads on a hoop around an electrified glass cylinder were explained as follows:

> If the Electrical Matter be emitted in Physical Lines, every where diverging from the Center of that Circle in which the Attrition is made ... towards the Circumference of the same Circle; then by the Rarefaction of the Medium contiguous to the Glass, and the necessary Pressure of the more remote and dense Medium, into the Plane of that same Circle, with Directions contrary to those in which the Effluvia are emitted: by this means (I say) the Threads may be regularly directed to the Center of that Circle, in whose Plane the Hoop to which they are fix'd is plac'd.

Similar explanations were given for the other results obtained with the threads.[46]

---

[46] It should be emphasized that although, in the passage just quoted, HAUKSBEE wrote of "the Electrical Matter", this does not imply that he had at this stage adopted a NEWTONIAN-like view of a universal electrical matter; the whole discussion in the appendix was limited to the effects obtainable with glass, and so he had no cause to indicate specifically that he was speaking of the electrical matter of glass. In earlier passages in the book, when discussing the electricity of other substances, while he was never very explicit, he did seem to be clinging to his earlier concept (as outlined in the *Phil. Trans.*) of each substance having its own characteristic effluvia.

It now becomes apparent why HAUKSBEE rejected his earlier ideas about the solidity of the matter which sustained the threads, and rephrased his account so that it described a fluid. Air, we are now told, is involved, and air is palpably a fluid. The theory sounds quite plausible; at any rate, it seems to cope adequately with most of the electrical phenomena known at the time (although we may note the germ of what turned out in the end to be one of the fatal weaknesses of later effluvium theories, an inability to explain *alternate* attractions and repulsions of light bodies). Everything rests, however, on Proposition V, that the external air is necessary for the effluvia to be able to act. BOYLE had found otherwise. It is perhaps significant that the experiment which led HAUKSBEE to such a vital conclusion was not described in detail by him. Whereas he described all the other experiments he had carried out in some detail in the body of the text, no mention was made of this one until it was included, as if it were an afterthought, in the appendix. Even here, the account he gave was very meagre: all he bothered to say was

> I can give what degree of Friction soever is necessary, in an exhausted Receiver. The contiguous Air ... being removed, the Electrical force seem'd to be quite gone; and continued so to be, till the presence of the Air was restor'd.[47]

Again, a seemingly obvious experiment for him to have tried would have been to include a glass cylinder and a surrounding hoop and threads in an exhausted receiver. He did not do this: instead, he simply asserted that

> *if* the semi-circular Hoop of Threads were plac'd in Vacuo, that Property of their regular Directions to a Center *would be* quite lost, even tho' the Globe or Cylinder were full of Air.[48] (My emphasis)

The suspicion inevitably forms that he had made up his mind about the role of the air before he carried out the appropriate experimental test, and that when he actually carried out the test, he did so in a very cursory fashion. We must not be too hard on him for this, however; even if he had not prejudged the outcome of the experiment, it is quite likely that he would have obtained the same result, for at the pressures he could achieve in his receiver, the residual air would be quite conducting. This being the case, the charge might well have been leaking off as fast as it was generated; the fact that he observed a vivid light when the rubbing took place *in vacuo* suggests that such a leakage was indeed taking place. Rather than criticism, HAUKSBEE deserves a great deal of credit for his efforts to build a theory which was firmly based on experiment; he consistently avoided the temptation to which many early electricians succumbed, to build vast speculative schemes on a minimum of evidence.

Epilogue

For a period of almost twenty years following HAUKSBEE's death, no further work on electricity of any importance was published.[49] Beginning in 1731, however, a number of important new dicoveries were announced by STEPHEN GRAY,[50]

---

[47] HAUKSBEE, *Physico-Mechanical Experiments*, p. 189.

[48] *Ibid*, p. 191.

[49] In this period, in fact, only one paper to do with electricity appeared in print, a minor piece by STEPHEN GRAY (*Phil. Trans.*, **31**, (1720), 104—107).

[50] All of GRAY's discoveries were announced in the *Phil. Trans.*, as follows: *Phil. Trans.*, **37**, (1731), 18—44; **37**, (1732), 227—230; **37**, (1732), 285—291; **37**, (1732), 397—407; **39**, (1735), 16—24; **39**, (1735), 166—170; **39**, (1736), 220; **39**, (1736), 400—403.

and GRAY's work in turn inspired a brilliant series of investigations in Paris by CHARLES-FRANÇOIS DE CISTERNAI DUFAY.[51] It was the discoveries made by these two men which led eventually to the overthrow of HAUKSBEE's theory.

GRAY's most important discovery was that certain materials would transmit the power to attract and repel light objects over long distances, provided they were suitably insulated. Soon afterwards, DUFAY established that any substance which could be electrified by friction (*i.e.* any "electric") would serve as an insulator, and that any substance which could not be electrified in this way (*i.e.* any "non-electric") would act as a conducting line for the electrical power. These results immediately raised a major theoretical difficulty, for the obvious inference to be drawn from them was that non-electric materials were so constituted that the electric matter could flow freely through them, while the constitution of electric materials was such that the electric matter could not pass freely through them. How was this to be reconciled with Hauksbee's earlier work which seemed to show precisely the opposite, namely that substances such as glass and wax, both electrics, were transparent to the electrical power, while non-electrics like the metals were not?

Such a blatant contradiction necessarily had a debilitating effect on HAUKSBEE's theory as a whole. That it played no larger role than this in the eventual demise of the theory is apparent, however, because the theory which replaced HAUKSBEE's was no better able to cope with it than HAUKSBEE's had been. Of far more immediate importance for the fate of the theory were the experiments of both GRAY and DUFAY relating to what we have already seen to have been the weakest point in HAUKSBEE's work, the question of the persistence of the electrical attraction *in vacuo*. It was GRAY who first reported that he had found that bodies would attract *in vacuo* almost as strongly as *in pleno*. He did no more than report his result, however; he did not concern himself at all with its theoretical implications. DUFAY, who had already revealed himself as generally in sympathy with HAUKSBEE's ideas, recognized the significance GRAY's result would have for the theory if it was correct, and so he immediately investigated the problem for himself. Much to his consternation, his results at least partly confirmed GRAY's. His full report was as follows:

> Le verre & les corps semblables frottés dans le vuide n'acquiérent que très-peu d'électricité, quoiqu'ils la conservent dans le vuide, s'ils ont été précédemment frottés dans l'air libre, & les corps, dont l'électricité est de la nature de celle que nous avons appellée *résineuse*, acquiérent cette vertu étant frottés dans le vuide, de même qu'ils feroient dans l'air libre.[52]

In other words, he confirmed HAUKSBEE's single result, but at the same time he obtained a number of other results which were quite at variance with the requirements of the theory.

---

[51] DUFAY reported his results in the *Mémoires* of the Académie Royale des Sciences, as follows: (1733), pp. 23—35; (1733), pp. 73—84; (1733), pp. 233—254; (1733), pp. 457—476; (1734), pp. 341—361; (1734), pp. 503—526; (1737), pp. 86—100; (1737), pp. 307—325. The contents of the first four memoirs were summarised in a letter to the Royal Society which was published in the *Phil. Trans.* (*Phil. Trans.*, **38**, (1733), 258—266).

[52] DUFAY, *Mémoires*, (1734), p. 355.

Equally puzzling from a theoretical standpoint was his discovery that when the air within his glass tube was strongly compressed, the electricity of the tube was considerably weakened, whereas according to the theory, it should have been strengthened. When he first obtained this result, DUFAY attributed it to the admission of grease and moisture into the tube during the compression, but when he still obtained the same result after he had convinced himself that he had overcome such technical problems, he could do no more than confess himself greatly astonished.[53] He made no attempt to construct a new theory which was less at variance with his results, and it was left to his one-time assistant, the abbé NOLLET, to do so in a paper read to the Académie Royale des Sciences in 1745.[54] In the new theory, NOLLET retained many of the features of HAUKSBEE'S theory, and in particular he retained the concept of a double stream of matter, one flowing out from the electrified body, the other flowing in towards it. However, to avoid the difficulties raised by DUFAY'S results, he denied that the inward-flowing stream was a current of air; instead, he postulated a current of the same matter as the outward-flowing stream, a current having its origin in the various objects surrounding the electrified body.

Conclusion

Historians of science have usually assumed that the science of electricity developed in the period prior to FRANKLIN, or at least prior to NOLLET, in what amounted to a theoretical vacuum. It has been my aim in this paper to demonstrate the falsity of that assumption. I have shown, I hope, that HAUKSBEE'S important researches were guided throughout by strong theoretical considerations, and I have indicated that DUFAY'S even more important studies were guided by exactly the same considerations. Nor was their theory in any sense a stagnant one. As it was developed by HAUKSBEE, it could give a fairly adequate explanation of almost all the known electrical phenomena; it even enabled him to predict the outcome of experiments such as the one involving the rubbing of a globe while it was positioned near a second, exhausted, globe. With the discovery of so many new phenomena in the 1730's, the theory turned out to be no longer adequate, but it is not at all surprising that it was a few years before the full extent of its inadequacies was appreciated, nor is it surprising that a strong continuity is evident between it and the theory which eventually replaced it. In the meantime, the theory continued to serve a useful function by suggesting new lines of research to its adherents. The theory functioned, then, in the same way as any other scientific theory, and it deserves a more serious treatment than it has usually received. This paper, I hope, can serve as a beginning.

---

[53] *Ibid*, pp. 360—361.

[54] JEAN ANTOINE NOLLET, "Conjectures sur les causes de l'Electricité des Corps," *Histoire de l'Académie Royale des Sciences*, (1745), Mémoires, pp. 107—151.

IV

'NEWTONIANISM' AND THE THEORY OF THE MAGNET

Voltaire's famous remark that in Paris he had left the universe a plenum, and in London found it mostly empty space, records in the author's usual sparkling style a widely held view of early eighteenth century scientific life. At this period, the French, so it is said, remained faithful to Descartes's vision of a science in which all natural events were to be explained as purely mechanical interactions between particles of omnipresent matter. The English preferred Newton's modified version of the ancient doctrine of atoms and the void. In Paris, particles acted on each other only by contact. In London, Newton and his followers tried instead to reduce the various natural phenomena to forces which acted, more often than not, at a distance. Not until the late 1730s and early 1740s, we are told—that is, a full half century or more after the first appearance of the *Principia*—did the power of Newton's instrumentalist approach finally compel the French to abandon their old-fashioned Cartesian ways.

By itself, Voltaire's comment might be dismissed as a piece of mere journalistic exaggeration. But in fact there is abundant evidence in the French scientific writings of the age to confirm what he says. Like Voltaire, French scientists, too, were clearly conscious of an intellectual gulf separating them at this period from their fellows across the Channel. It is therefore not surprising that modern commentators, of whom Pierre Brunet has been one of the most influential,[1] have fashioned their histories around the same theme. It has become a commonplace that there was a marked division between Cartesian and Newtonian modes of thought in early eighteenth century science, a division which was eventually resolved some time in the 1740s in an overwhelming victory for the Newtonian point of view.

In recent years, as this general outline has become ever more thoroughly entrenched in the literature, its boundaries have been extended beyond those envisaged by its originators. Brunet, for example, even though he chose an extremely general title for his work, in fact confined his attention to eighteenth century French attitudes towards Newtonian celestial mechanics and the philosophical implications of the unexplained attractions that lay at the heart of the Englishman's great work. In doing so, he was faithfully reflecting the central concerns of contemporary authors such as

Voltaire. More recently, however, especially since the publication twenty years ago of I. Bernard Cohen's well known study, *Franklin and Newton,*[2] it has come to be widely recognized that Newton's work had as great an influence on experimental physics as it did on planetary astronomy, but that here the chief inspiration came from the master's second great book, his *Opticks,* rather than from the mathematically much more rigorous *Principia.* The effect of this has been to widen the denotation of the term 'Newtonian', to give it an application extending to all the physical sciences, and no longer restricted to dynamics and celestial mechanics alone. More recently still, in the wake of Kuhn's *Structure of scientific revolutions,*[3] it has been all too easy to see 'Newtonianism' as an all-encompassing Kuhnian paradigm, and, forgetting the actual limitations of Brunet's coverage, to see his study as delineating an epic struggle between two paradigms of extremely general application. To explain the thinking of an eighteenth century scientist, it has increasingly come to be thought sufficient to characterize him as a Newtonian or Cartesian, as the case may be—or even as a Leibnizian if, as happens all too infrequently, we concern ourselves momentarily with developments beyond the Rhine. In other words, it has come to be freely assumed, if not acknowledged in so many words, that once we have discovered to which of these groups a man belongs, we can confidently predict which approach he will adopt to any particular scientific problem.

This is, in my opinion, to ask too much of our historical categories. It is also to run dangerously far ahead of the evidence. One way in which it does so is in its failure adequately to reconcile the Cohen-inspired extension of the meaning of the term 'Newtonian' with the story traditionally told of the battle between French Cartesians and Newtonians, the story told in such detail by Brunet. In describing a tradition of experimental Newtonianism inspired by the *Opticks,* Cohen confined his attention to the British scientific community of the day (including its colonial offshoots in North America) and, to a lesser extent, the Dutch. Given this, suppose we grant Brunet's conclusion, that Newton's celestial mechanics, and his doctrine of universal attraction, came to be widely accepted in France after 1740. Following Cohen, let us likewise grant that at this period we can clearly discern, in Britain and her colonies, and perhaps in Holland, a tradition of experimental Newtonianism inspired by the *Opticks.* It by no means follows that this latter brand of Newtonianism was also taken up by the French. Whether French experimental physics after 1740 was dominated by Newton, in the same way as French astronomy came to be, is an entirely different question from the one that Brunet set out to answer, and to answer it, a very different body of evidence needs to be studied.

Brunet himself took up the question in a work published several years before the one mentioned above.[4] In discussing the rise of the experimental method in France, however, he carefully refrained from identifying experimentalism generally with experimental Newtonianism in particular. On the contrary, he was well aware not only that the Dutch professors Boerhaave, 'sGravesande and Musschenbroek, who played so large a part in his story, drew their inspiration from other sources besides Newton, but also that the leading French experimental physicists of the age—men such as Du Fay, Nollet, and Brisson—could in no way be described as 'Newtonian'. If anything, the latter group would have to be called anti-Newtonian. To cite only one example, Nollet, in his famous textbook of physics, carefully re-wrote Newton's laws of motion in order to avoid any possible imputation that they might refer to forces acting at a distance, and to make it clear that whatever the astronomers might do, physicists were concerned with contact action alone.[5] If Nollet were a relatively minor figure in the French science of the period, this would scarcely be worth noticing; and there has, indeed, been a tendency among English-language historians of science to treat the man as a discredited, lonely, and slightly foolish relic of an earlier age, after the dramatic successes of Benjamin Franklin's rival 'Newtonian' theory of electricity in the early 1750s. Franklin's own autobiography has been the most frequently cited source here: the Abbé, Franklin tells us, "lived to see himself the last of his sect—except Mr. B—— [i.e., Brisson] his *élève* and immediate disciple".[6] Franklin, however, was deluding himself. I have shown elsewhere[7] that though Franklin's doctrines attracted an initial flurry of interest in France, they thereafter made very little headway for a period of twenty years or more. Contrary to what Franklin reported, Nollet died in 1770 as he had lived, the acknowledged master of French experimental physics.

It thus becomes clear that French experimental physics did not become unequivocally Newtonian at the same time as French astronomy and rational mechanics fell under the Englishman's spell. Hence, instead of simply assuming that Brunet's chronology applies not merely to the introduction of Newtonian celestial mechanics into France, but to the fate of Newton's doctrines generally, instead of speaking simply of *the* introduction of Newtonianism into France, we need to concede, if we wish to continue speaking in these terms, that the word can be understood in various ways, and to specify more carefully which kind of Newtonianism it is that we are talking about in any given context.

There is, of course, no reason to restrict this moral to the French case. Such a diversity of viewpoints is included under the 'Newtonian' umbrella these days that using the term in any context without qualification at best tells us very little, and is sometimes, as in the French example just discussed, positively misleading.

Remarks such as these are not altogether new.[8] In what follows, I shall pursue a somewhat different line of criticism, directed particularly at the habit of regarding designations such as 'Cartesian' or 'Newtonian' as sufficient to *explain* the thinking of eighteenth century scientists. Any history worthy of the name must aim to explain, and therefore is obliged to deal in generalizations which have some explanatory import. Otherwise it will never rise above the level of mere antiquarianism and isolated individual biography. But in history as in science, generalizations need to be tested; and when tested, they sometimes prove to be of more limited utility than was thought at first. To classify someone as, say, a Newtonian, ought, if the classification has any explanatory power, to provide us with some understanding of his intellectual predilections in cases outside those upon which our classification was based in the first place. Stating the point in slightly stronger terms, it ought to enable us to predict, if only in general terms, the likely direction of our man's thinking in as yet unstudied cases; and it is these predictions that we can then submit to test, by investigating how he in fact tackled the subjects in question.

It is the result of just such a test that I wish to describe in this paper. To avoid difficulties such as those mentioned already concerning possibly doubtful applications of our classificatory scheme, I shall confine my attention to a period and a group of scientists so unambiguously Newtonian that if the term cannot be applied to them, it surely cannot be applied to anyone. This group is Newton's immediate circle of disciples in early eighteenth century London. The subject of our inquiry is the hitherto little explored question of the attitude of members of the group to the phenomena of magnetism.

Given the usual connotations of the word 'Newtonian', as they have been spelled out by a number of authors in recent years,[9] there is no question but that we would expect members of our group to reject the typically cumbersome mechanism of circulating streams of a certain kind of subtle matter put forward by Descartes,[10] and to try instead to reduce this particular set of phenomena, like all others, to the action of some kind of force, acting in this instance between particles of iron or loadstone as the case may be. True, we would expect our Newtonians to pay lip service to the need eventually to find the unknown cause or causes that lay behind this force. But we would in this context expect them to emphasize the word "unknown", as they did in the gravitational case, and to concern themselves in practice not with discovering this cause, but with determining the way in which the force in question gave rise to the various observed effects. Our expectations in this case are all the stronger because the approach I have outlined seems to be the one suggested by Newton himself on a number of occasions in his published writings, most notably in the famous opening lines of the thirtyfirst Query in the *Opticks,* where gravity, magnetism and electricity were presented as "Instances shew[ing] the Tenor and Course of Nature", to support his suggestion that "the small Particles of Bodies [have] certain Powers, Virtues, or Forces, by which they act at a distance . . . upon one another for producing a great Part of the Phaenomena of Nature".[11]

Yet appearances can sometimes be deceiving. Though the evidence presented below is sparse, it is sufficient to show that, in regard to magnetism, Newton's thinking remained throughout almost all his life very close to that of his supposed arch-rival, Descartes! It also shows that most of our early 'Newtonian' group adopted the same approach. What, then, becomes of our traditional historical categories? From one point of view, perhaps, it is reassuring thus to find our Newtonians agreeing once again with Newton, even if in such a totally unexpected way. But in every other respect, such a conclusion is profoundly upsetting for the classification to which we have become accustomed. Only if 'Newtonian' meant simply "someone who agreed with Newton in every particular" would there be no difficulty. But apart from the probability that we would then be dealing with a null set, this is clearly *not* the meaning that authors who have used the term have had in mind. Instead, they have intended by it to encapsulate a particular approach to scientific questions—an approach based upon the concept of 'force' and standing in sharp contrast with that of, above all, the so-called Cartesians. If we find, however, that Newtonians can sometimes behave in orthodox Cartesian ways that are the very antithesis of what is normally expected of them, how strongly can we maintain such a sharp Cartesian-Newtonian dichotomy? We must surely conclude that this has failed the test proposed for it. And this in turn must force us to conclude more generally that the term 'Newtonian' does not really have anything like the unambiguous and very broad applicability and explanatory power that have been claimed for it. As a descriptive label within a carefully restricted context, it is undoubtedly still very useful. But that is all the weight it will bear.

What, then, is the evidence upon which these heretical assertions are based? What, precisely, did Newton and his followers have to say on the subject of magnetism? Here it must first be said that, as in so many cases, Newton's published writings do not reveal anything like the whole story, and in fact in this instance they are very far from making his opinion clear. At times his choice of phrase seems to imply that magnetic forces act, like gravity, in some unexplained fashion at a distance, but elsewhere he ascribes them to the mechanical action of certain subtle material effluvia. Nowhere, however, does he provide a systematic exposition of his views.

For the most part, when Newton refers to magnetism, he mentions it merely in passing, as a force that is already known to act in the world.

Very early in the *Principia*, for example, he uses well-known facts about the loadstone to illustrate his definitions concerning forces in general. The fact that "the magnetic force is greater in one loadstone and less in another, according to their sizes and strength of intensity" is used to illustrate the proportionality of any central force to "the efficacy of the cause that propagates it from the centre, through the spaces round about".[12] Likewise, in order to explain what he means by the "accelerative quantity" of a central force, Newton calls upon the fact that "the force of the same loadstone is greater at a less distance, and less at a greater".[13] A little further on, he illustrates the equality of action and reaction with an experiment using loadstone and iron.[14] When he needs an example of a force whose action extends only to neighbouring particles, he cites the fact that the attractive power of a magnet is interrupted by a nearby iron plate.[15] Later, magnetic and electric attractions are brought forward as examples showing that "all attraction towards the whole arises from the attractions towards the several parts".[16] In the *Opticks*, too, in Query 29, magnetism provides a useful analogy, this time for a point Newton wishes to make in connection with the anomalous refraction of light by Iceland crystal.[17] On other occasions in Newton's writings, however, he mentions magnetism only to distinguish between the magnetic force and his new force of gravity. In opposition to a large number of seventeenth century authors, he insists that the two are quite different, even to the extent that they seem to vary according to different laws.[18]

None of these remarks in fact commits Newton in any way to a particular theory of magnetic action, but his discussing the magnetic force so frequently in parallel with the force of gravity might well have given his readers the impression that he accorded them both the same causal status. For much of his life, Newton was convinced that the cause of gravity was non-mechanical, and his followers were well aware of this. The suspicion that he held a similar view about the cause of magnetism was therefore perhaps inevitable. Furthermore, the passage from the beginning of the thirtyfirst Query in the *Opticks,* quoted above, would have done nothing to dispel this idea. Admittedly, Newton immediately qualified what he said on this occasion by adding, "What I call Attraction may be perform'd by impulse, or by some other means unknown to me. I use that Word here to signify only in general any Force by which Bodies tend towards one another, whatsoever be the Cause". Yet later in the same Query, having mentioned gravity, fermentation and cohesion as examples of "certain active Principles", "general Laws of Nature, by which the Things themselves are form'd", Newton added "magnetick and electrick Attractions" to the list of "manifest Qualities" whose causes yet remained occult and unknown, but were very possibly, he seemed to imply, non-mechanical.[19]

It would however be a mistake to accept any such implication so far as magnetism is concerned. As indicated above, it seems for most of his life Newton actually held, as did all but the Aristotelians among his contemporaries, that magnetic phenomena were caused by mechanical processes, namely circulating streams of subtle material effluvia which passed axially through a magnet, emerged from one pole, and returned to the other through the surrounding air before renewing their course. Even if we do not do so in other cases, with respect to magnetism we ought to take literally Newton's caveat that what he called an attraction might be caused by mechanical impulse.

Newton's belief at the outset of his career in magnetic circulations rather like those described by Descartes is plainly revealed by an as yet unpublished manuscript dating from 1666 or 1667, now preserved in the Cambridge University Library. Here, Newton actually sketches the likely patterns of flow in various circumstances![19a] Similar views are set forth in a document published by A. R. and M. B. Hall under the heading "De Aere et Aethere", and dated by them to a period a few years later, namely 1673-75. Newton, arguing the existence of various kinds of matter subtler than air, at one point in this paper remarks without further elaboration: "I believe everyone who sees iron filings arranged into curved lines like meridians by effluvia circulating from pole to pole of the [load-]stone will acknowledge that these magnetic effluvia are of this kind".[20] The same doctrines appear again in other writings from this period, namely the "Hypothesis explaining the Properties of Light" that Newton sent to the Royal Society in late 1675, and another still unpublished manuscript in the Cambridge University Library. In the former of these, Newton twice refers casually to "Magnetic effluvia" in a way that indicates that he simply takes their existence for granted.[21] However, this document, like "De Aere et Aethere", was not published at the time; only when it was included by Thomas Birch in his *History of the Royal Society of London* in 1757 did its contents enter the public domain. Those of the second document have still not done so. In it, Newton discusses magnetism at greater length than anywhere else. Throughout, he assumes the existence in the vicinity of a magnet of peculiar "streams": indeed, he assumes (though in terms somewhat different from those used by Descartes) two separate and "unsociable" streams entering a magnet at its two opposite poles and passing through it in opposite directions.[22]

Testimony that Newton still held mechanistic views about magnetism during the 1690s comes from the pen of the mathematician David Gregory. In May 1694, while visiting Newton, Gregory recorded, amongst other things, his host's view that the magnetic virtue "seems to be produced by mechanical means".[23] Another casual allusion to "magnetick Effluvia" in the first edition of the *Opticks,* published ten years later, tells the same story.[24] Finally, in one of the famous 'aether' Queries that Newton added to the second English edition (1717) of this work, he uses the existence and activity of the magnetic effluvia, which he here again takes as beyond dispute, to justify certain assumptions he is making about the aether: "If any one would ask how a Medium can be so rare", he suggests, "let him tell me . . . how the Effluvia of a Magnet can be so rare and subtile, as to pass through a Plate of Glass without any Resistance or Diminution of their Force, and yet so potent as to turn a magnetick Needle beyond the Glass".[25]

Several scholars have argued in recent years that Newton's views concerning the possibility of finding mechanical explanations for the various forces he had introduced into his science changed during his lifetime. More precisely, they have seen him as much more sanguine about the aether as an explanatory mechanism in the early and late periods of his life than he was in between.[26] It by no means follows, however, that Newton's views about the cause of magnetism varied in the same sort of way, since in his thinking he seldom conflated the "magnetic effluvia" with his more general explanatory mechanisms. (Magnetism is, for example, a conspicuous absentee from the list set out in the "Scholium Generale" added to the 1713 edition of the *Principia* of phenomena which might be explained in terms of the "certain most subtle spirit" introduced at that point.[27]) If Newton's faith that the magnetic force was caused mechanically ever wavered, it did so only once, and that very briefly. In a draft "Conclusion" to his *Opticks* which he drew up during the early 1690s, Newton seems definitely to put the "attractive vertue" of the particles of a magnet in the same category as those other primitive unexplained forces, gravity and electricity: "The particles of bodies have certain spheres of activity", he here asserts, "within which they attract or shun one another. For ye attractive vertue of the whole magnet is composed of ye attractive vertues of all its particles & the like is to be understood of the attractive vertues of electrical & gravitating bodies".[28] Yet we have Gregory's testimony that within two years of Newton's composing this document, his conviction that the attractive power of the magnet was brought about by mechanical means was fully restored; and in the light of his subsequent allusions to "magnetic effluvia" in the various published versions of the *Opticks,* there seems no reason to doubt that thereafter he always continued in this belief. It is less certain that he continued to believe in the same circulatory mechanism as before, but what we know of the views of his closest disciples during this period suggests that he did.

Confirmation that we have understood Newton's scattered remarks about magnetism correctly, and a first indication of the views of his 'Newtonian' supporters on this question, come from the writings of one of Sir Isaac's earliest advocates, Dr Samuel Clarke. Clarke produced one of the best known of all expositions of Newtonian natural philosophy, in the unusual guise of notes appended to his series of Latin editions of Jacques Rohault's *Traité de physique,* a work that was itself thoroughly Cartesian in outlook.[29] He was, we know, in close contact with Newton. He it was who prepared the Latin edition of the *Opticks* and saw it through the press on behalf of the master. Later, he acted as spokesman for the Newtonian position in a famous exchange of letters with Leibniz, and it has been established that he did so in close consultation with Sir Isaac himself.[30] He would therefore have been well able to ensure that in his editions of Rohault's book, as elsewhere, the positions he adopted faithfully reflected the master's views.

In one of his notes, in responding to a typical attack by Rohault on the admissibility of the very notion of attraction in science, Clarke quoted extensively from the thirtyfirst Query of the *Opticks,* and included both the opening passage, in which magnetism is cited as an example of a force acting at a distance, and the passage, noted above, in which Newton mentions magnetism in close conjunction with his idea of "active principles" underlying various classes of natural phenomena.[31] Much more significant than this for our purposes, however, is Clarke's failure to respond at all to the whole chapter that Rohault devoted elsewhere in his book to an exposition of the Cartesian theory of magnetism. Far from attempting to rebut Descartes's theory, Clarke added but three notes to the entire chapter, and these merely unimportant glosses.[32] Given that elsewhere in the book Clarke seized every possible opportunity to advocate the Newtonian viewpoint, his failure to comment here is most conspicuous, and suggests that he accepted at least the general outline of the Cartesian account. And this in turn suggests that Newton did likewise.

Another influential early Newtonian was John Keill, lecturer in natural philosophy and later Savilian Professor of Astronomy in the University of Oxford. Like Clarke, Keill was in close personal contact with Newton, being deeply involved, in particular, in defending Newton's rights in his famous dispute with Leibniz over the invention of the calculus. In his *Introduction to natural philosophy* (London, 1720)—though not in the original Latin version of this—Keill revealed his acceptance of traditional ideas about magnetism by alluding in a manner very similar to Newton's to "magnetick Effluvia" that freely pervaded the pores of dense matter.[33]

John Harris, editor of the popular *Lexicon technicum* which appeared in two bulky volumes in 1704 (the same year as the first edition of Newton's *Opticks*) and 1710, was another early convert to Newtonianism. Geoffrey Bowles has pointed out that the chief influence on the 1704 volume was Robert Boyle, but by the time the second volume appeared, six years later, Harris had become a dyed-in-the-wool Newtonian.[34] It is therefore noteworthy that, so far as magnetism was concerned, Harris made no attempt in 1710 to retract his earlier opinion that "the Operation of the *Magnet* depends on the Flux of some fine Particles which go out at one Pole, then round about and in again at the other".[35] Apparently he saw no need to do so.

The case of the astronomer Edmond Halley, who worked closely with Newton for many years, is even more striking. In a very interesting paper that he read to the Royal Society in May 1716, Halley attempted to explain the aurora borealis in terms of magnetic effluvia emerging from the north pole of the Earth and either shining faintly by themselves or causing nearby matter to glow as a result of friction. That these effluvia circulated through and around the Earth he simply took for granted. The only aspect of Descartes's theory which he regarded as hypothetical was the Frenchman's notion that the effluvia consisted of "certain skrewed or *striate Particles*, adapted to the Pores they are to enter".[36]

Yet not all British scientists after Newton were prepared to commit themselves so openly to the traditional theory. Most authors of articles in the *Philosophical transactions* dealing with magnetism—men such as William Derham, the elder Hauksbee, Brook Taylor and George Graham—studiously avoided the question of what might be causing the phenomena they were studying. Rather, they used theoretically neutral terms like "magnetic virtue", "polarity", and "attractive virtue", and without elaborating further, confined themselves exclusively to the practicalities of magnetic research. An exception was Servington Savery, who did at least refer in passing to the cause of the "invisible Force" by which one loadstone acted on another. Less confident, however, than Halley, he described it as "unknown".[37]

Another exception was the arch-Newtonian J. T. Desaguliers, who alluded at one point to the Frenchman Du Fay's thoroughly Cartesian idea that magnetization involved "Threads or Beards fix'd at one End . . . laid all one way" in the iron.[38] Unfortunately, Desaguliers failed to indicate how these were supposed to account for the phenomena in question, or—of particular interest to us here—his opinion of Du Fay's hypothesis. Nor did he discuss the theory of magnetism at all in either the printed summary of his immensely popular course of lectures on experimental philosophy, or the treatise that grew out of the lectures.[39] In the absence of a definite lead from the master (and it must be remembered that some of the most important evidence described above concerning Newton's real views could not be cited by his followers), Desaguliers, like most British writers of the period, preferred not to commit himself in print. Yet the fact that he deliberately drew attention to Du Fay's ideas—he was, so far as I am aware, the first Englishman to mention them in print—and that while doing so he stressed the conformity of his experiments with those ideas, is highly suggestive. It would seem that, at the very least, Desaguliers thought Du Fay was on the right track. And that would indicate that Desaguliers, like Du Fay, accepted the basic Cartesian idea that magnetism was brought about by a circulating subtle fluid of some kind.

One British author who was prepared to commit himself in public was the instrument maker and itinerant scientific lecturer Benjamin Martin. In a brief discussion in his popular and professedly Newtonian work, *Philosophia Britannica,* Martin, too, revealed himself an orthodox Cartesian on this particular subject. Having defined magnetic poles as "points which emit the Magnetic Virtue", he went on to describe the traditional experiment in which iron filings were scattered on a sheet of paper held over a magnet as "a very curious Method of rendering visible the Directions which the Magnetic Effluvia take in going out of the Stone".[40]

Britain's foremost magnetic researcher in the first half of the eighteenth century was undoubtedly Dr Gowin Knight. He, like Martin, approached the theory of his subject in a traditional way, setting out his views at some length in a paper read to the Royal Society as late as 1747. "The magnetic Matter of a Loadstone", he maintained, "moves in a Stream from one Pole to the other internally, and is then carried back in curve Lines externally, till it arrives again at the Pole where it first entered, to be again admitted". As proof, he, like Martin, cited the patterns formed by iron filings on a sheet of paper held above a magnet. The Society, bastion of Newtonianism though it was, was far from expressing contempt for such ideas. On the contrary, less than three weeks before he presented his "valuable contribution",[41] Knight was presented with the Society's Copley Medal for 1747 as a token of the esteem in which his magnetic researches were held, and among the experiments which the President singled out on that occasion for special mention in his presentation address were some in which Knight displayed his ability to "turn into such a direction as he pleases the position of [a magnet's] Poles, and the stream of the Magnetic matter which is supposed to flow through them".[42] Like Knight's earlier papers, his latest contribution was promptly published in the *Philosophical transactions,* and the good doctor was encouraged to pursue his investigations further.[43]

Opinions such as Knight's clearly retained a wide measure of credibility in England, as they did elsewhere, even in the late 1740s. Should any doubts remain, we need only consult the article on magnetism in Ephraim Chambers's famous *Cyclopaedia,* the fifth edition of which appeared at about this time: "The opinion that principally prevails among the moderns", we read, "is that of Des Cartes".[44] There seem, indeed, to have been very few dissenters. As noted earlier, Savery may have been one, though this is far from proven. William Whiston certainly was, at least on some points, because in a work published in 1721 he argued somewhat egotistically that, since he could not imagine a mechanism to explain the magnetic attraction, its cause must be non-mechanical, like that of gravity. Even Whiston's position was equivocal, however, since he also maintained that the *directive* power of a magnet was due to the usual "magnetic effluvia" circulating continually around it.[45] Beyond this, to find a dissenting voice we must cross to Holland, where Musschenbroek argued loud and long against the circulation theory. His arguments were by no means as conclusive as he said they were, however, and in any event he was unable to devise a better explanation for the phenomena of magnetism than the one he rejected. Convinced that the traditional theory was false, he could do nothing better by way of an alternative than advocate a wait-and-see attitude, a suspension of belief in all hypotheses until more information could be obtained.[46] Without even the beginnings of a mathematical theory of magnetism akin to that which Newton had provided in the gravitational case, such a position was hardly likely to win him many followers. And so indeed it proved. In the magnetic case, even the most convinced Newtonians continued to uphold orthodox Cartesian opinions throughout the first half of the eighteenth century.

What conclusions are we to draw from this curious story? Here, we need do little more than recall what was said earlier. If, as we can now see to have been the case, members of our 'Newtonian' test group almost to a man adopted an approach towards the phenomena of magnetism the very opposite of that we expected of them, any confidence we may have had in the explanatory power of the term 'Newtonian' is necessarily undermined. Never again, I would suggest, can we extrapolate with any assurance from previously studied patterns of supposedly 'Newtonian' behaviour to as yet unstudied instances. Early eighteenth century English scientists, we conclude, did not, after all, invariably approach scientific problems from a doctrinaire standpoint dictated by their image of a Newtonian world in which the various natural phenomena were brought about by particles exerting unexplained forces of various kinds on each other across an intervening void. At least in certain cases such as the one studied here, they were much more open-minded than this, and were prepared to treat the case on its merits. Not for them the dogmatism of the paradigm or the Lakatosian research programme! The triumph of Newtonianism in certain areas did not carry with it the automatic rejection elsewhere of everything Cartesian.

REFERENCES

1. Brunet, *L'introduction des théories de Newton en France au XVIIIe siècle* (Paris, 1931).
2. I. B. Cohen, *Franklin and Newton* (Philadelphia, 1956).
3. T. S. Kuhn, *The structure of scientific revolutions* (Chicago, 1962; 2nd ed. 1970).
4. Brunet, *Les physiciens hollandais et la méthode expérimentale en France au XVIIIe siècle* (Paris, 1926).
5. J. A. Nollet, *Leçons de physique expérimentale*, tome i (Paris, 1743), 207, 260-1.
6. B. Franklin, *The autobiography, and selections from his other writings*, ed. H. W. Schneider (New York, 1952), 153.
7. R. W. Home, "Electricity in France in the post-Franklin era", *Actes du XIVe Congrès international d'histoire des sciences, Tokyo-Kyoto, 1974* (Tokyo, 1974-5), ii, 269-72.
8. *Cf.* Thomas L. Hankins, *Jean d'Alembert: Science and the Enlightenment* (Oxford, 1970), 4; and P. M. Heimann, "Newtonian natural philosophy and the scientific revolution", *History of science*, xi (1973), 1-7.
9. Cohen, *Franklin and Newton*, 179-82; Robert E. Schofield, *Mechanism and materialism: British natural philosophy in an Age of Reason* (Princeton, N.J., 1970), *passim*; Arnold Thackray, *Atoms and powers: An essay on Newtonian matter-theory and the development of chemistry* (Cambridge, Mass., 1970), chs ii, iii.
10. Descartes, *Principia philosophiae* (1644), pt IV, §§ 133-83; in his *Oeuvres*, ed. C. Adam and P. Tannery (Paris, 1897-1913), viii, 275-311. I have also used the French edition (1647), published in *Oeuvres*, vol. ix.
11. Newton, *Opticks, or a treatise of the reflections, refractions, inflections & colours of light* (New York, 1952; based on the 4th ed., London, 1730), 375-6.
12. Newton, *Mathematical principles of natural philosophy*, trans. Andrew Motte, rev. Florian Cajori (Berkeley/Los Angeles, 1934), 4.
13. *Ibid.*
14. *Ibid.*, 25-26.
15. *Ibid.*, 302.
16. *Ibid.*, 415.
17. Newton, *Opticks*, 373-4.
18. E.g., Newton, *Mathematical principles*, 414. Also *The correspondence of Isaac Newton*, ed. H. W. Turnbull *et al.* (Cambridge, 1959-), ii, 341-2; 360-2.
19. Newton, *Opticks*, 401.
19a. Cambridge University Library, Add MS. 3974, ff. 1-3. I am grateful to Dr D. T. Whiteside for advice concerning the dating of this item.
20. A. R. and M. B. Hall, *Unpublished scientific papers of Isaac Newton* (Cambridge, 1962), 220 (English trans., p. 228). R. S. Westfall argues (*Force in Newton's physics* (London, 1971), 409-10) that this document ought to be given a somewhat later dating, namely 1679, than that provided by the Halls. Joan L. Hawes has cited the passage quoted to support her view, similar so far as magnetism is concerned to that developed in this paper, that "when Newton writes of the forces of magnetism and electricity, he envisages a kind of spirit or effluvium which emanates from

the electric or magnetic body, performs its actions, and returns to the body" (Hawes, "Newton and the 'Electrical Attraction Unexcited'", *Annals of science*, xxiv (1968), 121-30, p. 122).

21. I. B. Cohen, ed., *Isaac Newton's papers and letters on natural philosophy* (Cambridge, 1958), 180, 183.
22. Cambridge University Library, Add. MS. 3970.3, f.473-4. I am indebted for this reference to Professor Westfall, who summarizes Newton's account in his *Force in Newton's physics* (ref. 20), 332.
23. Newton, *Correspondence* (ref. 18), iii, 335 (English trans., p. 338). Gregory himself seems to have been less certain than Newton at this point: *cf.* his "Notae in Newtoni Principia", Royal Society of London, MS. 210, f.79 (insert).
24. Newton, *Opticks* (1st ed., 1704), book II, 69. *Cf.* reprint ed. (New York, 1952), 267.
25. Newton, *Opticks* (reprint ed.), 353.
26. Henry Guerlac, "Newton's optical aether: His draft of a proposed addition to his *Opticks*", *Notes and records of the Royal Society of London*, xxii (1967), 45-57; J. E. McGuire, "Force, active principles, and Newton's invisible realm", *Ambix*, xv (1968), 154-208; Westfall, *Force in Newton's physics* (ref. 20), ch. VI (esp. 363-77, 391-5).
27. Newton, *Mathematical principles*, 547. Magnetism is included in a similar list drawn up by Newton on another occasion (Cambridge University Library, Add. MS. 3970, f. 241; quoted by Joan L. Hawes, "Newton's two electricities", *Annals of science*, xxvii (1971), 95-103, p. 97).
28. Cambridge University Library, Add. MS. 3970.3, f. 338r-338v; quoted by I. Bernard Cohen, "Hypotheses in Newton's philosophy", *Physics*, viii (1966), 163-83, pp. 179-80. Dr Whiteside kindly drew my attention to the significance of this document.
29. Rohault's *Traité* was first published in 1671. The first edition of Clarke's Latin translation appeared in 1697, and his notes were steadily expanded in succeeding editions. I have used the English translation of 1723, *Rohault's system of natural philosophy, illustrated with Dr. Samuel Clarke's notes taken mostly out of Sir Isaac Newton's philosophy* (London, 1723; reprinted New York, 1969). On Clarke's notes, see M. A. Hoskin, "'Mining All Within': Clarke's notes to Rohault's *Traité de physique*", *The Thomist*, xxiv (1962), 353-63.
30. A. Koyré and I. B. Cohen, "Newton and the Leibniz-Clarke correspondence", *Archives internationales d'histoire des sciences*, xv (1962), 63-126.
31. *Rohault's system* (ref. 29), i, 54-55.
32. *Ibid.*, ii, 163-87.
33. Keill, *Introduction to natural philosophy* (London, 1745), 67. I have used the 1745 edition of this work, but Schofield, *Mechanism and materialism* (ref. 9), 29, indicates that a similar passage appears in the earlier edition. The original Latin version was published in 1702; I have not consulted later editions of this to see whether the passage in question was incorporated in them.
34. Geoffrey Bowles, "John Harris and the powers of matter", *Ambix*, xxii (1975), 21-38.
35. Harris, *Lexicon technicum: or, an universal English dictionary of arts and sciences* (London, 1704), art. "Magnet". *Cf.* vol. ii (London, 1710), art. "Magnetism".

36. Halley, "An account of the late surprizing appearance of the lights seen in the air . . . with an attempt to explain the principal phaenomena thereof", *Philosophical transactions of the Royal Society of London,* xxix (1716), 406-28. See also J. Morton Briggs, Jr, "Aurora and Enlightenment: Eighteenth-century explanations of the aurora borealis", *Isis,* lviii (1967), 491-503, p. 492.
37. Savery, "Magnetical observations and experiments", *Philosophical transactions,* xxxvi (1729-30), 295-340.
38. Desaguliers, "An account of some magnetical experiments", *Philosophical transactions,* xl (1737-8), 385-7. *Cf.* Du Fay, *Histoire de l'Académie royale des Sciences [de Paris], avec les mémoires . . . ,* 1728, Mém. 355-69; 1730, Mém. 142-57; 1731, Mém. 417-32.
39. Desaguliers, *Physico-mechanical lectures: or, an account of what is explain'd and demonstrated in the course of mechanical and experimental philosophy* (London, 1717); and also his *Course of experimental philosophy* (2 vols, London, 1734-44).
40. Martin, *Philosophia Britannica: or, a new and comprehensive system of the Newtonian philosophy . . .* (Reading, 1747); 2nd ed. (London, 1759), i, 42-44.
41. Royal Society of London; Journal Book Copy, xix, 388.
42. *Ibid.,* 363. Extracts from the Journal Book are quoted by permission of the Society.
43. Gowin Knight, "Some further experiments relating to the general phenomena of magnetism", *Philosophical transactions,* xliv (1746-7), 665-72. Knight elaborated his ideas, and incorporated them into an all-inclusive theory of matter, in his well known *Attempt to demonstrate that all the phaenomena in nature may be explained by two simple active principles, attraction and repulsion . . . ,* published a few months later (London, 1748). Schofield has discussed Knight's views about magnetism in his *Mechanism and materialism* (ref. 9), 175-6, 179-81, but has failed to perceive either their Cartesian origins or the fact that such ideas were widely current in Knight's day.
44. Chambers, *Cyclopaedia: or, an universal dictionary of arts and sciences* (5th ed., London, 1741-3), art. "Magnetism".
45. Whiston, *The longitude and latitude found by the inclinatory or dipping needle . . .* (London, 1721), 11.
46. Musschenbroek, "De viribus magneticis", *Philosophical transactions,* xxxiii (1724-5), 370-8; *idem, Dissertatio physica experimentalis de magnete* (Leiden, 1729), 4-5, 57-67; *idem, Epitome elementorum physico-mathematicorum in usus academicos* (Leiden, 1726). In the case of the last of these works, I have used John Colson's English translation of the second edition, *The elements of natural philosophy* (London, 1744), and also Sigaud de la Fond's French rendering of the greatly-expanded final (1762) version of the Latin text, *Cours de physique expérimentale* (Paris, 1769). Musschenbroek took care to bring his text up to date from one edition to the next, but his opinions concerning the cause of magnetism seem not to have changed very much with the years.

V

Out of a Newtonian Straitjacket: Alternative Approaches to Eighteenth-Century Physical Science

Any history worthy of the name must deal in generalisations. Otherwise it will never rise above the level of mere antiquarianism and isolated individual biography. Yet an incautious reifying of our generalising concepts is as fraught with danger as an obstinate refusal to use such concepts at all. The literature of the history of science, like that of the history of ideas in general, is crowded, however, with 'isms'. Terms such as Aristotelianism, Platonism, mechanism, vitalism, Cartesianism, Newtonianism, are constantly employed in a manner that suggests that the things for which they stand have an unambiguous existence and an explanatory power of their own. But was Newton's work, for example, the climax of the seventeenth-century search for a truly 'mechanist' philosophy of nature, or was it, on the contrary, a decisive rejection of the very possibility of such a mechanist philosophy? Both positions have been maintained before now—and, I would argue, with equal validity. The answer one gives will depend on what one means by the term 'mechanist', and there are various ways in which this can, with perfect propriety, be used.

More to the point for my purposes here, however, is not so much the fact that very different concepts and widely differing implications may underlie a given 'ism', but that any such label is ultimately no more than a tool the historian uses in the process of unifying and systematising his knowledge. There is no ideal Platonic 'mechanist' to whose image individual scientists of the past approximated more or less closely; no more is there an ideal Newtonian, Cartesian, or whatever. On the contrary, individual historians must recognise that they make their own choices about how to use such terms, and also recognise that others may use them differently, or even use a different set. The test of whether the particular choice of analytical categories is a good one will be the extent to which it leads to an enhanced understanding of the historical period in question.

Discussions of eighteenth-century science are as replete with 'isms' as those of the science of any other period. The picture to which we have long since become accustomed depicts the science of the early part of the century as still generally Cartesian in outlook. The Newtonian alternative is seen as gradually gaining ground, first in Britain at the beginning of the century, then in Holland a decade or two later, and eventually in France (and, by implication, in the rest of Europe) some time in the late 1730s or early 1740s. For the remainder of the century, we are told, and well into the next, Newtonianism reigned supreme, with its influence spreading beyond science proper into every field of human intellectual endeavour.

I do not intend to concern myself here with any of the wider, extra-scientific ramifications of Newtonianism in eighteenth-century thought. Rather, I shall confine my attention to the science, and indeed the physical science alone, of the period. So far as this is concerned, I shall draw attention to certain difficulties in the received account. I shall argue that, if the traditional Cartesian-Newtonian dichotomy provides a reasonably satisfactory basis for understanding developments in the first part of the century, it becomes less and less helpful as the century proceeds. Finally, I shall suggest an alternative analysis of what happened in the second half of the century.

One of the most frequently cited studies of the triumph of Newtonianism is Pierre Brunet's classic *L'introduction des théories de Newton en France au XVIII siècle.* This work, published in 1931, was envisaged by its author as merely the first part of something much more extensive, and so it is perhaps unfair to criticise it for its rather limited horizons; yet it turned out to be the only part of the projected larger work ever to appear, and in fact it has been as a complete study in its own right that it has made its influence felt. In any event, it is clear that when Brunet wrote of 'les théories de Newton', he had almost exclusively in mind Newton's contributions to celestial mechanics as set out in the famous *Principia mathematica philosophiae naturalis* (1687) together with the philosophical implications of the unexplained attractions that lay at the heart of that great work. Brunet's book was concerned with the first stages of the French acceptance of Newtonian gravitational theory in place of the older

Cartesian doctrines of aetherial vortices. His theme is Voltaire's—Voltaire, who left the universe a plenum in Paris, and in London found it mostly empty space. Voltaire bulks large in Brunet's story, as does Maupertuis, who, like Voltaire, spent some time in England in the 1720s and whose *Discours sur la figure des astres* (1732) is often described as the first work to publicise the Newtonian point of view in France.

Brunet's is an outstanding piece of work, at once scholarly, thorough, and extremely penetrating in its analyses. It charted an expanse of territory scarcely even broached before then. Yet it is also, unquestionably, a child of its time, a pioneering effort from the days when the serious study of the history of science was in its infancy. Rupert Hall has made the point in a recent article that, to Brunet, 'the manuscript archives and correspondence collections remained closed',[1] his work being based instead on the published sources alone. Hall's own work as editor of the final volumes of Newton's correspondence has made him familiar with this previously untapped material, and has brought to light a network of lesser Newtonian influences in France in the earlier years of the century.

Hall also alludes to Brunet's 'special view of the problem' which 'led him to attach little importance even to some of the evidence in print'.[2] Brunet's emphasis on the ideas set out in the *Principia* and their reception in France is typical of an older generation in the history of science. Today, no one writing a book with Brunet's title would dare to limit his discussion to Newton's impact on celestial mechanics. Especially since the publication twenty years ago of I. Bernard Cohen's well known study, *Franklin and Newton*,[3] it has been widely recognised that Newton's work had as great an influence on experimental physics as it did on planetary astronomy, but that here the chief inspiration came from the master's second great book, his *Opticks*, rather than from the mathematically much more rigorous *Principia*. Before this, there was for many years a tendency to downgrade the importance of the *Opticks*, largely on the positivistic and antihistorical ground that, following the labours of Young and Fresnel at the beginning of the nineteenth century, Newton's optical studies could be seen to have been based on a false theory of the nature of light, and for this reason, to have held back the progress of optical science for a hundred years. Fortunately, the history of science has now long since outgrown the habit of making judgments on such an invidious basis. In the case of Newton's *Opticks*, Cohen

[1] A. R. Hall, 'Newton in France: a New View', *History of Science*, XIII (1975), 235.
[2] Ibid.
[3] Philadelphia, 1956.

has shown that the book gave rise to a vigorous tradition of experimental inquiry into precisely those properties of inorganic matter—its optical, thermal, chemical, electrical and magnetic properties—that had been bypassed in the idealised formalism of the *Principia*, where undifferentiated point masses were supposed to exert forces on each other in some unexplained manner at a distance.

In thus reinstating Newton's *Opticks*, however, Cohen in no way challenged the orthodox account, outlined above, concerning the reception of Newton's ideas in the various nations of western Europe. Rather, while widening the denotation of the term 'Newtonian', to give it a meaning that related to all the physical sciences and was no longer restricted to dynamics and celestial mechanics alone, Cohen still confined his analysis of this wider Newtonianism to the British scientific community of the day (including its colonial offshoots in North America) and, to a lesser extent, the Dutch. So, too, by and large, have more recent authors such as Arnold Thackray and R. E. Schofield, who have enriched our understanding of eighteenth-century experimental Newtonianism in a number of important ways.[4] Thackray's book includes, to be sure, a brief section entitled 'Newtonianism in France', and this ends with Newtonianism 'established' during the 1730s, as it was in Brunet's book.[5] Upon closer examination, however, we find that Thackray is not extending Cohen's *Opticks*-inspired experimental Newtonianism to France after all; on the contrary, his analysis is based upon those same few *Principia*-oriented works that Brunet drew upon, the books by Maupertuis, Voltaire and Madame du Châtelet. Elsewhere in his book, in fact, in another brief section on 'The French Situation', Thackray recognises that, at least so far as chemistry was concerned, 'the situation in France was by no means so simple' as he had suggested earlier, and that the ideas of the German scientist Stahl, though totally opposed to those of Newton in almost every respect, were at least as influential as his.[6]

Thackray's latter point is an important one, but he fails to trace out its wider implications for our understanding of Newtonianism in general. Let us grant that Newton's celestial mechanics, and his doctrine of universal attraction, came to be widely accepted in France after 1740. Let us likewise grant that at this period we can clearly discern, in Britain and her colonies, and perhaps in Holland, a tradition of experimental Newtonianism inspired by the *Opticks*.

[4] A. Thackray, *Atoms and Powers: An Essay on Newtonian Matter-Theory and the Development of Chemistry*, Cambridge, Mass., 1970; R. E. Schofield, *Mechanism and Materialism: British Natural Philosophy in an Age of Reason*, Princeton, N.J., 1970.

[5] Thackray, *Atoms and Powers*, pp. 95-101.

[6] Ibid., pp. 192-8.

It by no means follows that this latter brand of Newtonianism was also taken up by the French. Whether French experimental physics after 1740 was dominated by Newton, in the same way as French astronomy came to be, is an entirely different question from the one that Brunet set out to answer, and to answer it, a very different body of evidence needs to be studied.

Brunet himself took up the question in a work published several years before the one mentioned above.[7] In discussing the rise of 'la méthode expérimentale' in France, however, he carefully refrained from identifying experimentalism generally with experimental Newtonianism in particular. On the contrary, he was well aware not only that the Dutch professors Boerhaave, 'sGravesande and Musschenbroek, who played so large a part in his story, drew their inspiration from other sources besides Newton, but also that the leading French experimental physicists of the age—men such as Du Fay, Nollet, and Brisson—could not in any meaningful sense be described as 'Newtonian'. If anything, indeed, the latter group was positively anti-Newtonian in outlook. To cite only one example, Nollet, in his famous textbook of physics, carefully re-wrote Newton's laws of motion in order to avoid any possible imputation that they might refer to forces acting at a distance, and to make it clear that, whatever the astronomers might do, physicists were concerned with contact action alone.[8] If Nollet were a relatively minor figure in the French science of the period, this would scarcely be worth noticing; and there has, indeed, been a tendency among English-language historians of science to treat the man as a discredited, lonely and slightly foolish relic of an earlier age, after the dramatic successes of Benjamin Franklin's rival 'Newtonian' theory of electricity in the early 1750s. Franklin's own autobiography has been the most frequently cited source here: the Abbé, Franklin tells us, 'lived to see himself the last of his sect—except Mr B.—[i.e. Brisson] his *élève* and immediate disciple'.[9] Franklin, however, was deluding himself. I have shown elsewhere[10] that though Franklin's doctrines attracted an initial flurry of interest in France, they thereafter made very little headway for a period of twenty years or more. Contrary to what Franklin reported, Nollet died in 1770 as he had lived—the acknowledged master of French experimental physics.

[7] P. Brunet, *Les Physiciens hollandais et la méthode expérimentale en France au XVIIIe siècle*, Paris, 1926.

[8] See J.A. Nollet, *Leçons de physique expérimentale*, Tome I, Paris, 1743, 207, 260-1.

[9] B. Franklin, *The Autobiography, and Selections from His Other Writings*, ed. H. W. Schneider, New York, 1952, p. 153.

[10] R. W. Home, 'Electricity in France in the Post-Franklin Era', *Actes du XIVe congrès international d'histoire des sciences, Tokyo-Kyoto, 1974*, Tokyo, 1974-5, II, 269-72.

There is an important lesson to be learned from this story. Thomas Hankins has drawn attention in his excellent study of d'Alembert to the vast diversity of outlook that is included these days under the 'Newtonian' umbrella. 'One begins to wonder', Hankins writes, 'if the denotation "Newtonian" serves any real purpose.'[11] But our Nollet example takes us a step further than this; it shows that too free a use of the terms is not merely unhelpful, it can be positively misleading. It makes it clear that French experimental physics did *not* become unequivocally Newtonian at the same time as French astronomy and rational mechanics fell under the Englishman's spell. Hence, instead of speaking simply of *the* introduction of Newtonianism into France, we need, if we wish to continue speaking in these terms, to qualify more carefully what kind of Newtonianism it is that we are talking about in any given context.

From my own work on eighteenth-century ideas about magnetism, I have been forcefully reminded of a difficulty related to this one. When we label a period 'Newtonian', we are, I have suggested, inevitably tempted to think of it as Newtonian in every respect. When one looks at eighteenth-century writings about magnetism, however, one quickly becomes aware that until well into the 1760s, or perhaps even later, almost everyone, including the most ardent Newtonians, simply took for granted the broad outlines of the Cartesian theory of magnetism.[12] In other words, they supposed that any magnetised body was surrounded by a circulating stream of subtle aetherial matter passing axially through the body and returning through the external air. Magnetic effects were held to result from interactions between such streams and samples of iron they encountered, and were brought about by contact action alone, not by any Newtonian-style forces acting at a distance between magnetic particles or poles. (I might add that what I have said applies even to Newton himself. However, his published references to magnetism were so scanty that his eighteenth-century readers could not possibly have learned from them what his real views were. If anything, they suggested that he put magnetic forces in the same category as gravity, but in fact he did not do so.) The lesson of this story is plain, and it renders our usual modes of classification even more suspect. From it we learn that even if we are careful to specify what kind of Newtonianism we are talking about, we still cannot be confident that the scientists we class as 'Newtonians' will consistently adopt a

[11] T. L. Hankins, *Jean d'Alembert: Science and the Enlightenment*, Oxford, 1970, p. 4.

[12] The evidence upon which this assertion is based is set out in my forthcoming study, *Aepinus's Essay on the Theory of Electricity and Magnetism*.

Newtonian approach to every question that arises in their physical investigations.

There are other respects, too, in which the usual analysis of eighteenth-century science lets us down. To limit discussion to the followers of Descartes and Newton is to ignore a third major stream, the Leibnizians. The influence of the great German philosopher was most strongly felt, of course, east of the Rhine, being disseminated particularly in the voluminous writings of Christian Wolff. The 'new philosophy' enshrined in Wolff's works swept through the German universities like wild fire in the 1730s and 1740s, quickly supplanting the traditional Aristotelian approach that had been steadfastly adhered to until then. Thereafter it continued to provide the cultural *milieu* in which virtually all German scientists were educated, until it in turn was displaced towards the end of the century by the teachings of Immanuel Kant. But Leibnizianism made its presence felt in the scientific centres of western Europe as well. For example, the conversion to Leibniz's cause of one of the earliest French Newtonians, Madame du Châtelet, has been noted,[13] as have the distinctly Leibnizian overtones of the biological theory developed by another erstwhile spokesman for the Newtonian position, Maupertuis.[14]

It is with points such as these in mind that the suggestion has recently been made that we should interpret eighteenth-century science, not as a straightforward struggle between Cartesians and Newtonians, but as a three-way contest, including the followers of Leibniz too.[15] Yet increasing the number of our categories in this way carries with it its own special dangers. If we recognise the Leibnizians as a separate group, should we not do the same for the followers of Stahl? Furthermore, if we are going to distinguish experimental Newtonianism from its mathematically oriented cousin, we must surely acknowledge a similar dichotomy at least within the ranks of the Cartesians, and very likely within the Leibnizian school as well.[16] And why stop there? If increasing precision is the aim we should refine our categories further, and thus further increase their number. The resulting more differentiated taxonomy may well reflect more

[13] See P. Hazard, trans. J. L. May, *European Thought in the Eighteenth Century*, London, 1954, p. 304.

[14] Ibid.

[15] Y. Elkana 'Newtonianism in the Eighteenth Century', *British Journal for the Philosophy of Science*, XXII (1971), 297-306.

[16] That there were in the eighteenth century two separate schools of Cartesians has been pointed out by Elkana (ibid.). So far as the Leibnizians are concerned, we may distinguish (though perhaps less sharply) between the work of mathematicians such as Jacob Hermann and Johann Bernoulli I on the one hand, and experimental treatises such as the *Elementa physica* of G. E. Hamberger, on the other.

accurately the complexity of our period than do the simple patterns we normally read about. It will also, however, defeat the historian's primary aim in distinguishing schools, namely a generality of description that enhances historical understanding. It will tend to that very particularism our initial categories were designed to avoid.

A perennial difficulty in categorisation is what to do with the major figures of an era. The reason for this is obvious. The people in question are regarded as major figures because they add something of their own and diverge at some point from the standard patterns of thought of their day. This is why they stand out. Hence the very individuality that makes such a person noteworthy at the same time ensures that he will be hard to fit into any standard category.

Historians of science have sometimes taken advantage of this individuality to add to the dramatic impact of the story they are telling. Hankins, for example, draws attention to the rationalist and *a priori* character of d'Alembert's mechanics in a most effective way by arguing that this renowned exponent of the Newtonian point of view adopted an essentially Cartesian approach in his own major scientific works. Arnold Thackray has adopted similar tactics in his treatment of the great atomic theorist, John Dalton. Dalton has traditionally been portrayed as a staunch Newtonian in his attitude to science. Thackray, however, insists that Dalton's atomism was strictly anti-Newtonian in character, and he thereby successfully draws our attention to some important but previously under-rated eighteenth-century developments in matter theory. To point up an historical account in this way is, of course, a perfectly valid rhetorical device. The possibility of doing so ought not, however, to surprise us. Both d'Alembert and Dalton are major figures in their own right in the history of science, and hence we should not expect them to follow slavishly any text-book tradition, no matter how great the prestige of its ancestry. And this in turn should remind us that, if we grant that neither d'Alembert nor Dalton was an orthodox Newtonian, we ought not then to swing too far in the opposite direction and conclude that they were wholly anti-Newtonian in outlook. The truth, as usual, lies somewhere in between. As both Hankins and Thackray recognise, their subjects did draw upon the Newtonian heritage. They also drew upon various other traditions current in their day, however. As well, they contributed something of their own particular genius, and thereby finished up with a distinctively d'Alembertian or Daltonian mix.

The eighteenth-century figure who has perhaps been most ill-used by our habit of classifying people into 'isms' is Leonhard Euler. In part, no doubt, this is because no adequate study of Euler's manifold contributions to eighteenth-century science has yet been written. But in part it must be attributed to a too rigid use of our ordinary historical categories. Euler was arguably the greatest scientist of the eighteenth century. Only for special and largely rhetorical purposes, therefore, and only then with very substantial reservations, ought we to expect to be able to classify him meaningfully as Newtonian, or Cartesian, or Leibnizian. Yet in recent years, all three of these labels from the past have in one form or another been attached to him.[17] The fact that reputable historians approaching the man from different points of view have rated him so differently is, I submit, more than sufficient to establish my point. In matters intellectual, Euler was nobody's disciple, but was very much his own man. And the same could be said, to a greater or lesser extent, of each of the major thinkers of the period.

This brings me to my final criticism of our present tendency to discuss the history of science almost exclusively in terms of competing intellectual 'schools'. If we use terms such as 'Newtonian', 'Cartesian' or 'Leibnizian' to describe later eighteenth-century or early nineteenth-century science, there is some danger this may be taken to imply that the scientists whose work is being considered were preoccupied with the same issues with which people had been concerned many years earlier. But such was not the case. These terms have a clear meaning and relevance, and their most obvious application, in discussions of, say the Leibniz-Clarke correspondence, or Fontenelle's fulminations against the doctrine of attraction. But most of the points at issue in these early debates were by the mid-century either settled, or agreed by both sides to lie beyond the proper bounds of scientific inquiry. Where a decision had been made, the verdict had gone most often, though not always, to Newton, and this is why the succeeding age in science can with some justification be regarded as Newtonian. But its Newtonianism was tempered by infusions from various Cartesian, Leibnizian and other sources; and more importantly for my purposes here, no longer did the practising scientists of the period feel constrained constantly to expound and justify their philosophical position, or adopt a perpetual attitude of confrontation in their writings. Inevitably, their work was influenced by their familiarity with Newton's extraordinary achievements. Yet instead of worrying about past disputes, they looked forward to the resolution of a new and quite different set of scientific problems. We ought to follow their lead: that is, we ought to assess this later period

[17] E.g. Hankins, *Jean d'Alembert* p. 35; V. Boss, *Newton and Russia: The Early Influence, 1698-1796*, Cambridge, Mass., 1972, p. 136; Y. Elkana, The independence of Euler's thought has recently been stressed by Ronald Calinger, in 'Euler's "Letters to a Princess of Germany" as an Expression of his Mature Scientific Outlook', *Archive for History of Exact Sciences*, XV (1976), 211-33.

on its own merits, rather than in terms of the intellectual concerns of a previous age.

In seeking an alternative and more illuminating analysis of eighteenth-century physics, it is, I believe, more profitable to look at the actual practice of the scientists in question than to linger unduly upon their prefatory methodological pronouncements. When we do this, we soon discover two very different approaches to the subject. This new dichotomy is not based on the differing ideologies of competing intellectual schools—indeed, it cuts right across the usual Newtonian-Cartesian division—but reflects instead two different views on the role that mathematics ought to play in the practice of physics. On the one hand, we find a relatively small group of outstanding mathematicians—men such as the older Bernoullis, Clairaut, d'Alembert, Euler and Lagrange—who welded the Newtonian doctrine of attraction and Leibniz's version of the infinitesimal calculus into what came to be known as rational mechanics. Their approach, almost wholly mathematical and largely divorced from the realm of experiment, contrasts sharply with that of the practitioners of what the French called *physique expérimentale*, to whom the higher reaches of mathematics were a closed book, and in whose writings even the most elementary mathematics appeared only very rarely.

The dichotomy I am suggesting is not altogether new. It parallels to some extent the two strands of Newtonian influence discerned by Cohen, one inspired by the *Principia* and the other by the *Opticks*. More recently, Gillmor has pointed to the existence of the two traditions, and has argued that they became united in the work of the man who forms the subject of his book, Charles Augustin Coulomb.[18] And Kuhn has developed the general theme more fully within the context of a much broader over-view of the history of the physical sciences.[19] In addition, Silliman has given a clear statement of the dichotomy in the course of an important article on the work of the great optical theorist Augustin Fresnel.[20] Nowhere, however, has it yet been completely developed. Cohen's discussion is by far the most extensive, but it is concentrated almost exclusively on the experimental tradition, and then mainly on its British exponents. Gillmor is concerned only with what he sees as the end of the story, as is Silliman—though from a somewhat different point of view;

[18] See C. S. Gillmor, *Coulomb and the Evolution of Physics and Engineering in Eighteenth-Century France*, Princeton, 1971, *passim*.

[19] See T. S. Kuhn, 'Tradition mathématique et tradition expérimentale dans le développement de la physique', *Annales: économies, sociétés, civilisations*, XXX (1975), 975-98.

[20] See Robert H. Silliman, 'Fresnel and the Emergence of Physics as a Discipline', *Historical Studies in the Physical Sciences*, IV (1975), 137-62.

while Kuhn himself admits that he has done little more than outline a program for future research.[21]

D'Alembert was perhaps a little more extreme than most eighteenth-century mathematicians in his disdain for experimental inquiry, but for the most part he typifies the rational mechanics tradition very well. The astronomer Lalande once complained of him that he had 'never held a prism in his hand', while Daniel Bernoulli, who was considerably more experimentally minded than his famous father and uncle, felt that d'Alembert's researches on the winds were so lacking in experimental foundation that 'after one has read his paper, one knows no more about the winds than before'.[22] Because d'Alembert regarded mathematical demonstrations more highly than anything that depended on observation, he preferred the theoretically based lunar tables of his friend Pierre-Charles Le Monnier to the observationally based but much more accurate tables constructed by Tobias Mayer.[23] D'Alembert wanted to ground the science of mechanics—which was, he felt, better viewed as a branch of mathematics than as a branch of physics—upon purely *a priori* principles. He confessed himself unable to accept a merely contingent truth as a fundamental principle in his mechanics because this 'would ruin the certainty of Mechanics, and reduce it to being no more than an experimental Science'.[24] Even more typical of the rational mechanics tradition is the fact that whenever d'Alembert tackled a new problem, it quickly became apparent that his real interest lay in generating interesting new differential equations and exploring their mathematical complexities, not in possible applications of the mathematics to actual physical situations. It was of this that Daniel Bernoulli complained in regard to d'Alembert's work on the winds. The same charge—lack of empirical underpinning and control—could equally well have been levelled at most of his other works. It is the chief characteristic of eighteenth-century rational mechanics, in the sense that I am using the term here.

On the other side, mathematics was entirely foreign to *physique expérimentale*. Here, the subjects of inquiry were surface tension and capillarity, heat, magnetism, electricity and (in the work of people like Stephen Hales) parts of what we regard today as chemistry and plant and animal physiology. Eighteenth-century experimentalists frequently paid lip service to the desirability of obtaining quantitative data. In practice, they seldom progressed beyond furnishing endless tables of numerical results, the most conspicuous feature of which

[21] This is a subject I intend to take up shortly elsewhere.
[22] See Hankins, *Jean d'Alembert*, p. 3.
[23] See ibid., p. 37.
[24] J. d'Alembert, *Traité de dynamique*, 1743; rpt. Bruxelles, 1967, p. xii.

was the total absence of any concept of 'significant figures'. They had no idea how to progress from their tables to genuine functional relationships between physical variables.

Probably the first work in which an attempt was made consistently to apply mathematical techniques to the subject matter of *physique expérimentale* was Aepinus's *Tentamen theoriae electricitatis et magnetismi*, published in St Petersburg in 1759. The reactions to it of the leading electrical and magnetic investigators of the day is most revealing. As is well known, Aepinus took Franklin's theory of electricity as his starting point. But unlike Franklin and Franklin's other followers, Aepinus perceived that the foundations of the theory, as the American had left them, led, if rigorously applied, to absurd conclusions. In his book, Aepinus showed plainly that this was so, and also how, with the addition of a further postulate, the theory could be rendered mathematically coherent. With the aid of some elementary mathematics, he then proceeded to deduce some new, and in some cases most unexpected, consequences from the theory. More importantly, with the aid of some highly ingenious experiments he succeeded in confirming these predictions. In addition, he developed a theory of magnetism analogous to his newly improved theory of electricity, and showed that it, too, led consistently to verifiable predictions. His book was, in all, a most impressive piece of work, and later in the century it was hailed by a committee of the Paris Academy of Sciences, headed by Laplace, as constituting 'an epoch in the history of the sciences'.[25]

And yet the reception Aepinus's book was given when it first appeared was almost totally uncomprehending. To those investigators, mainly French, who still adhered to the theory of electricity espoused by the Abbé Nollet, Aepinus's work was wrongly based from the start. But the Franklinists of the day, whom one might expect to be more sympathetic to his approach, equally missed the point of what he was doing. They were happy to let Aepinus persuade them—up to a point at least—in cases where he was able to marshal direct experimental backing for his position; but they invariably considered each proposition or group of propositions in isolation. They failed to see that, in Aepinus's hands, these formed part of a coherent and mathematically consistent whole. Henry Cavendish was the sole exception in this regard. He did appreciate the rigour and systematic character of Aepinus's analysis, since these were features

[25] Extract from the *Registres* of the Academy, published in R. J. Haüy, *Exposition raisonnée de la théorie de l'électricité et du magnétisme, d'après les principes de Mr Æpinus*, Paris, 1787, p. [xxix].

he constantly strove for in his own work.[26] But his work fell on equally stony ground. The reaction to the additional and somewhat paradoxical postulate that Aepinus introduced into Franklin's theory is particularly illuminating. Aepinus showed that unless one supposed that particles of ordinary matter mutually repelled each other, it followed from Franklin's theory that two uncharged bodies ought to exert electrical forces on each other. (He also showed that, in the form in which his new postulate entered into the theory, it did not conflict with Newton's law of gravity.) Most Franklinists of the day, however, unused to requiring of their theories the degree of rigour demanded of Franklin's theory by Aepinus, ignored his argument altogether. Failing to perceive that Aepinus was exposing and then rectifying serious weaknesses in the received account, they tended to dismiss his treatise as an unnecessarily elaborate gloss on Franklin's work. This was certainly the attitude of the Harvard professor John Winthrop, for example, who complained that Aepinus had 'the right German prolixity, even to a degree of tediousness', before concluding that 'As to Electricity, I don't find that he has advanced much, if at all, beyond our excellent Countryman . . . & I scarcely recollect any thing he has said on this subject, but what I had learn'd before, & with greater pleasure, from Dr Franklin'.[27]

Others reacted with even more hostility to Aepinus's use of mathematical arguments in what had been, until then, the exclusive preserve of the experimentalists. Benjamin Wilson, the leading English electrical investigator of the day, wrote to him in patronising terms that 'The introducing of algebra in experimental philosophy, is very much laid aside with us, as few people understand it; and those who do, rather cho[o]se to avoid that close kind of attention: tho' I make no doubt but I dar[e] say you had a very good reason for making use of that method'.[28]

As a final illustration of my point, let me draw attention to an interesting comment, from the pen of the Franeker professor Jan Hendrik Van Swinden, on the structure of Aepinus's work as a whole. Van Swinden wished to attack all those, but particularly Aepinus and the Turinese academician Gian Francesco Cigna, who had claimed that worthwhile analogies could be drawn between various electrical and magnetic phenomena. In the course of this attack, he complained that Aepinus's arguments were much more difficult to rebut than those of Cigna. The latter's claims could, he

[26] See R. McCormmach, 'Henry Cavendish: a Study of Rational Empiricism in Eighteenth-Century Natural Philosophy', *Isis*, LX (1969), 293-306.

[27] J. Winthrop to E. Stiles, 21 February 1764: Stiles Papers, Beinecke Rare Book and Manuscript Library, Yale University.

[28] B. Wilson to F.U.T. Aepinus, n.d. (draft: British Library. Add. MS. 30094, f. 91).

said, be isolated and countered one by one, whereas in Aepinus's case, his principles were only suppositions: 'they cannot be refuted directly . . . it is only by examining the consequences, more or less remote, that this physicist has drawn from them, that one can evaluate them'.[29] To Van Swinden, Aepinus's approach was unsatisfactory because it smacked too much of the old *esprit de système* that experimental science had long since outgrown: 'I acknowledge', he said, 'that I am not attached to systems in my work on this subject, but to the facts alone'.[30] Yet Aepinus had in fact provided, in the course of his analysis, a great deal of experimental backing for his position. Hence Van Swinden's criticism amounts to a rejection of the *way* in which Aepinus used experimental evidence to support this theory. Van Swinden, like others in the eighteenth-century tradition of experimental physics, wanted to use only well attested experimental facts as his starting point in constructing his theories. Aepinus allowed himself much greater flexibility than this, requiring instead, as his empirical control, that his theories led to experimentally confirmable consequences.

The new analysis of eighteenth-century physics that I am proposing provides, I believe, a better understanding of the fate of Aepinus's book than do the traditional analyses of the period into different intellectual 'schools'. The mathematics that Aepinus used in his book was so elementary that it was of no interest whatever to eighteenth-century mathematicians working within the rational mechanics tradition. Yet, at the same time, it was, together with his insistence that physical theories be subjected to the mathematician's stricter standards of rigour and consistency, sufficient to render the work unintelligible to most experimentalists of the age. In other words, the reason why so little attention was paid to it at first was that it fell squarely between the two stools. It simply had no audience.

Let me conclude with a brief sketch of subsequent developments, as I see them. I have remarked already that Aepinus's book was the first work in which mathematics was applied consistently and profitably to the subject matter of *physique expérimentale.* Henry Cavendish's famous paper on the theory of electricity, published in the *Philosphical Transactions* of the Royal Society of London in 1771, was another early example of this approach. In the 1780s, works written in a similar style began to appear in increasing numbers, at first exclusively in Paris. Leading figures in the development of this new kind of physics were Coulomb, Laplace,

[29] J. H. Van Swinden, *Recueil de mémoires sur l'analogie de l'électricité et du magnétisme, couronnés & publies par l'Académie de Bavière*, La Haye, 1784, I, 157.
[30] Ibid., I, 246.

Monge, and the crystallographer Haüy. In contrast to rational mechanicians such as d'Alembert, for whom the physical situation provided little more than an excuse for a piece of pure mathematics, the new breed of mathematical physicist carefully restrained his mathematical imagination with a series of strict empirical controls. It was to this group, rather than to old-style electrical investigators such as Jean Baptiste Le Roy and Joseph Sigaud de la Fond, that Aepinus's book appealed. It was they who first realised the significance of what he had attempted, and who on this account hailed his book as marking an epoch in the history of the sciences.

The establishment of the École polytechnique in 1794 gave the new style of mathematical physics a secure institutional base. Within a few years, the first wave of outstanding young graduates from this school—men such as Biot, Poisson, Gay-Lussac, Malus and Fresnel—together with older men such as Laplace and Joseph Fourier who were associated with the school for a time, transformed the whole of traditional physics into the highly mathematised form it has retained ever since. In the years that followed, the subject grew in spectacular fashion. If the present paper has a moral, however, it is that to study these further developments, we shall inevitably need a different set of categories again from the one I have proposed here. Each age, in other words, must be treated on its own merits: categories suitable for one inevitably lead to distortions if applied to another.

Leonhard Euler's 'Anti-Newtonian' Theory of Light

Summary

Leonhard Euler was the leading eighteenth-century critic of Isaac Newton's projectile theory of light. Euler's main criticisms of Newton's views are surveyed, and also his alternative account according to which light is a wave motion propagated through the aether. Important changes are identified as having occurred between 1744 and 1746 in Euler's thinking about the way in which a wave such as he supposed light to be is propagated through a medium. Paradoxically, in view of Euler's overtly anti-Newtonian stand, these amount to Euler abandoning his early, Malebranchian notions about the physical basis of wave propagation, in favour of the ideas set out by Newton in Book II of his *Principia*.

Contents

1. Introduction

There have been many attempts to characterize Isaac Newton's influence on the science of the age that succeeded him. The story of science in the eighteenth century has frequently been represented as the inevitable triumph of a new and distinctively Newtonian style of thought over the older orthodoxies of Aristotle and Descartes. Newton's victory is generally held to have been complete by the early 1740s, whereafter his views are supposed to have reigned largely unchallenged for many years.[1] In recent times, however, a number of authors, myself included, have argued that it is misleading to see eighteenth-century science in such simple terms as these.[2] In the present paper, I wish to develop this point further by showing that labels like 'Newtonian' and 'anti-

[1] Pierre Brunet, *L'introduction des théories de Newton en France au XVIIIe siècle avant 1738* (Paris, 1931); I. B. Cohen, *Franklin and Newton: An Inquiry into Speculative Newtonian Experimental Science and Franklin's Work in Electricity as an Example thereof* (Philadelphia, 1956); Robert E. Schofield, *Mechanism and Materialism: British Natural Philosophy in an Age of Reason* (Princeton, 1970); Idem, 'An Evolutionary Taxonomy of Eighteenth-century Newtonianisms', *Studies in Eighteenth-Century Culture*, 7(1978), 175–192; Arnold Thackray, *Atoms and Powers: An Essay on Newtonian Matter-Theory and the Development of Chemistry* (Cambridge, Mass., 1970).

[2] Roger Hahn, *Laplace as a Newtonian Scientist* (Los Angeles, 1967); P. M. Heimann, 'Newtonian Natural Philosophy and the Scientific Revolution', *History of Science*, 11(1973), 1–7; R. W. Home, '"Newtonianism" and the Theory of the Magnet', *History of Science*, 15 (1977), 252–266; Idem, 'Out of a Newtonian Straightjacket: Alternative Approaches to Eighteenth-century Physical Science', in *Studies in the Eighteenth Century, IV*, edited by R. F. Brissenden and J. C. Eade (Canberra, 1979), pp. 235–49; Simon Schaffer, 'Natural Philosophy', in *The Ferment of Knowledge: Studies in the Historiography of Eighteenth-Century Science*, edited by G. S. Rousseau and Roy Porter (Cambridge, 1980), pp. 55–91.

Newtonian' cannot be applied unproblematically even in such a seemingly straightforward case as Leonhard Euler's widely publicized opposition to Newton's theory of light.

My concern, however, is not merely to point out the inadequacies of a currently popular way of characterizing eighteenth-century science. Though Euler's writings on the nature of light have never been studied systematically by historians, both the ideas he espoused and their subsequent fate are—or so I shall argue—of considerable interest in their own right. Furthermore, by paying proper heed to them, we arrive at a picture of eighteenth-century physics, and of eighteenth-century optics in particular, that differs in a number of significant respects from the standard account.

Leonhard Euler was one of the leading figures of eighteenth-century science. Chiefly remembered as a mathematician, he also wrote extensively on virtually all aspects of the science of physics. The sheer volume of his work is astonishing. The massive edition of his *Opera omnia* that has been appearing steadily since the early years of this century now extends to over seventy volumes.[3] His equally massive correspondence, which includes important exchanges of letters with many of the major figures among his scientific contemporaries, is a fundamental if still under-utilized resource for the historian of eighteenth-century science.[4] Through his letters and publications, and through the powers of patronage he enjoyed by virtue of the leading positions he long occupied in the Academies of Science in St Petersburg and Berlin, he wielded immense influence within the scientific community of his age.

As David Speiser has noted,[5] in matters of physics, Newton provided the starting point for most of Euler's work. In Euler's *Mechanica* (1735), one of his principal objectives was to translate the first two Books of Newton's *Principia* into the language of the calculus. Both here and in his many publications on mathematical astronomy, Euler accepted wholeheartedly, at least from a methodological point of view, Newton's highly contentious habit of reducing the motions of bodies to the actions of unexplained forces exerted on them by other bodies at a distance. In his optical

[3] *Leonhardi Euleri Opera omnia*, edited by F. Rudio *et al.* (imprint varies, 1911–).

[4] A detailed calendar of Euler's correspondence has been published as vol. 1, Series 4A of the *Opera omnia*. Sections of the correspondence itself have also been published as follows:

(*a*) *Opera omnia*, Ser. 4A, vol. 5 (correspondence with d'Alembert, Clairaut and Lagrange) and vol. 6 (correspondence with Maupertuis and Frederick II).

(*b*) *Correspondance mathématique et physique de quelques célèbres géomètres du XVIIIe siècle*, 2 vols, edited by P. H. Fuss (St Petersburg, 1843; reprinted New York, 1968); correspondence with Goldbach and Johann I Bernoulli and Daniel Bernoulli.

(*c*) G. Eneström, 'Der Briefwechsel zwischen Leonhard Euler and Johann I. Bernoulli', *Bibliotheca mathematica*, series 3, 4 (1903), 344–88; 5 (1904), 248–91; 6 (1905), 16–87. Also 'Der Briefwechsel zwischen Leonhard Euler and Daniel Bernoulli', ibid., 7 (1906–7), 126–56.

(*d*) *Eulers und Johann Heinrich Lamberts Briefwechsel*, edited by Karl Bopp (Abhandlungen der Preussischen Akademie der Wissenschaften, 1924, 2).

(*e*) *Leonard Eiler: pisma k uchenym*, edited by V. I. Smirnov *et al.* (Moscow/Leningrad, 1963); letters to Bailly, Bulfinger, Bonnet, Ehler, Wolff and others.

(*f*) *Leonhard Euler und Christian Goldbach: Briefwechsel 1729–1764*, edited by A. P. Yushkevich and E. Winter (Berlin, 1965).

(*g*) *Die Berliner und die Petersburger Akademie der Wissenschaften im Briefwechsel Leonhard Eulers*, edited by A. P. Yushkevich and E. Winter, 3 vols (Berlin, 1956–76).

(*h*) *The Euler–Mayer Correspondence (1751–1755): A New Perspective on Eighteenth-Century Advances in Lunar Theory*, edited by Eric G. Forbes (London, 1971); also in the original German, in *Istoriko-astronomicheskia issledovaniya*, 5 (1959), 271–444; 10 (1969), 285–310.

[5] D. Speiser. *Leonhardi Euleri Opera omnia*, series 3, vol. v (introduction to Euler's optical writings), p. LIII; see also Speiser's essay in *Leonhard Euler 1707–1783: Beiträge zu Leben und Werk*, edited by J. J. Burckhardt *et al.* (Basel: Birkhäuser, 1983), pp. 215–28.

writings, too, Euler cited Newton's work more frequently than that of any other investigator. In contrast to his attitude to the *Principia*, however, when Euler cited Newton's *Opticks*, he usually did so in order not to uphold but to oppose Newton's views.

Euler's opposition to Newton's theory of light is well known. Historians have tended, however, to treat his views as highly idiosyncratic, and have seldom paid them serious attention. Euler and Benjamin Franklin have frequently been portrayed as the only two major figures of eighteenth-century science to reject Newton's corpuscular or projectile theory of light and to uphold instead the idea that light was an undulation propagated through an all-encompassing aetherial medium. Neither the roots of their dissatisfaction with the projectile theory nor the details of the alternative they proposed have been systematically discussed. This neglect has been part of a wider neglect of eighteenth-century developments in optics. The period has tended to be depicted by historians as one devoid of either challenge or interest. No major new discoveries were made, so it is said, nor was optical theory advanced to any significant extent. Only recently has this attitude begun to change and, as it has done so, attention has begun at last to be paid to Euler's optical work. In particular, Casper Hakfoort has made an extensive study of Euler's theory of light and its reception in eighteenth-century Germany in which he has shown that, at least in some parts of Europe, the Newtonian view did not hold undisputed sway.[6] Also, Geoffrey Cantor has indicated that Euler's ideas were frequently cited by eighteenth-century British writers on optics, even if in the main they generally disagreed with them; and he has also linked Euler's work with later developments in optical theory by showing that Thomas Young drew heavily upon Euler's ideas in presenting his own well-known version of the wave theory in the years around 1800.[7]

I would go somewhat further than this and say that, east of the Rhine, Euler's became possibly the dominant influence in optical thinking during the second half of the eighteenth century. Hakfoort has already documented the support provided for Euler's position by several influential German writers of the period, notably J. C. P. Erxleben in his widely used textbook of general physics,[8] J. H. Lambert,[9] and G. S. Klügel, who prepared the German edition of Joseph Priestley's *History... of vision, light and colours*, and added no fewer than four pages of his own to Priestley's brief and inadequate account of Euler's opinions.[10] Hakfoort estimates, indeed, that between 1750 and 1790, about two-thirds of German authors publishing on optics accepted Euler's views.[11] Euler's correspondence, now largely preserved in Leningrad, confirms that there was wide-ranging support for his views from members of the German scientific community. He was thus very far from being the isolated figure, so far as his theory of light was concerned, that he appears in the general literature on the subject.

The myth that he was isolated is partly a result, I would suggest, of the long-standing dominance of the history of science literature by English and French-language

[6] C. Hakfoort, *Optica in de eeuw van Euler: Opvattingen over de natuur van het licht, 1700–1795* (Amsterdam: Rodopi, 1986); also 'Nicolas Béguelin and his Search for a Crucial Experiment on the Nature of Light (1772)', *Annals of Science*, 39 (1982), 297–310.

[7] G. Cantor, *Optics after Newton: Theories of Light in Britain and Ireland, 1704–1840* (Manchester, 1983), especially chapter 5.

[8] J. C. P. Erxleben, *Anfangsgründe der Naturlehre*, 6th edn (Göttingen, 1794), p. 313.

[9] J. H. Lambert, *Photometria* (Augsburg, 1760), pp. 2–3.

[10] J. Priestley, *Geschichte und gegenwärtiger Zustand der Optik*, translated by G. S. Klügel (Leipzig, 1775–76), II, 261–4. Despite his publicizing Euler's views in this way, Klügel himself remained agnostic so far as optical theory was concerned.

[11] Hakfoort (footnote 6), chapter 5.

scholars who in their studies of eighteenth-century science have failed to take sufficient note of developments in countries beyond the Atlantic seaboard. Recent work on the history of electricity and magnetism, including some of my own, has shown how misleading any account is that neglects developments in Germany, Sweden and St Petersburg.[12] The same will be found to apply, I believe, to the history of optics, once the subject is sufficiently attended to. In fact, as indicated already, Euler's writings on the nature of light richly repay a closer scrutiny. To be sure, we do not find here any sudden or revolutionary advance in scientific comprehension, but instead a gradual strengthening of understanding in a situation where there was no definitive argument for or against the theory in question. Yet this manner of progressing is, I believe, much more representative of the way science develops than the supposed revolutionary leap, and thus no less worthy of study; while the final outcome in this case, the elucidation of what is involved in wave propagation in general, demands our attention because it was one of the major achievements of eighteenth-century science and, indeed, a fundamental step in the evolution of modern physical science. Finally, as I suggested at the outset, the story raises interesting questions about some still commonly used historical labels.

The principal sources so far as Euler's optics is concerned are, of course, his published writings, most conveniently available in the superb *Opera Omnia*. In addition, sections of his correspondence have been published, together with a calendar of the whole;[13] but there are many other letters in the Leningrad branch of the Archives Department of the Soviet Academy of Sciences, along with other important unpublished papers of Euler's. In what follows, I shall be drawing upon some of this as yet untapped material. From it one can begin, at least, to reconstruct the course of development of Euler's ideas about the nature of light. I do not propose here, however, to explore all the ramifications of Euler's thinking on this subject. Rather, I shall focus upon just one component of the story—one that will, I hope, shed some light of its own on eighteenth-century styles of physical thought. The particular question I shall address is Euler's conception of the underlying physical process involved in the propagation of a wave through an elastic aether.

2. Euler's critique of Newton

Euler went to considerable lengths on several occasions to emphasize the anti-Newtonian character of his theory of light. The most definitive statement of his position appeared in his essay 'Nova theoria lucis et colorum', published in 1746, in the first chapter of which he took up and systematically refuted the principal arguments presented by Newton in favour of the projectile theory and against the vibrations-in-a-medium alternative. As Euler observed, one of Newton's strongest arguments was his denial of the very existence of an aetherial medium capable of transmitting pulses in the manner proposed by those whose view he was opposing. As Euler put it:

> The chief argument for proving the true emanation of light from shining bodies is usually based upon the incongruity of a space everywhere full... If light is

[12] J. L. Heilbron, 'Franklin, Haller and Franklinist History', *Isis*, 68 (1977), 539–49; idem, *Electricity in the 17th and 18th Centuries: A Study of Early Modern Physics* (Berkeley, 1979); Willem D. Hackmann, *Electricity from Glass: The History of the Frictional Electrical Machine, 1600–1850* (Alphen aan den Rijn, 1978); *Aepinus's Essay on the Theory of Electricity and Magnetism*, edited by R. W. Home (Princeton, 1979).

[13] See footnote 2.

> propagated to us from the sun and the fixed stars in the same way as sound, the universal space between us and the sun and the fixed stars must be replete with some subtle matter. This matter would interpose some resistance to the motion of the planets and comets; so since Newton was unable to discover any retardation in these bodies, he was forced to conclude that the heavens were devoid of all resistance and were therefore empty.[14]

But, argued Euler, Newton and his followers were inconsistent here, for even as they denied the existence of the aether they postulated streams of light corpuscles flowing from the sun and fixed stars in all directions:

> The universal space must be continuously brim full of effluvia of this sort, since it is impossible to conceive of a point anywhere which is not penetrated ceaselessly by an infinite multitude of rays from the sun and the fixed stars.[15]

The choice was not, therefore, as Newton would have had it, between a space that is effectively empty and one that is full, but between a space filled with a chaos of light corpuscles travelling at enormous speeds in every direction, and a space filled with the more tranquil and orderly medium proposed by the vibration theorists.

Euler also took up Newton's other great objection to the vibration theory, namely that it was incompatible with the observed rectilinear propagation of light and the formation of shadows. According to the vibration theory, the propagation of light through the aether was closely analogous to that of sound through air. Yet sound appears to be propagated in all directions from a source and to bend around objects in its path. As Newton had put it:

> sounds...are heard through a mountain interposed; and, if they come into a chamber through the window, dilate themselves into all the parts of the room, and are heard in every corner; and not as reflected from the opposite walls, but directly propagated from the window, as far as our sense can judge.[16]

Euler's response was to deny the validity of the analogy between such situations and the casting of a shadow by an opaque body interposed in a beam of light. Much of the sound entering the room did so, he said, not through the hole but through the surrounding walls; only if the walls were totally sound-proof—something he regarded as impossible of achievement—could one discover whether the kind of objection proposed by Newton held good. Euler was confident that, could such an experiment be arranged,

> sound would arouse the sense of hearing only in the direction from which it had come, and...the result would be completely similar to that which is usually observed when a ray of light is introduced through a hole into a camera obscura... For we see that, in the open air, sound is propagated from a sonorous body on all sides in straight lines and is never deflected to the sides.[17]

[14] Euler, 'Nova theoria lucis et colorum', in *Leonhardi Euleri Opera omnia*, series 3, vol. v, 1–45 (pp. 2–3). I am currently preparing an English edition of this work for publication.

[15] Ibid., p. 3.

[16] Isaac Newton, *Mathematical Principles of Natural Philosophy*, translated by Andrew Motte, revised by Florian Cajori (Berkeley/Los Angeles, 1934), p. 370; cf. Newton, *Opticks* (New York, 1952), p. 363, for an elaboration of the 'mountain interposed' argument.

[17] Euler (footnote 14), p. 5.

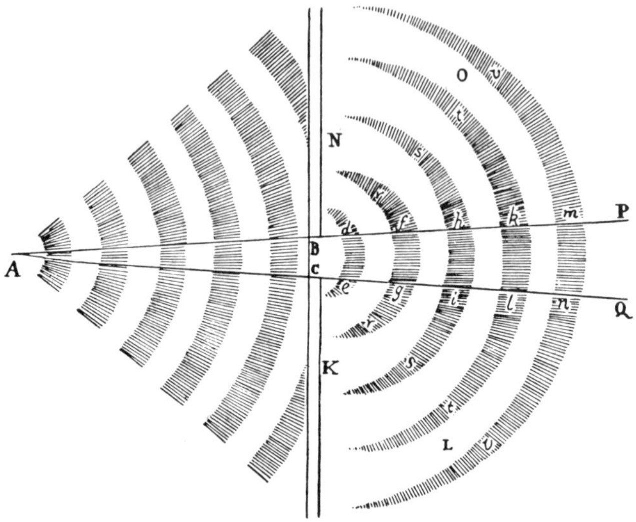

Figure 1. Newton on the passage of a wave motion through a hole (*Mathematical Principles of Natural Philosophy*, p. 369).

Euler did not respond directly to Newton's remark about sounds being 'heard through a mountain interposed', but he doubtless would have maintained that the mountain itself would have been transmitting the sound to the hearer on the other side.

Yet Euler's approach at this point merely evades and does not at all overcome Newton's objection, the basis of which is set out as Proposition XLII, Book II of the *Principia*: 'All motion propagated through a fluid diverges from a rectilinear progress into the unmoved spaces'.[18] Newton's accompanying diagram (Figure 1) makes clear the implication he wished to draw from the proposition: when a wave emanating from a source A passes through a hole BC, the hole itself should function as a new centre as the wave spreads into the medium beyond it, and hence no shadow should be cast. The heart of the objection, in other words, was not the success or otherwise of a hypothetical experiment in a sound-proofed room, but a seeming contradiction between the known behaviour of fluids and what was needed in order to generate a viable account of light as a wave motion in an all-encompassing medium. No more than any other fluid-medium theorist could Euler ignore this problem, and later in his essay he did in fact address it more directly.

Newton's notion that a section of an advancing wavefront, such as that passing through the hole BC in his illustration, could act as centre for a new hemispherical wave, was stated much more generally by Huygens in his *Traité de lumière* (1690) when he proposed his famous construction whereby every point in the wavefront is treated as

[18] Newton, *Mathematical Principles* (footnote 16), p. 369.

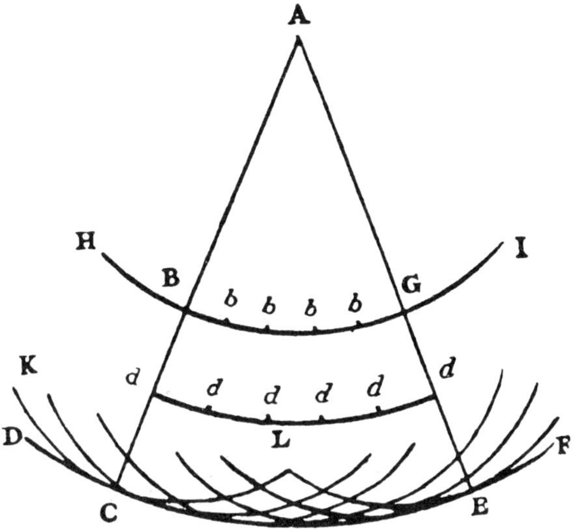

Figure 2. Huygens on the formation of shadows (*Treatise on Light*, p. 21).

the source of a new wave. Huygens did not, however, accept Newton's conclusion concerning the impossibility of shadows, but argued that (Figure 2):

> if, for example, there were an opening BG, limited by opaque bodies BH, GI, the wave of light which issues from the point A will always be terminated by the straight lines AC, AE..., the parts of the partial waves which spread outside the space ACE being too feeble to produce light there.[19]

To be sure, Huygens' argument was illicit, as was apparent to, for example, Johann II Bernoulli.[20] Nevertheless, his sustaining it at all is a salutory reminder of the deep conceptual difficulties confronting those who were trying at this period to construct an adequate theory of wave motions in elastic media. In particular, it reminds us that Newton's position on this matter was by no means unassailable in the eighteenth century. (It was, of course, eventually overthrown by Fresnel.)

Euler does not explicitly refer to Huygens, but he must have known his work either directly or through his friend, Bernoulli; and at two different points in his essay he formulates arguments that work to undercut Huygens' construction as well as that of Newton. 'Once a pulse has been formed', he says,

> it moves forward in a straight line if the medium is uniform ... The transmission of pulses arises from an agitation of the particles of the elastic medium in which the

[19] Christiaan Huygens, *Treatise on Light*, translated by Silvanus P. Thompson (New York, 1962), p. 21.

[20] Jean Bernoulli, 'Recherches physiques et géométriques sur la question: comment se fait la propagation de la lumière' (Paris, 1736), in *Recueil des pièces qui ont remporté les prix de l'Académie Royale des Sciences*, 3 (1734–37), separately paginated. Huygens lacked both the notion of periodicity—his theory involved the propagation of irregular pulses rather than of waves, properly so-called—and a principle of interference, so was unable to develop a well-formed theory of destructive interference that would account for the formation of shadows.

> pulse is situated. *Since the agitation at all times tends in a definite direction, it gives the pulse a motion towards that same region.* Therefore we completely eliminate the previously mentioned difficulty of Newton's...[21]

There is, he notes, an added difficulty: 'According to [Newton's] reasoning a pulse should also revert backwards if it could deviate sideways; but this is opposed to experience as well as to theory'.[22] Later, he repeats and elaborates the main argument:

> Since the pulses are everywhere progressing at the same speed they will be curved into arcs of circles whose centre is at the point *A* [Figure 3]... The direction of the rays must be estimated from the position of the pulses, such that if the pulse at *G* affects the organ of sight, the ray of light is judged to have arrived along the direction normal to the pulse *GG*. Therefore rays exist in the aether only in as far as straight lines normal to the pulses are conceived there; but *because the pulses act and progress along this direction, from that the effect of the rays is perceived and understood.*[23]

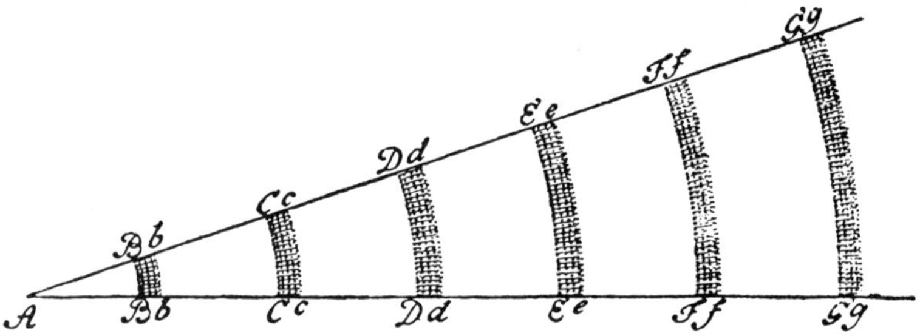

Figure 3. Euler on the spreading of waves ('Nova theoria', p. 22).

Euler also draws attention to a number of 'enormous difficulties' confronting Newton's projectile theory. The three arguments he presents are very familiar. First, if light rays are composed of streams of material corpuscles, either the mass of the source must in a finite time be sensibly diminished, or the density of the rays must be improbably small. Euler calculates that, in order for the Sun's mass not to have sensibly altered in the past five thousand years (as it appears not to have done), the solar rays in the vicinity of the Earth must be 10^{18} times less dense than the Sun itself. Secondly, Euler says, 'it seems absolutely inexplicable how two or more rays from different regions meeting each other at such incredible speed do not disturb each other's motion'. Finally,

> since rays are observed to be capable of passing in all directions through pellucid bodies, these bodies would have to be perforated in straight lines in every direction, so that there could be no straight line conceived of in them which was not at the same time a passage of this kind. Hence the matter could not find any

[21] Euler, 'Nova theoria' (footnote 14), p. 18 (my emphasis). The sentence in italics amounts to a denial of the validity of Huygens' construction.

[22] Ibid.

[23] Ibid., pp. 22–3 (my emphasis).

place where it might be located to constitute these bodies and it could not in any way cohere with itself; for however the matter of these bodies is conceived of as being disposed, it will be completely impossible for passages to exist and spread out in absolutely all directions.[24]

3. Euler's alternative

The extent to which Euler cast his theory of light in an anti-Newtonian mould is thus very evident. Throughout his essay and, indeed, elsewhere in his writings on optics, he constantly presented himself in the guise of challenger to prevailing orthodox opinion as laid down by the great Sir Isaac. Nevertheless—or so I wish to argue—even in the context of his theory of light, it is not so straightforward as it might at first appear to classify Euler unambiguously as an anti-Newtonian. In particular, problems arise when we turn to the details of his account of the way a wave is propagated through a medium.

So far as this question is concerned, the most significant source is a paper that Euler presented to the Berlin Academy of Sciences on 6 February 1744. A summary appeared in the *Histoire* of the Academy for that year,[25] but the paper was never published in full. The text survives, however, under the title 'Pensées sur la lumière et les couleurs', in the archives in Leningrad.[26] In this work, Euler set out for the first time in public most of the themes that were to characterize the later essay 'Nova theoria lucis et colorum' that is usually regarded as the definitive statement of his ideas. It appears, however, that during the two years that separated these two presentations, Euler's views on the underlying physical mechanism altered considerably; and it is upon this change that I intend to focus.

The key to Euler's theory of light was, as we have seen, his idea that a close analogy existed between the propagation of light and of sound. Light required a different medium from sound, of course, but that presented no difficulty since Euler simply assumed the existence of an all-encompassing elastic aether that could act as a carrier: there is, indeed, a hint in the paper we are considering that it was his prior belief in the existence of the aether that led him to oppose Newton's theory of light in the first place.[27] Euler's understanding of how sound was propagated through air therefore became his model for explaining, in detail, the transmission of light through the aether. That this could give rise to any kind of question is at first glance surprising because Newton, years before, had given an apparently unexceptionable account in Section VIII, Book II of his *Principia* of what was going on in the propagation of a sound wave. Yet Euler's account initially differs significantly from Newton's; only later does he come to adopt the Englishman's approach.

According to Newton, when a body is set trembling while immersed in an elastic medium,

> The parts of the tremulous body, alternately going and returning, do in going urge and drive before them those parts of the medium that lie nearest, and by that impulse compress and condense them; and in returning suffer those compressed parts to recede again, and expand themselves. Therefore the parts of the medium

[24] Ibid., pp. 6–7.

[25] *Mémoires de l'Académie des Sciences et des Belles-Lettres* [*de Berlin*], 1 (1745), Histoire, pp. 17–24.

[26] Archives Department of the Academy of Sciences of the U.S.S.R., Leningrad Branch, fond 136, op. 1, no. 243.

[27] Loc. cit., § 5.

> that lie nearest to the tremulous body move to and fro by turns, in like manner as the parts of the tremulous body itself do; ... these parts, being agitated by like tremors, will in their turn agitate others next to themselves; and these others, agitated in like manner, will agitate those that lie beyond them, and so on *in infinitum.* And in the same manner as the first parts of the medium were condensed in going, and relaxed in returning, so will the other parts be condensed every time they go, and expand themselves every time they return.[28]

As the discussion proceeds, Newton makes it clear that the 'parts' (*partes*) of which he speaks are to be understood not as the ultimate corpuscles of which the medium is composed, but as small but finite physical elements of it. He in fact determines the elastic restoring force acting on a 'physical short line' (*lineola physica*) or 'linear part' (*pars linearis*) of the medium and shows it to be proportional to the displacement of that part from its mean position, so that the motion of the part will be simple harmonic.[29] At a crucial stage in the argument, he assumes that the elastic restoring force of each part is inversely proportional to its compression; that is, he assumes that Boyle's law applies. This of course leaves the source of the restoring force unspecified; but everyone knows that, elsewhere in the *Principia*, Newton showed that Boyle's law and the elasticity determined by it could be accounted for by assuming that the particles of which the medium was composed mutually repelled each other.[30]

Euler's approach, in his 1744 presentation, is completely different. Here, each corpuscle of the medium is conceived of as an elastic globule, and the compressions and rarefactions are thought of as occurring in the individual corpuscles themselves, whole corpuscles not being bodily displaced. When a vibrating body presses on the first molecule in a line, Euler explains, this is 'reduced into a smaller space'; it will therefore 'have a greater spring [*ressort*] than the others, and consequently it will restore itself to its natural state. By this restitution it will compress the molecule next to it, and this, as it restores itself, the third, and so on'.[31] Euler's account here is, in fact, entirely consonant with the description of the structure of the air that he had given in his influential essay *De sono*, published some years earlier, in 1727. There, too, it was a matter of an array of elastic globules transmitting an impulse by being squeezed and then, as they expanded by virtue of their innate elasticity, squeezing in turn those next to them:

> If we imagine, in a series of airy globules, one more compressed than the others, it will, *sui juris*, dilate by impelling the surrounding globules in all directions and producing compression in them... Sound arises, when the same globule suffers repeated compressions by an outside force, finite intervals having been interposed; that is, to excite sound it is required that the same globule alternately contracts and relaxes.[32]

[28] Newton, *Mathematical Principles* (footnote 16), p. 371.

[29] Ibid., pp. 375–8. Cf. *Isaac Newton's Philosophiae Naturalis Principia Mathematica: the third edition (1726) with variant readings*, edited by A. Koyré and I. B. Cohen (Cambridge, 1972), I, 515–6.

[30] Newton, *Mathematical Principles* (footnote 16), pp. 300–1.

[31] Euler (footnote 26), § 4: '[cette molecule] aura un ressort plus grand que les autres, et ... par consequent elle se remettra dans son état naturel. Par cette restitution elle comprimera la molecule voisine, et celle cy en se remettant la troisieme, et ainsi de suite ...'.

[32] Euler, *Opera omnia*, series 3, vol. 1, p. 184.

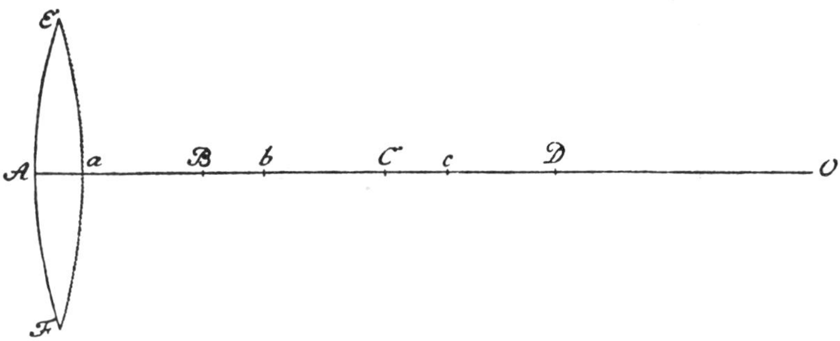

Figure 4. Euler on the generation of pulses ('Nova theoria', p. 10).

The contrast with the ideas Euler expounded in the 'Nova theoria' of 1746 is remarkable. In this later work, his account is indistinguishable from Newton's. He supposes a vibrating cord *EAF* (Figure 4) fixed at *E* and *F* and initially drawn back to *A*:

> When the cord is released at the point *A*, as it moves forward to *EaF*, the point *A* will move the adjacent particle of the medium forward to *a* and will condense the particles of the medium lying beyond *a*, because the particles that were previously distributed along the line *AO* are now reduced to the smaller space *aO*... If the cord remains in the position *EaF* or recedes, since the particle *a* has a greater elasticity than the following ones, it will expand and will condense these by pushing them... When the particle *B* is pushed, however, it will not only compress, it will also move forward through the space *Bb*...[33]

Euler does not stop to tell us whether the elastic 'particles' (*particulae*) of which he speaks are the actual molecules of the medium, as in his previous analysis, or are, like Newton's, small but finite elements of it. The point no longer matters to him (though in fact it seems certain that he is now following Newton's usage[34]). The crucial thing is that Euler now shares with Newton the notion that these particles undergo a simple harmonic oscillation about some median position; and on this basis he is able to give his first detailed mathematical analysis of the propagation of a wave. However, he evidently regards Newton's use of Boyle's law as no more than an artifice. That the oscillations the particles undergo is simple harmonic is, for him, no more than a simplifying assumption.

Euler's discussion here makes clearer the basis of his confidence that rectilinear propagation and the formation of shadows poses no threat to his theory. To be sure, light radiates in every direction from a source, so that, as shown in Figure 3, the advancing wavefront is spherically symmetrical about the source. But for Euler this does not lead to any Huygens-like analysis of the propagation process. As Figure 4 suggests and the accompanying discussion makes plain, Euler conceives of the longitudinal wave generated by a vibrator as a *linear* effect. His entire mathematical analysis is founded on this view of things. Most surprisingly in view of the fact that he

[33] Euler, 'Nova theoria' (footnote 14), p. 10.
[34] Ibid., p. 14.

talks constantly about the varying density (and hence elastic force) of the medium as a pulse or wave passes through it, for Euler it is *only* the actual generation of that pulse or wave that is a three-dimensional phenomenon; the subsequent process of propagation through the medium, in any direction that one cares to name, is not:

> to excite pulses in the aether, the aether must be struck somewhere by some force and be disturbed from its state of equilibrium at that point. For the removal of equilibrium will give rise to a pulse which will propagate itself further in any direction whatsoever, *and doubtless in the same fashion in all directions as we have shown ought to happen in one.*[35]

The difference between Euler's physical picture of wave propagation in an elastic medium as set out in his *De sono* and that presented in the 'Nova theoria lucis et colorum' has been remarked by Clifford Truesdell,[36] but he has not pursued the point. The discovery of the unpublished 'Pensées' enables us to pin down much more precisely this important shift in Euler's thinking which dates, we can now see, from after the presentation of the 'Pensées' to the Berlin Academy in early 1744, but before the publication of the 'Nova theoria' in 1746. It would be of great interest to know more about the circumstances surrounding the shift, and in this regard a systematic study of Euler's other papers and especially his correspondence from the period we have now identified might prove richly rewarding.

Until now, Euler's early theory of elastic globules has been known only in connection with his ideas on the propagation of sound. By linking the theory also with his ideas about light, the 'Pensées' suggest an unexpected source, namely the work of Malebranche. He, it will be recalled, had substituted for the rigid spherical globules that made up the bulk of Descartes' cosmos an array of tiny vortices. These were, by their very nature, elastic, and were therefore able to transmit 'pressure vibrations' (*vibrations de pression*) that in Malebranche's view constituted light. Following the analogy with sound, he had also proposed that the different colours of light derived from pressure vibrations of different frequencies.[37] There are thus striking parallels between his opinions and those of Euler as revealed in the 'Pensées'; and while here, as elsewhere, mere parallelism does not establish a filial connection, it may well be significant that, except for Newton, the author who figures most prominently in Euler's exposition as an authority on optical theory is Malebranche's acolyte, Dourtous de Mairan.[38] Perhaps more conclusive is the fact that Euler's friend, Johann II Bernoulli, who likewise invoked a medium composed of elastic globules, explicitly cited Malebranche's work in this regard.[39]

Whether or not a link with Malebranche can be sustained, however, there can be no mistaking the generally Cartesian flavour of Euler's 'Pensées'. There is thus a rich irony here, because what we have discovered is that the key to Euler's developing a theory of light strong enough to rival Newton's, at least in some parts of Europe, was his replacing this earlier, Cartesian style of thinking on the subject by an unmistakably Newtonian account of a wave motion in an elastic medium! Once again the

[35] Ibid., p. 18 (my emphasis).
[36] C. Truesdell, 'The Theory of Aerial Sound, 1687–1788', Editor's Introduction, Part 2, in *Leonhardi Euleri Opera omnia*, series 2, vol. 13, pp. XXV, XXXII.
[37] *Oeuvres de Malebranche, tome 3: Recherche de la vérité*, edited by G. Rodis-Lewis (Paris, 1964), pp. 257 ff.
[38] Euler (footnote 26), § 3; 'Nova theoria' (footnote 14), § 59 (*Opera omnia*, series 3, vol. 5, p. 21).
[39] Bernoulli (footnote 20), p. 9.

insufficiency of analysing eighteenth-century science in terms of a supposed dichotomy between Cartesian and Newtonian approaches to the natural world becomes glaringly apparent. In truth, the situation was very much more complicated than that.

Acknowledgments

Research for this paper was supported by a grant under the Australian Research Grants Scheme. Staff of the Leningrad Branch of the Archives Department of the Academy of Sciences of the U.S.S.R. provided generous hospitality and assistance. A shorter version of the paper was presented at a conference on 'Newton and Halley' at the William Andrews Clark Memorial Library, Los Angeles, in August 1985.

VII

THE NOTION OF EXPERIMENTAL PHYSICS IN EARLY EIGHTEENTH-CENTURY FRANCE*

One of the most characteristic features of the science of physics, as we know it today, is its highly mathematical form. Physical theories are almost always presented in the guise of extended pieces of mathematical analysis linked to experience through a number of mathematically expressed quantitative functional relationships between physical variables. During the past century and a half, this manner of proceeding has proved immensely successful, to the point where one can now all too easily assume that any worthwhile physics is necessarily quantitative and highly mathematical in its approach. It is therefore important to emphasize that physics has not always been done this way, and that our modern mathematical form of the subject is a relatively recent development whose beginnings are, except in a few special cases, scarcely to be traced back beyond the early years of last century. To be sure, some of the issues that arise whenever one attempts to bring mathematics to bear on the study of natural events had been considered and resolved somewhat earlier in such fields as astronomy, mechanics and geometrical optics. Yet even the very possibility of extending the lessons that had been learned in those cases to the generality of physical phenomena was by no means established, let alone the manner in which, in practice, this might be done.

When this point is recognized, there at once emerges a major task for the historian, namely, to trace the origins of the new mathematical version of physics, as it emerged in the early years of the nineteenth century, from its historical roots in the preceding age. This, however, would be a very large task indeed. In the present paper, my aim is the more modest one of delineating some seldom remarked features of the eighteenth-century background out of which nineteenth-century mathematical physics evolved. It is generally agreed that the new mathematical approach was a largely French invention, the achievement of a remarkable group of investigators based on the Paris Academy of Sci-

J. C. Pitt (ed.), Change and Progress in Modern Science, 107–131.
© 1985 by D. Reidel Publishing Company

ences (or the First Class of the Institut de France, as it was called during the Consulate and Empire) and the infant Ecole Polytechnique. Here, I shall be concerned with the situation in France in the period preceding this development, and more particularly with an influential tradition of experimental physics that emerged there during the first half of the eighteenth century and that came eventually to be associated particularly with the name of the Abbé Jean-Antoine Nollet (1700-1770). I shall show, among other things, that this was a tradition in which not only did mathematical and quantitative considerations play no part, they were actually ruled out, being held to be inappropriate to the subject-matter in question.

The historical importance of this tradition has generally been overlooked in the literature of our subject. Its practitioners do not have major scientific discoveries to their credit, and as a result, few of their names are recognized today. Furthermore, even when they are cited, it is not always sympathetically. Nollet in particular has had a bad press among historians of science, especially those of the English-speaking world. When he is remembered today – if he is remembered at all – it is chiefly as a remarkably successful vulgariser of science. Nollet, we recall, was the man who carried electrical demonstrating to the very pinnacle of French society, the court of Louis XV himself, and amused the King by causing a long line of monks to leap as one into the air when a battery of Leyden jars was discharged through them. His actual scientific work, and especially the theory of electricity he announced in 1745 and defended in a number of subsequent publications, have tended to be dismissed rather scornfully as unworthy of serious attention.

Both Nollet himself, however, and the intellectual tradition he represents, deserve better than this. In the context of his time, Nollet made a major contribution to the science of electricity, and his ideas dominated French discussions of the subject for a generation and more at a time when it was expanding very rapidly indeed.[1] More importantly, his significance for our story is by no means limited to his work in electricity. For a period of thirty years in the middle of the eighteenth century, Nollet was France's leading experimental physicist. His work inspired an entire generation of French investigators, and also led to an important reorientation of physics teaching in universities and colleges throughout the country. Moreover, his influence extended well beyond the confines of France, being felt particularly strongly in Italy, Spain and parts of Germany. Throughout much of Europe his name was for many years almost synonymous, indeed, with the words "physique expérimentale".[2]

There has long been a tendency – especially, once again, among English-speaking authors – to regard French experimental physics in the first part of the eighteenth century as something of an intellectual backwater. The general French caution about embracing the new approach of Newton and his followers has tended to be regarded, with hindsight, as mere blind conservatism prompted by an undue fondness for the *a priori* doctrines of Descartes. That the situation was in fact more complicated than this has recently begun to be recognized.[3] Yet the basic picture of a doctrinaire Cartesianism being slowly overwhelmed by the power of the Newtonian method has as yet scarcely been disturbed. This general outline was first developed by eighteenth-century writers such as Voltaire in discussing the fortunes during the first part of the century of Newtonian celestial mechanics. There are dangers in extending it uncritically to other areas of eighteenth-century science: in fact, our habit of viewing eighteenth-century science as a whole entirely from the stand-point of a debate between Newtonians and Cartesians is often unhelpful, and can sometimes be positively misleading.[4] Nowhere is this more apparent than in the field of general physics, because early eighteenth-century French physics was by no means uncritically Cartesian in outlook.

Nollet – to return to him as our example – was very much an eclectic in his approach. "I do not present myself here under the auspices of any philosopher", he proclaimed in the preface to his immensely popular *Leçons de physique expérimentale*, "it is neither the physics of Descartes, nor that of Newton, nor that of Leibniz, that I have resolved to follow particularly". Though he rendered honor, he said, to great men of all nations,

> I admit nothing on their word, unless it bears the stamp of experiment or is demonstrated according to the rules [i.e. of logic]: in matters of Physics, one ought not to be a slave to authority, and still less to one's own prejudices; one ought to recognize the truth wherever it shows itself, and not affect to be a

Newtonian in Paris and a Cartesian in London. (Nollet 1743-1764, I, pp. xviii-xix.)

True to his promise, in the body of his book Nollet drew in turn from each of the major intellectual traditions of the day. In so far as he rejected forces acting at a distance as a basis for his physics – he even went so far as to re-word Newton's third law of motion to read "Reaction is equal to compression", so as to avoid any suggestion that there might be situations other than the direct contact of two bodies to which it might apply (Nollet 1743-1764, I, p. 261) - Nollet would have to be classed as anti-Newtonian in outlook. He was firmly persuaded that gravity was brought about by a circulating vortex of aetherial matter, and he was generally content with the standard Cartesian practice of invoking "invisible fluids" as mechanical agents to account for all those physical phenomena where no more obvious mechanical cause could be discerned. On the other hand, he recognized that as yet no satisfactory detailed explanation of gravity had been provided, and so far as aetherial mechanisms more generally were concerned, he conceded that some authors had allowed their imaginations too free a rein by far. (Nollet 1743-1764, II, pp. 79-80, 455) He clearly had no sympathy whatsoever for Descartes' efforts to construct an *a priori* foundation for physics. When it came to the science of motion, it was to Newton that he turned. He took over in its entirety, for example, Newton's concept of inertia, and cited as he did so the Englishman's experimental demonstration, using pendulums, of the proportionality of inertia and weight. (Nollet 1743-1764, I, pp. 179-180) He even accepted Newton's conception of a universal attraction, together with his general analysis of the planetary motions. Nollet insisted, however, that this attraction should be regarded merely as a fact, and not "a principle whose immediate cause is none other than the will of the Creator". On this point, he claimed the support of Newton himself, in opposition to various latter-day authors. (Nollet 1743-1764, II, pp. 411-419; VI, pp. 148-158) Finally, while Nollet simply took over large sections of the *Opticks*, including the whole of Newton's discussions of colored rays, he at the same time firmly rejected the corpuscular theory of light that Newton advocated, in favor of a modified version of Descartes' transmission theory. (Nollet 1743-1764, V, passim)

It is in fact impossible to class Nollet and his fellow French experimental physicists meaningfully as either Cartesian or Newtonian, or to fit them at all comfortably into the story traditionally told of a life-and-death struggle between Newtonian and Cartesian doctrines in eighteenth-century France. Generally in sympathy with the mechanistic spirit of Descartes' physics, they were at the same time well aware of the magnitude of Newton's achievement, and drew freely upon the Englishman's work wherever they felt it to be well-founded. In other words, in matters intellectual, they generally adopted a bracingly eclectic approach. Furthermore, in their doctrinaire experimentalism their attitude towards their subject was shared by many eighteenth-century physicists outside France as well, by whom it continued to be widely accepted into the early decades of the following century even after it had been supplanted in France itself. Their style of thinking provides the historical backdrop against which the new mathematical physics constructed by men such as Coulomb, Laplace, Fourier and Fresnel needs to be set. In this paper, then, I shall try to bring out some of the more important features of their approach and of the way in which it evolved.

* * * * * * * * * *

As traditionally studied in universities throughout Europe, physics was not an experimental science. Nor was its subject matter the same as it is today. In both style and content the pattern was set instead by the Aristotelian treatises *Physica, De caelo, De generatione et corruptione, Meteorologica* and *De anima*, the study of which constituted a major part of the traditional undergraduate philosophy course. Consistent with the original meaning of the Greek word φυσις, from which the subject took its name, "physics" dealt with "nature" in general. Traditionally, its primary concern was with very broad questions such as the nature of body, or the conditions determining natural change in general, rather than with particular natural effects.

In France, this conception of the subject still predominated in the early years of the eighteenth century, notwithstanding the well known changes that had taken place in intellectual outlook during the preceding period of the "scientific revolution". In particular, physics continued to be seen, even by the "moderns", as an explanatory science, not a merely descriptive one.[5] Indeed, physics was commonly distinguished from natural history on precisely this basis; that is, natural history was seen as resting content with descriptions of phenomena while physics ascended to their causes. Furthermore, "physics" continued to embrace not only the topics it includes today but also questions that we now assign to the chemical, biological and human sciences and, to some extent, to astronomy. Only very gradually during the course of the succeeding century did the scope of the discipline gradually become confined within the boundaries it enjoys today.

An important change of emphasis can, however, be detected at this period, with ever-increasing amounts of attention being paid to questions of special physics (*physique particulière*) – the branch of the science concerned with "the examination and discussion of the particular bodies or corporeal individuals this universe contains" – at the expense of the more traditional general physics (*physique générale*) that dealt with "the nature, essence and properties appropriate to all bodies".[6] The change is apparent, for example, in Rohault's famous *Traité de physique* (1671), where Cartesian physics is expounded without the metaphysical underpinnings upon which Descartes himself had placed so much emphasis, and is presented instead as an experimentally oriented science chiefly distinguished by its mode of explanation in terms of corpuscular mechanisms in a universally disseminated aether.[7] Likewise, as the Cartesian philosophy at last began to penetrate the curriculum of the University of Paris in the late 1690s, a particular complaint levelled against the reformers by their more traditionally-minded colleagues was that they devoted undue attention to *physique particulière* at the expense of metaphysics and moral philosophy. (Freudenthal 1980, pp. 267-295) A comparison between late seventeenth-century and early eighteenth-century university texts – between, for example, Jean Du Hamel's *Philosophia universalis* (5 vols., 1705) and Pierre Le Monnier's *Cursus philosophicus ad scholarum usum accommodatus* (6 vols., 1750) – tells the same story, with a massive increase in attention to questions of *physique particulière* being evident in the later works.

The idea of experimental investigation was simply inappropriate to old-style Aristotelian "general" physics. With the increasing attention being paid to "special" physics, however, empirical considerations began to play a more and more important role. To some extent this was so even among Descartes' immediate followers, for the master, despite his overwhelmingly *a priori* outlook, had nevertheless allowed experiment a definite place in his science. Rohault, in particular, who was unquestionably the leading Cartesian of his generation, placed great emphasis on experimental inquiry.

Throughout the second half of the 17th century, however, Descartes' supporters remained outside the main institutions of French intellectual life. As is well known, Cartesians were formally excluded, along with Jesuits,' from the early Académie Royale des Sciences, where the experimental approach found its most secure institutional base. One of the Academy's two meetings each week was regularly devoted to experimental research in physics, while the other was given over to mathematical pursuits. To be sure, here, too, "physics" was understood more broadly than it is today, with most of the experimental work done being concerned with questions we would today regard as anatomical, chemical or botanical rather than physical. But men such as Mariotte and Perrault also took up for experimental investigation questions of an unarguably physical kind. At the same time, they adopted a highly sceptical attitude towards scientific theory far removed from the claims of both peripatetic and Cartesian natural philosophy. For both Perrault and Mariotte, and with them the entire *physicien* group in the early Academy, physics was a pragmatic science no longer concerned with either certainty or the ultimate causes of things.

One can, in fact, distinguish at this period three separate groups in France that were concerned in one way or another with "physics", as this subject was traditionally defined. In effect, these groups constituted, for a few years, three separate research traditions, each with its own values, methodological outlook and ontology and each forming, as well, a relatively distinct social group. First, there were the University professors, who took as their task the propagation of traditional learning, not the discovery of new knowledge. Their method was verbal and expository; their theory, Aristotelian; their objective, knowledge that was certain; their primary concern, the elucidation of causes. Secondly, there was the *physicien* group in the Académie Royale des Sciences, committed above all to the discovery of new knowledge by means of experiment, generally inclined to mechanistic theories to explain their observations but fundamentally sceptical about all theory and less and less concerned with the discovery of causes. Finally, there were the Cartesians, whose objective was, as with the professors, certainty and a knowledge of causes, but whose method was experimental and whose theory was programmatically mechanistic. The separation between these groups began to break down in the 1690s. In effect, we are concerned in this paper with the regrouping of forces that subsequently occurred.

The Académie des Sciences underwent a major restructuring in 1699, one consequence of which was that Cartesians were admitted to the membership for the first time. In due course this had the effect of somewhat diluting the scepticism of the early years. Yet even after its reform, the Academy remained an exclusive and, intellectually speaking, a very inward-looking body. As a result, even though its members did some important work, except for Malebranche they had surprisingly little impact on intellectual patterns in French society more generally. In particular, their empirical outlook had little influence outside the Academy's walls, so that empiricism had to fight its battles anew at the more popular level of the university curriculum and the public lecture, with little assistance from the King's academicians. From this process it emerged in a rather different form that involved, in particular, a more naive view of the relationship between experiment and theory in natural science than had been espoused by, for example, Mariotte. Eventually, with the appointment of Nollet to the Academy, this view, developed outside the Academy's walls, glorying now in the name of *physique expérimentale* but limited in its application to a narrower range of phenomena than included heretofore in the domain of "physics", became the new orthodoxy even within the ranks of that institution.

The change began, however, in the universities where, as I have mentioned, Cartesianism began to make significant inroads into the curriculum in the late 1680s.[8] Though the overall structure of the courses was in fact unaffected by this development, so that general philosophical principles continued to be presented first and then applied in turn to the various traditional domains of special physics – the heavens, and then inanimate and animate nature – increasing attention did begin to be paid to these latter sections of the course. A factor that contributed greatly to the success of the new approach at this period was a very popular course of demonstration experiments, similar in some respects to Rohault's earlier courses for the general public, but now introduced into the University of Paris for the first time. The lecturer was Pierre Polinière, and he offered his course of lectures at the invitation of one of the chief instigators of the Cartesian invasion of the university, Guillaume Dagoumer.

The importance of the demonstration experiment for the history of eighteenth-century physics has recently been stressed anew by Heilbron (1979). In regard to the history of French experimental physics in particular, Pierre Brunet argued many years ago that the example of the Dutch professors 'sGravesande, Boerhaave and Musschenbroek was crucial. (Brunet 1926) Yet in my opinion Brunet has seriously under-rated the significance of Polinière's work, which has yet to be studied with the attention it deserves. Though the Dutch undoubtedly exerted a major influence on French physics, especially on Nollet, I believe that Polinière represents a perhaps equally important native source for a tradition of experimental philosophy.

Polinière commenced his studies at the University of Caen before moving to Paris, where he applied himself particularly to the study of mathematics under the guidance of the celebrated Pierre Varignon. Ultimately this bore fruit in his first publication, a bulky textbook of elementary mathematics that appeared in 1704. (Polinière 1704) Polinière's real love, however, was always physics, not mathematics. He taught himself from books how to repeat the leading experiments of the day, to such good effect that, as has been noted already, he was commissioned by Dagoumer in the mid-1690s to give a course of experimental demonstrations for students at the Collège d'Harcourt. This proved so successful that he was subsequently engaged to repeat the course each year in several of the other colleges as well. He published his lectures in 1709 under the title *Expériences de physique*, and the work proved popular enough to go through two further editions during his lifetime, in 1718 and 1728. A posthumous fourth edition based on Polinière's own manuscript amendments appeared in 1734, a few months after his death, and a fifth, almost identical with the fourth, in 1741. Polinière was of a retiring disposition, and spent most of his life on his estate in Normandy, moving to Paris for only a few months each year in order to give his lectures. Despite this, the lectures achieved a great success in Parisian society as well as in the University, to the extent that the young King Louis XV himself attended the course in 1722.[9]

A consistent emphasis on the dependence of physical theory upon experiment was one of the hallmarks of the experimental physics of Nollet and others later in the century. Something of an evolution towards the views eventually espoused by Nollet and many of his contemporaries can already be discerned, however, in successive editions of Polinière's book.

When Polinière first began to give his course of lectures, it was clearly intended merely to supplement the systematic course of lectures on natural philosophy being taught by his sponsor, Dagoumer. This subsidiary status is reflected in the first published version, in which experiment is accorded a distinctly inferior role in physics to theory. The professors of Paris, we are told, have taken advantage of their students' enthusiasm for Polinière's experiments "to teach them in an agreeable manner what is most curious and useful in Physics, and have applied [the experiments] very ingeniously to the principles they have established in their writings, in order to confirm these and make their truth better understood". (Polinière 1709, preface) By the time the second edition of his book was published nine years later, Polinière had grown more confident, and no longer restricted experiment to the role of merely confirming principles already established in the theoretical writings of the learned doctors. Experiment, he now indicated, could of itself bring students to a knowledge of true physical principles. The professors, he wrote, "advancing step by step have used these experiments to lead their scholars by the hand, as it were, into the truth of the principles they have established by their writings". "These experiments", he went on,

> can even be considered as the principles and fundamentals of an exact Physics (*une Physique exacte*). I bring in the proximate causes of these effects when I believe I understand them, without troubling myself too much about seeking the remote and general causes on which these proximate causes depend. Because these remote causes often not being well known to us, we are in danger of propounding fables. (Polinière 1718, preface)

It has been pointed out elsewhere that in the 1718 edition of his book, Poliniere substantially altered his discussion of light and colours to include an account of Newton's work. (Corson 1968, pp. 402-413) Though he presented Newton's ideas only hesitantly – "I do not pretend that these novelties are incontestable and without difficulty", he said, "I hope only that they will contribute to further deepen research to which the objects we see every day invite us" (Polinière 1718, p. 465) – his readiness to present them as fully as he did at a time when they remained little known in France reveals his determination to keep his presentation as up to date as possible. So, too, does his prompt inclusion in the 1734 edition of an account of the important magnetic researches communicated to the Paris Academy of Sciences shortly before by Du Fay (Polinière 1734, I, pp. 364-366)[10] But what most appealed to him in both cases, it is clear, was the way in which important theoretical advances here so obviously depended upon prior experimental investigation. By the third (1728) edition of his book, Polinière's confidence in himself and in the experimental method he had embraced was complete. Physics, he now maintained, was concerned with "a knowledge of the effects our senses reveal to us, and the study of their causes". (Polinière 1728, p. [viii]) Experiment when properly ordered could lead directly to the fundamental principles of the science, as he believed his own work showed. The proper ordering, however, was important:

> There is here no accumulation of useless, confused and amusing observations of different kinds, without design or consequences. Here are materials investigated with care, prepared and arranged methodically, and which can be considered as the fundamentals of an exact Physics.

Unlike conclusions reached on a merely verbal basis, he said, "the truth I announce will be recognized in all times and places, will always be novel despite its age, and will always gain the admiration of those who begin to have a knowledge of it". (Polinière 1728, pp. [i]-[ii]) Experiment, he maintained, was "the shortest and surest way to know the essence and properties of bodies, and even the economy of the Universe". It is by this means, in short, that we come to know the "immutable laws ... that [God] has imposed on nature".

To insist upon the priority of experiment in physical inquiry was, of course, by the eighteenth century scarcely original. Bacon, whom Polinière cited approvingly from the outset (Polinière 1709, p. 251), had been dead for almost a hundred years, while the Royal Society of London and the Paris Academy of Sciences had long since been successfully established. Nevertheless, I must insist that, in its own time and circumstances, Polinière's stance was significant, as is the evolution in his thinking that is revealed in successive editions of his treatise. Poliniere's lectures constituted the first major infiltration of an experimental outlook into the French educational system. As he himself tells us, once his book was published, the example of the Parisian masters in utilizing his work was imitated in a number of other provincial and Paris colleges. (Polinière 1728, pp. [iii]-[iv]) Initially presented as merely supplementary to the traditional and largely theoretical physics of the schools, experiment became in the third and subsequent editions of Polinière's book the *only* dependable way of arriving at a true physics. In other words, by 1728 Polinière was propounding a complete and thorough-going empiricism, independent of both Aristotle and Descartes; and though many of the explanations he offered for particular effects were modelled after those provided earlier by Descartes, his commitment to them was free of the *esprit de système* that had marked even Rohault's work of a generation earlier.[11]

Of particular interest in the later editions of Poliniere's book is his claim that the principles of physics are directly discoverable by experiment. Such a view was implicit in Bacon's inductivism, but has now been generally discredited. In the eighteenth century, however, it was maintained by many other authors after Polinière. In particular, it was embraced enthusiastically by Nollet (who during his student days in Paris in the early 1720s may well have attended Polinière's lectures), and, following Nollet, by numbers of lesser writers. It became, in fact, for almost a century the leading feature of the experimentalist tradition in physics.

This brings me, then, to Nollet himself. Nollet's commitment to the experimental approach doubtless derived, in part at least, from his personal circumstances of life. Having been born of peasant stock, he had no private income, and so was faced with the need to make a career for himself. Endowed with both a keen brain and remarkable manual dexterity, he found in the physics lecture-demonstration a way of making a living that fitted perfectly with his own talents and interests. He announced his first course of lectures in 1734, so soon after Polinière's death that the timing seems more than coincidental. In these, and later in the published *Leçons de physique expérimentale* that grew out of them, he sought systematically to develop general physical principles upon experimentally demonstrable foundations. Not only was this an excellent pedagogical technique – there is nothing quite so convincing as evidence laid before one's very eyes – it also used Nollet's personal skills to the utmost in the design of spectacular yet meaningful demonstrations.

Already by the time he initiated his course, Nollet had acquired powerful patrons in the Academy of Sciences in the persons of Du Fay and Réaumur, with both of whom he had worked as a resesarch assistant. His lectures, which were aimed at upper-class Parisian society rather than at university undergraduates, were even more successful than Polinière's. Together with the published *Leçons*, they gained enormous public prestige for *physique expérimentale* which culminated in 1753 in the King's creating and naming Nollet as first incumbent of a new chair of experimental physics in the University's Collège de Navarre. This was the first such chair in any French university, and it set the pattern for others established in the course of the next two or three decades.

Nollet, however, went beyond mere pedagogy in his reliance on experiment. Like Polinière, and in a manner similar to Bacon, he maintained that the general principles of any worthwhile physical theory *had* to be straightforward generalizations from experience. "The aim of experimental physics", he announced, "is to know the phenomena of nature, and to show the causes of these *by proofs of fact*." (my emphasis) (Nollet 1753, p. 9) He roundly condemned "all those hypotheses which a too bold imagination takes pleasure in fabricating and multiplying on its own authority to form a body of doctrine, before knowing how they square with the facts that Observation and Experiment can furnish." (Nollet 1753, p. 34) On the other hand, like Bacon, he allowed the *controlled* use of hypotheses a place in his science: "I dare to say", he wrote, "that one can and should allow some, if one is content to conceive of possibilities, in order to submit them to experiment, and learn in that way what reality they can have." (Nollet 1753, p.

33) Yet he insisted that the hypothetical status of such propositions should never be forgotten, and stated plainly that "observation and experiment are the surest means, I would say practically the only ones, that a scholar who seeks to extend the progress of physics can employ." (Nollet 1753, p. 24).

Unlike Polinière, Nollet became a member – indeed, in time, a leading member – of the Paris Academy of Sciences. He was a practising experimental physicist in his own right, and not merely an experimentally oriented physics teacher.[12] His research publications exhibit the same Baconian philosophy, however, as his pedagogical works. When he first presented his new theory of electricity at a meeting of the Academy in April 1745, for example, he acknowledged that what he was proposing was a "system", but insisted that his, unlike certain others, was built upon experimentally established truths: "in forming it", he said, "the imagination has worked only on what has been furnished by experiment, and I make bold to say that it would be a mistake to take it for an assemblage of mere possibilities, or speculations devoid of proof." (Nollet 1745, Mém. p. 108) A few months later, in reporting to a friend that Bose of Wittenberg had advanced ideas very similar to his own, he remarked: "I am not surprised – I flatter myself that the same ideas will come to anyone with a little experience in physics who considers all the electrical phenomena at leisure."[13]

A closer look at the arguments Nollet offered in support of his electrical theory brings out his position even more clearly. Already convinced of the inadequacy of resorting in this instance to unexplained forces acting at a distance, Nollet described a series of experiments which, he said, made manifest the existence of some matter in motion in the region around an electrified body. These included the effect known as the "electric wind" – the cobweb-like sensation experienced when an electrified rod is held near the skin – and the crackling sound, characteristic smell, and luminous discharges associated with a well rubbed tube. "What but a substance", he asked, "is tangible, makes itself heard, has a smell, and is visible? Do not all these features incontestably indicate a matter?" (Nollet 1745, Mém. p. 112) As to the nature of this matter, Nollet's starting point was Herman Boerhaave's conclusion, also reached on what purported to be wholly empirical grounds, that the world was filled with an imponderable elastic all-pervading subtle fiery medium. Nollet had long been of the belief that it was this substance that gave rise, under appropriate circumstances, to the various phenomena of electricity. In the absence of positive proof, however, he had refrained from making his views known. The recent German experiments in which electrical discharges were used to ignite alcohol and other readily inflammable materials gave him the "proof" he had been waiting for, and prompted the presentation to the Academy of his initial paper on the subject.[14]

The naivety of Nollet's empiricism is perhaps most evident of all, however, in his discussion of the means by which the supposed all-pervading Fire brought about the various electrical effects. Fundamental to his entire argument here was the assumption that the various luminous discharges associated with electrified bodies consisted simply of streaming fiery matter rendered visible as a result of being concentrated into a more confined space than usual. The various motions of this matter could thus, Nollet assumed, be ascertained by direct inspection. The explanations of other effects such as the attraction and subsequent repulsion of nearby light objects were to be sought in terms of the motions thus discovered. As is well known, he concluded that any electrified body was surrounded by a double stream of subtle fiery matter, an "effluent" stream flowing out from the body and an "affluent" one flowing inward from the bodies surrounding it.

Nollet's attitude here had a number of interesting consequences. To begin with, the chief desideratum for a researcher who adopted his views became the ever more detailed study of common and well known effects, and not the "torturing" of nature to obtain data about new and unusual situations. One or two seemingly contrary results obtained in special situations could not, according to Nollet, overthrow an understanding gained from multiple observations of more common states of affairs. It was precisely on this basis that he rejected certain difficulties raised by Jallabert against his theory: "it is not by one or two proofs that electricity must be judged", he wrote, "but by the same experiments repeated a thousand times."[15]

It also becomes easy to understand Nollet's bafflement when confronted soon afterwards by the writings of Benjamin Franklin and his supporters. So far as Nollet was concerned, his opponents, in attacking his theory, were denying not a set of abstract propositions but experimentally proven facts. For example, misunderstanding Franklin's doctrine of "plus" and "minus" as an attempt to replace his own theory of effluences and affluences by a theory based on unidirectional streams of electrical fire, Nollet cited again his observation of luminous brushes directed from neighboring objects towards a rubbed glass globe, and could not comprehend how, in the light of this, Franklin's supporters could deny the existence of a second, inward-moving stream. Likewise, to combat Franklin's notion that glass was totally impermeable to electrical fire, he contrived some striking experiments in which luminous discharges appeared inside an evacuated glass chamber when this was electrified on the outside. To his astonishment, however, his opponents remained unconvinced. So far as Nollet was concerned, to continue to assert the impenetrability of glass in these circumstances was tantamount to denying the plain testimony of one's own senses.[16]

To modern eyes – and this brings me back to my starting point – one of the most striking features of Nollet's writings on physics is the complete absence from them of quantitative considerations and mathematical analysis. Nor is this a feature peculiar to Nollet's work. Polinière's treatise is equally non-mathematical, even though, as we have seen, he was a competent mathematician, at least at an elementary level. Economic and social factors clearly played a role here. Both Polinière and Nollet designed their courses of lectures for lay audiences, and would not have been able to assume even the most rudimentary prior knowledge of mathematics. In this respect, the French texts were very similar to those published in Holland and England at about the same time. Even in such a mathematical branch of the science as mechanics, there was everywhere a tendency to offer experimental demonstrations rather than mathematical proofs.

The avoidance of mathematical forms of argument extended, however, even to the research publications of Nollet and many of his fellow eighteenth-century experimental physicists. Some paid lip service, it is true, to the ultimate goal of a truly mathematical physics. But even then they seldom meant by this phrase what we mean by it today. By "mathematical" they usually meant merely "rigorous": they had no conception of a physics in which quantitative functional relationships were manipulated by means of formal mathematical arguments. Indeed, for those who shared Nollet's views of the proper relationship between experiment and theory, such a physics was strictly impossible, since direct observation of nature could never yield precise mathematical relationships between physical variables.

Nollet's published remarks about mathematics were in fact somewhat equivocal. In his 1753 inaugural lecture, he remarked that a knowledge of languages could be useful for a budding physicist. This included, he said, a knowledge of mathematics:

> A language which it is indispensible to learn is that of algebra and geometry; these two sciences have been successfully introduced into physics; everywhere where they can be applied, they bring the exactitude and precision that is peculiar to them; they spread illumination in the mind, they make it reason aright; with their help, it proceeds more rapidly, more surely, and can go further; it is essential if one is to follow authors who travel by the light of these torches. (Nollet 1753, pp. 17-18)

Later in the same work he further emphasized the usefulness, even the necessity, of mathematics for the physicist:

> As it is to be hoped that Beginners who seek to instruct themselves by reading Works on Physics understand the Geometrical and Algebraic expressions that are very commonly used in these today, we should also regard it as necessary for those who want to extend the progress of Physics to possess enough of these two Sciences to assist them in their researches, and to evaluate their discoveries: doubtless there will be many occasions when one is reduced to regretting being unable to make use of them; but even then the geometrical spirit will prevent one from departing from the truth by following false trails, and will make one see the *approximately's* more correctly. Everywhere, moreover, combinations, measurement, calculation will indicate in advance what can be expected from an investigation, will open new views, and will prevent false appearances being taken for reality. (Nollet 1753, pp. 42-43)

Yet this kind of usefulness for mathematical reasoning remains a far cry from the much more crucial role it has since come to play in physics. So far as Nollet was concerned, the principal function of mathematics in physics seems to have been to prevent one from straying into false paths of reasoning. Furthermore, Nollet proceeded immediately to a fierce denunciation of those who had sought to make physics a mere plaything for mathematicians:

> Having recommended in good faith the application of Geometry to Physics, having likewise recognized that the study of Nature only commenced with that happy union to make real progress, would I dare say that it is dangerous for a Physicist to develop too great a taste for Geometry? Living examples will not fail to be opposed to me, who will prove beyond possibility of reply that one can be at the same time an excellent Geometer and a very skillful Physicist; but are these good models always imitated? For a few wise Geniuses, who have not by excelling in an exact Science lost the taste for a study where one almost never finds either precision or complete certainty, and who only have recourse to calculation and Geometrical expressions when the importance of the questions and the nature and necessity of the proofs demands it: how many do we not see who can no longer descend from the high speculations to which they have risen, who disdain everything which is beneath! How many others who could without difficulty reach the common level, amuse themselves by rendering in Algebraic characters truths which would lose none of their value if they were expressed in a manner intelligible to everyone! Such writings, much appreciated, show clearly enough that the small amount of Physics in them has served as pretext for making a parade of a different Science. (Nollet 1753, pp. 43-44)

There is bitterness evident in these remarks, but there is also hidden in them a revealing hint concerning Nollet's attitude towards physics itself. Physics, Nollet says, is a science in which "one almost never finds either precision or certainty." To us, coming after Hume, the lack of complete certainty will come as no surprise; but the claim that physics is generally an *imprecise* science is not something we are used to. Nevertheless, it is entirely consistent with Nollet's naively empiricist approach to his subject. Strictly speaking, even in mechanics and optics, fields where both mathematical and physical styles of reasoning were seen to have roles to play, these were conceived by Nollet and his colleagues as distinct and separate. Thus for many thinkers at this period, mechanics done mathematically was a branch of mathematics; done with nuts and bolts, it was a branch of physics. And though many would have agreed with Nollet that a good mechanic needed a sufficient knowledge of both mathematics and physics, (Nollet 1743-1764, III, pp. 2-3) they would also have shared his belief, as set out in another letter to his friend Jallabert, that the exactitude of mathematics was, at bottom, simply inappropriate to physics. Commenting on the news that Daniel Bernoulli was trying to unite experiment to mathematical theory, Nollet stressed the need for "Geometrical exactitude not to be disdainful of lowering itself to the *à peu près* inseparable from experimental physics", adding that "I can well imagine that that would come hard to someone accustomed to rigorous demonstrations."[17]

What we have here, I believe, is an eighteenth-century reflection of a very traditional view. "The minute accuracy of mathematics is not to be demanded in all cases", Aristotle had written in the *Metaphysics*, "but only in the case of things which have no matter. Hence its method is not that of natural science; for presumably the whole of nature has matter."[18] In the seventeenth century, van Helmont wrote that "the Rules of the Mathematicks, or Learning by demonstration, do ill square to Nature. For Man doth not measure Nature; but she him."[19] Even Descartes had protested in a letter to Mersenne that "to require of me Geometrical demonstrations in a matter which depends on Physics is to demand that I achieve impossible things."[20] According to Joseph Priestley writing later in the eighteenth century, one of the great achievements of the scientific revolution had been to separate natural philosophy (i.e. in this context, physics) from the mere ratiocination that characterized mathematical reasoning:

> Before this period [following Descartes], we hardly meet with any production among the learned, but such as might have been executed

> without stirring from their closets, except in what respected mechanicks; so that the proper distinction between the natural philosopher and the mathematician was but little attended to. (Priestley 1772, I, p. 132).

As I indicated earlier, however, I believe that so far as Nollet was concerned, the experimental method he adopted was itself sufficient to rule out a quantified mathematical physics. With his over-riding emphasis on direct observation, he could never have reconciled himself to the notion of "experimental error" that is so crucial to the determination of quantitative experimental generalisations. For Nollet, and with him the whole school of eighteenth-century experimental physicists, deviations from mathematical exactness in the data could never be discarded as mere chance departures from a ruling norm. Rather, they were bound up with the irreducibly inexact nature of the subject itself. They were, in fact, of the very essence of a truly experimental physics.

Hence it was that when, in the first years of the following century, mathematics began for the first time to be applied successfully to a wide range of physical questions, those who remained faithful to the old-style *physique expérimentale* quickly found themselves outflanked. Though their predecessors had played a central role in reconstructing the subject on an experimental footing, the too-direct empiricism that I have described, that left no room for a theory expressed in mathematical terms, now in its turn also became outmoded. But that is another story.

NOTES

* Research for this paper has been supported by a grant from the Australian Research Grants Scheme.

[1] Cf. Home (1974, II, pp. 269-272); Home (1981, passim); Heilbron (1979, chaps. 11,15).
[2] Torlais (1954). For an account of Nollet's influence even as far away as distant Bohemia, see Smolka (1968, III B, pp. 131-135).
[3] Marsak (1959, pp. 51-60); Hall (1975, pp. 233-250); Guerlac (1977, pp. 479-490); Guerlac (1981, esp. chap. 5). For the more traditional view of French resistance to Newton, Cf. Brunet (1931).
[4] Home (1979a, pp. 235-249); Home (1977, pp. 252-266).
[5] Cf. Rohault (1671, I, p. 1): "This Word, *Physicks*, strictly speaking, and according to the Etymology of it, signifies no more than *Natural*; but we here use it to signify the Knowledge of natural Things, that is, that Knowledge which leads us to the Reasons and Causes of every Effect which Nature produces."
[6] Anonymous (1756, p. 582).
[7] On Rohault, see Schuster (1975, pp. 506-508); also Mouy (1934, pp. 108-138); and McClaughlin (1977, pp. 225-240).
[8] Brockliss (1981a, pp. 33-69); Brockliss (1981b, pp. 131-168); Freudenthal (1980).
[9] These biographical details are drawn from the anonymous "Abregé de la vie de M. Polinière" inserted in the posthumous editions of his *Expériences de physique* (1734, 1741), and from Polinière's own preface to his book.
[10] Cf. Du Fay (1728, Mém. pp. 355-369); Du Fay (1730, Mem. pp. 142-157); Du Fay (1731, Mém. pp. 417-432).
[11] Many of Polinière's detailed discussions have been summarized in Hanna (1972, pp. 13-39).
[12] Polinière's only piece of significant research was his study very early in his lecturing career of the phenomenon now known as electroluminescence. On this, see Corson (1968).
[13] Nollet to E. F. Du Tour, 10 December 1745 (Dibner Library, Smithsonian Institution): "je nen suis pas surpris – je me flatte, que les memes idées viendront a tout homme un peu initié en physique, qui aura vû a loisir tous les phenomenes Electriques".

[14] In a letter to the Genevan professor Jean Jallabert dated 19 July 1745, Nollet wrote as follows concerning the new German experiments: "Mais ce qui ma le plus occupé, c'est que jay aperçu dans ces nouveaux phenomenes des preuves assez fortes pour établir un systeme d'explication que je meditois depuis longtemps, mais que je n'osois hazarder sur des raisons foibles ou equivoques". (MS collection, Société d'Histoire et d'Archéologie de Geneve; quoted in Home (1979b, pp. 171-176).

[15] Nollet to Jallabert, 28 July 1747, MS collection, Société d'Histoire et d'Archéologie de Geneve: "Ce nest ny par une ny par deux epreuves quil faut juger de lelectricité, mais par les memes experiences repetées mille fois". Nollet's attitude is well described in Yamazaki (1976, pp. 37-64).

[16] For Nollet's response to Franklin's ideas, see Home (1981, chap. 7).

[17] Nollet to Jallabert, 10 December 1750, MS collection, Société d'Histoire et d'Archéologie de Genève: "Entre les mains dun tel homme cela doit bien aller, pourvû que l'exactitude Geometrique ne dedaigne pas s'abaisser jusquaux *a peu prés* inseparables, de la phys[ique] experimentale: Car je conçois bien que cela doit couter quelque chose a quelqu'un qui est accoutume aux demonstrations rigoureuses".

[18] Aristotle (330 BC, Bk. II, chap 2, 995a 15-18).

[19] Debus (1968, p. 24).

[20] Descartes to Mersenne, 27 May 1638; Descartes (1897-1913, II, pp. 141-142). The translation is from Sabra (1967, p. 23).

REFERENCES

Anonymous (1756), *Memoires de Trevoux 57*.

Aristotle (330BC), *Metaphysics*.

Brockliss, L. (1981a), "Aristotle, Descartes and the New Science: Natural Philosophy at the University of Paris, 1600-1740", *Annals of Science 38*.

(1981b), "Philosophy Teaching in France, 1600-1740", *History of Universities 1*.

Brunet, P. (1926), *Les physiciens hollandais et la méthode expérimentale en France au XVIIIe siecle*. Paris.

_________ (1931), *L'introduction des théories de Newton en France au XVIIIe siècle avant 1738*. Paris.

Corson, D. (1968), "Pierre Polinière, Francis Hauksbee and Electroluminescence: A Case of Simultaneous Discovery", *Isis* 59.

Debus, A. (1968), "Mathematics and Nature in the Chemical Texts of the Renaissance", *Ambix* 15.

Descartes, R. (1897-1913), *Oeuvres de Descartes*, eds. C. Adam and P. Tannery. Paris.

DuFay, C. (1728, 1730, 1731), *Histoire de L'Academie Royale des Sciences, avec les memoires*.

Freudenthal, G. (1980), "Littérature et sciences de la nature en France au debut du XVIIIe siècle: Pierre Polinière, l'introduction de l'enseignement de la physique expérimentale à l'Université de Paris et l'*Arret burlesque* de Boileau", *Revue de synthese 101*.

Guerlac, H. (1977), "The Newtonianism of Dortous de Mairan", *Essays and Papers in the History of Modern Science*. Baltimore: Johns Hopkins University Press.

_________ (1981), *Newton on the Continent*. Ithaca/London: Cornell University Press.

Hall, A. (1975), "Newton in France: A New View", *History of Science 13*.

Hanna, B. (1972), "Poliniere and the Teaching of Experimental Physics at Paris, 1700-1730", *Eighteenth Century Studies presented to Arthur M. Wilson*, ed. Peter Gay. N.H.: Hanover.

Heilbron, J. (1979), *Electricity in the 17th and 18th Centuries: A Study of Early Modern Physics*. Berkeley: University of California Press.

Home, R. (1974), "Electricity in France in the Post-Franklin Era", *Actes du XIVe Congrès International d'Histoire des Sciences, Tokyo-Kyoto, 1974* Tokyo.

_________ (1977), "'Newtonianism' and the Theory of the Magnet", *History of Science 15*.

_________ (1979a), "Out of a Newtonian Straitjacket: Alternative Approaches to Eighteenth-Century Physical Science", *Studies in the Eighteenth Century IV*, ed. R. Brissenden and J. Eade. Canberra: Australian National University Press.

_________ (1979b), "Nollet and Boerhaave: A Note on Eighteenth-Century Ideas about Electricity and Fire", *Annals of Science 36*.

_________ (1981), *The Effluvial Theory of Electricity*. New York: Arno Press.

Marsak, L. (1959), "Cartesianism in Fontenelle and French Science, 1686-1752", *Isis 50*.

McClaughlin, T. (1977), "Le concept de science chez Jacques Rohault", *Revue d'histoire des sciences 30*.

Mouy, P. (1934), *Le développement de la physique cartesienne, 1646-1712*. Paris.

Nollet, J. (1743-64), *Leçons de Physique expérimentale* 6 vols., Paris.

_________ (1745), "Conjectures sur les causes de l'électricite des corps", *Histoire de l'Académie*.

_________ (1753), *Oratio habita . . . cum primum Physicae Experimentalis cursum professor a Rege institutus auspicaretur*. Paris.

Polinière, P. (1704), *Elémens des mathématiques*. Paris.

_________ (1709), *Expériences de physique*. Paris.

_________ (1718), *Expériences de physique*. 2nd ed.

_________ (1728), *Expériences de physique*. 3rd ed.

_________ (1734, 1741), *Expériences de physique*. posthumous eds.

Priestley, J. (1772), *History and Present State of Discoveries relating to Vision, Light, and Colours*. London

Rohault (1671), *Rohault's System of Natural Philosophy, illustrated with Dr. Samuel Clarke's Notes*. trans. John Clarke. London, 1723, reprinted New York, 1969.

Sabra, A. (1967), *Theories of Light from Descartes to Newton*. London.

Schuster, J. (1975), Art. "Rohault", *Dictionary of Scientific Biography*. *XI*. New York.

Smolka, J. (1968), "L'abbé Nollet et la physique en Bohême", *Actes du XIIe Congrès International d'Histoire des Sciences, Paris 1968*. Paris 1971.

Torlais, J. (1954), *Un physicien au siècle des lumières: l'abbé Nollet, 1700-1770*. Paris: Sipuco.

Yamazaki, E. (1976), "L'abbé Nollet et Benjamin Franklin", *Japanese Studies in the History of Science 15*.

Nollet and Boerhaave: A Note on Eighteenth-Century Ideas about Electricity and Fire

History has not been kind to the Abbé Nollet. Greatly esteemed in his lifetime as the master of French experimental physics, and especially of the then new science of electricity, he is remembered today—when he is remembered at all—simply as a remarkably successful vulgarizer of science. Nollet, we recall, was the man who carried electrical demonstrating to the very pinnacle of French society, the court of Louis XV himself, and amused the King by causing a long line of monks to leap as one into the air when a battery of Leyden jars was discharged through them. But we tend to dismiss with, at best, a shrug and a knowing smile, his more serious efforts at constructing a theory of electricity adequate to account for the then known phenomena. His theory was from the outset, such gestures imply, both impossibly complex and hopelessly wrong, and it was a fortunate chance indeed that saw it overwhelmed within a few years of its first being made public by Benjamin Franklin's vastly better alternative.

Even more damaging for Nollet's subsequent reputation among historians of science has been his obduracy in refusing gracefully to concede the victory to his American rival, and in persisting right up until his death in 1770 with a forlorn defence of his supposedly by then thoroughly outmoded and discredited doctrines. His most frequently quoted epitaph has been that provided by none other than Franklin himself: the Abbé, Franklin tells us, 'lived to see himself the last of his sect—except Mr. B—[that is Mathurin-Jacques Brisson] his *élève* and immediate disciple'.[1]

However, Nollet deserves better than this. His ideas about electricity, when first made public, far from being dismissed as unworthy of serious attention, won immediate and wide acceptance. His subsequent writings were inevitably treated as important contributions to the subject. And Franklin was plainly deluding himself so far as at least the French reception of his new theory of electricity was concerned. Nollet's ideas, far from being eclipsed almost overnight by those of the American, continued to dominate French discussions of the subject for another twenty years and more, until well into the 1770s.[2] His doctrines therefore cannot be dismissed as

[1] B. Franklin, *The autobiography, and selections from his other writings* (ed. H. W. Schneider: New York, 1952), 153.

[2] R. W. Home, 'Electricity in France in post-Franklin era', in *Actes du XIVe Congrès International d'Histoire des Sciences, Tokyo-Kyoto, 1974* (Tokyo, 1974–75), vol. 2, 269–272. See also J. L. Heilbron, 'Franklin, Haller and Franklinist "History"', *Isis*, **58** (1977), 539–549.

having been so peculiarly his own as to be unworthy of serious historical study. On the contrary, their initial rapid acceptance shows that when they were first formulated, they were well attuned to the kinds of demands that scientists of the day were making of their theories. A study of Nollet's ideas can therefore aspire to do more than illuminate a curious cul-de-sac in the history of science. It can also draw our attention to some more general features of science in his day, features that had important consequences for the way the subject subsequently developed.

Traditionally, the principal phenomenon that electrical theorists had sought to explain was the attractive power acquired by various substances such as amber and glass when rubbed. The attraction had generally been attributed to the action of certain subtle effluvia elicited from the substances by the rubbing, but opinions had differed as to whether or not these were composed of 'the finer parts of the attrahent', and were thus specific to the particular kind of matter being rubbed. With the notable exceptions of Descartes and Newton, most seventeenth-century authors had thought that they were.[3] The remarkable discoveries of the early years of the eighteenth century, however, especially Stephen Gray's announcement in 1731 that the attractive power could be readily transferred from one kind of matter to another,[4] had lent powerful support to the view that, far from being specific, the effluvia associated with different electrified bodies consisted of streams of a single all-pervading subtle matter. This notion became a central doctrine of the theory developed by Nollet.

In other respects, the general outlines of Nollet's theory of electricity are well known.[5] He supposed that when an electrifiable body was rubbed, subtle matter which had previously lain hidden within its pores was agitated and squeezed out into the surrounding air, giving rise to a stream of matter flowing outwards from the body. To compensate for this loss, a second stream of similar matter set in towards the body from its surroundings. According to Nollet, it was the combined action of these two streams of matter that gave rise under appropriate circumstances to the various effects described as 'electrical'. In particular, nearby light objects would be attracted or repelled, he said, according to whether, in the region in which they found themselves, the incoming or the outgoing stream was the stronger. Also, to account for the then very recent discovery that sparks and flame could be generated when a finger, for example, was brought near an electrified body, Nollet argued that because the finger was a conductor, and hence allowed the streaming matter to enter and leave its pores with greater ease than did the surrounding air, there would be a concentration of both inward and outward-moving streams in the direction of the finger. If the rays composing these streams became sufficiently forceful, the particles of the two streams would collide head on with enough force to break open their surrounding envelopes of sulphureous matter and release the active fiery matter within.

Thus baldly stated, Nollet's theory sounds like yet another example of that unrestrained system-building, that multiplication of mechanical hypotheses, for

[3] The phrase quoted is Boyle's (*Experiments and notes about the mechanical origine or production of electricity* (London, 1675; reprinted Oxford, 1927), 6. Descartes's views are set out in his *Principia philosophiae* (1644): see C. Adam and P. Tannery (eds.), *Oeuvres de Descartes* (Paris, 1897–1913), vol. 8 312–314. For Newton's opinion, see his *Mathematical principles of natural philosophy* (trans. Andrew Motte, rev, Florian Cajori, Berkeley/Los Angeles, 1934) 547. See also my forthcoming article, 'Newton on electricity and the aether'.

[4] S. Gray, 'A Letter . . . containing several experiments concerning electricity', *Philosophical transactions of the Royal Society of London*, **37** (1731), 18–44.

[5] This account is based on Nollet's initial presentation of his ideas in the paper he read to the Académie Royale des Sciences on 28 April 1745: 'Conjectures sur les causes de l'électricité des corps', *Histoire de l'Académie Royale des Sciences, avec les Mémoires, (1745)*, Mém., 107–151. Nollet subsequently defended the same doctrines in his *Essai sur l'électricité des corps* (Paris, 1746), and in many later publications.

which the followers of Descartes have been so roundly and so justifiably condemned. This is, indeed, how historians of electricity have sometimes, though wrongly, represented it. Nollet himself would have angrily rejected any such characterization of his work. He always claimed that his ideas were developed independently of all party spirit, that in matters intellectual he was no more beholden to Descartes than he was to Newton.[6] Nollet was, in fact, a leading exponent of a peculiarly eighteenth-century mode of scientific investigation that is perhaps best referred to under its eighteenth-century French appellation as *physique expérimentale*. He claimed in a manner typical of those who advocated this approach that the basic premises of his theory, far from being hypothetical, were straightforward matters of fact. He argued, for example, that the outwardly directed streams of subtle matter could easily be seen whenever a glass tube was briskly rubbed in a darkened room. Likewise, he said, the brush of light seen on the end of, say, a finger brought near a glass globe mounted on one of the newly developed German electrifying machines, was merely a part of the inward-moving stream rendered visible by being condensed into a narrower space than usual.

I do not wish to enter here into the merits or drawbacks of the naively empiricist view of scientific theory-formation implicit in Nollet's claim. In any event, Nollet's approach has been well described in a recent paper by Yamazaki.[7] I intend to take as read the fact that Nollet adopted this position, and to focus instead upon certain assumptions that he made and certain arguments he presented concerning the nature of the subtle matter that lay at the heart of his theory. A considerable part of Nollet's initial presentation of the theory was given over to discussing this question; indeed, he regarded the resolution of it as one of his chief tasks.[8]

Nollet's approach to his self-imposed task was altogether characteristic. First, he argued that electricity (which was, be it noted, for him as for all his contemporaries an effect and not a cause) must be 'the action of a matter in motion between the electrified body and that on which it makes its mark'. This matter, he insisted, could plainly be seen, smelled, heard and felt upon the skin, and, when brought to bear on inflammable liquors or vapours, could set them alight. With this point established, the next move was obvious: 'I search Nature for some subtle fluid known on other grounds, or at least supposed and admitted by the majority of physicists, a fluid which has characteristics similar to those of the matter which brings about electricity'. If he found such a substance, he argued, could he not also legitimately attribute electrical effects to it? To suppose otherwise, he implied, would be to violate Ockham's razor and multiply entities without necessity.[9] After such a build-up, it comes as no surprise that Nollet had a suitable candidate ready to hand. The interesting thing is the particular substance he had in mind, for he here brings us face to face with one of the great unifying themes of eighteenth-century physics. 'These characteristics', Nollet announced, 'are those of fire properly so-called, of that element which is supposed for all sorts of reasons to be present everywhere, not only in fluid media and around solid bodies, but in their interior as well, and even in the pores of their integrant parts'. And he proceeded to spell out in detail a series of properties that the matter of fire and the matter of electricity had in common.

Following the pioneering efforts of the late Hélène Metzger,[10] there has developed among historians of science in recent years a growing recognition that the doctrine, that the world is everywhere pervaded by a subtle material Fire, occupied

[6] Nollet, *Leçons de physique expérimentale*, vol. 1 (Paris, 1743), xviii–xix.

[7] E. Yamazaki, 'L'Abbé Nollet et Benjamin Franklin', *Japanese studies in the history of science*, **15** (1976), 37–64.

[8] Nollet, 'Conjectures ...' (footnote 5), 113–123.

[9] *Ibid.*, p. 113.

[10] Metzger, *Newton, Stahl, Boerhaave et la doctrine chimique* (Paris, 1930).

an important position in eighteenth-century scientific thinking. This notion, unfamiliar to us, is usually associated with the name of the famous Dutch professor Herman Boerhaave because, although it was certainly not altogether original with him,[11] he did expound it systematically and at considerable length in his well-known text *Elementa chemiae* (Leyden, 1732), and it was chiefly from this source that it was taken up by later eighteenth-century authors.

The impact on Nollet's thinking of his personal contacts with the other leading Dutch authorities of the day, 'sGravesande and Musschenbroek, is well known.[12] While in Leyden he would also almost certainly have met Boerhaave. Neither in Nollet's paper on the theory of electricity, however, nor elsewhere that I am aware of, does he acknowledge that the Dutchman's doctrines concerning fire might have played a part in the development of his own ideas on the subject. Instead, he always presented these ideas as notions that were generally accepted in his day. On the other hand, in the long section on fire in his very popular *Leçons de physique expérimentale*, Nollet regularly used Boerhaave as his chief authority.[13] At the very least, therefore, we may be confident that the doctrines set out in the *Elementa chemiae* helped to consolidate the Frenchman's thinking.

Certainly the structure of the theory of electricity that Nollet made public in 1745 was intimately bound up with his acceptance of a theory of fire very similar to Boerhaave's. In particular, the fundamental mechanism underlying the theory, his notion that an outward flow of subtle matter from a rubbed body would automatically give rise to a compensating inwardly directed stream, depended entirely on his acceptance of the idea that all the bodies surrounding the one being rubbed—including, of course, the air—were themselves filled with subtle matter. Furthermore, this matter had to be highly elastic, as Boerhaave said was the case with fire, or else Nollet's mechanism would have involved appealing to a very crude form of action at a distance. Finally, Nollet's explanations of the production of sparks, and of the ignition by electrical means of alcohol and other inflammable materials, depended entirely on his identification of the subtle matter involved with fire.

There are good grounds for believing, indeed, that it was Nollet's inability to establish this particular point that had dissuaded him from publishing his ideas earlier. When, in 1745, he did eventually announce his theory, he said he had conceived it more than ten years before, but had not at that stage regarded it as sufficiently well established to warrant publication.[14] What had caused him to change his mind were certain experiments made newly possible by the powerful electrifying machines developed in Germany. With these, continued brush discharges (*aigrettes lumineuses*) and not merely transient sparks could be obtained. More striking still, it was from these machines that sparks strong enough to ignite alcohol and other inflammable liquids were obtained for the first time. Here was the evidence Nollet had been seeking. 'I saw in these new phenomena', he wrote soon afterwards to his friend, the Genevan physicist Jean Jallabert, 'proofs sufficiently strong to establish a system of explanation that I had been thinking about for a long time, but that I had not dared to risk with [only] feeble and equivocal justifications

[11] R. Love, 'Some sources of Herman Boerhaave's concept of fire', *Ambix*, (1972), 157–174.

[12] P. Brunet, *Les physiciens hollandais et la méthode expérimentale en France au XVIIIe siècle* (Paris, 1926), 108*ff*.

[13] Nollet, *Leçons*, vol. 4 (Paris, 1748), 154*ff*.

[14] Nollet, 'Conjectures ..., 107–108.

[for it].'[15] At last the way was open, it seemed, for a grand unification of the previously obscure and very puzzling phenomena of electricity with the more general and—or so it was believed—well established theory of fire. How welcome a prospect this must have seemed not just to Nollet but to many an embattled philosopher at the time!

If we are to understand the initial broad appeal of Nollet's doctrines, therefore, and also their surprising resilience in the face of the Franklinist challenge, it is essential that we recognize the strength of the support they drew from their intimate connections with the theory of fire expounded so effectively by Boerhaave. Several recent authors have been at pains to emphasise the Newtonianism of Boerhaave's outlook, and the importance of his doctrines concerning fire for the later evolution of eighteenth-century Newtonian natural philosophy. Of particular relevance here has been Cohen's demonstration that Boerhaave's teachings were an important source for Franklin's ideas on the subject of electricity.[16] Nollet, however, could in no sense be described as a Newtonian: and yet he depended even more than Franklin did upon the ideas that the Dutch professor elaborated in his book. The work of a major historical figure cannot be easily confined, it appears, to narrow party lines.[17]

[15] Nollet to Jallabert, 19 July 1745: 'Mais ce qui ma le plus occupé, c'est que jay aperçu dans ces nouveaux phenomenes des preuves assez fortes pour établir un systeme d'explication que je meditois depuis longtemps, mais que je n'osois hazarder sur des raisons foibles et equivoques' (Ms. collection, Société d'Histoire et d'Archéologie de Genève).

[16] I. B. Cohen, *Franklin and Newton* (Philadelphia, 1956), 214–234. See also Robert E. Schofield, *Mechanism and materialism* (Princeton, 1970), ch. VII; and Arnold Thackray, *Atoms and powers* (Cambridge, Mass., 1970), 106–113.

[17] This article is a slightly amended version of a paper first presented at the XVth International Congress of the History of Science, Edinburgh, August, 1977.

IX

FRANKLIN'S ELECTRICAL ATMOSPHERES

BENJAMIN FRANKLIN'S enunciation of his so-called 'one-fluid' theory of electricity in the period 1747–55 has often been hailed as an important milestone in the history of physics. So indeed it was, for, with the rapid and widespread acceptance of the new theory, the science of electricity became based for the first time on the view that the electrification of a body involved the accumulation of a 'charge' from elsewhere, rather than the excitation of matter already present in the body. Only a little more remotely, the triumph of Franklin's theory over the rival 'excitation' theory espoused by the Abbé Nollet paved the way for the more sophisticated fluid theories propounded by Aepinus, Cavendish, and Coulomb in the years that followed.

It has been generally recognized by historians that these later theories differed from Franklin's in a number of important ways, notably in their introduction of additional assumptions in order to render the foundations of the theory more consistent, and in their rejection of Franklin's notion of an atmosphere of electrical matter surrounding a positively charged body. In the latter case, however (and in the former too, for that matter), insufficient attention has been paid to the importance of the change, and to the reasons which prompted it. In Franklin's theory the atmospheres played a central role, particularly in explaining electrostatic attractions and repulsions, but at the same time, they gave rise to some awkward contradictions which seem to have been overlooked by historians. The mere presence of such contradictions tells us something about Franklin himself, and in seeking to discover the extent to which they were recognized, explicitly or implicitly, by his supporters, we can trace some interesting aspects of the transition to a complete reliance on actions at a distance in eighteenth-century electrical theory.

In this paper, then, I shall direct attention to the place of attractions and repulsions in Franklin's theory, and consequently to the role of his electrical atmospheres. I shall also discuss the attitude of some of the more important of his followers to this aspect of his theory, and I shall draw attention to some interesting modifications which they introduced into the theory at this point in an effort to render it more satisfactory.

The development of Franklin's theory

The phenomena which eighteenth-century 'electricians' sought to

An earlier version of this paper was read at a conference of the Australasian Association for the History and Philosophy of Science held in Melbourne in August 1968. Subsequent research was supported by a grant from the Australian Research Grants Committee.

explain fell into two broad categories. On the one hand, there were various effects—particularly attractions and repulsions—associated with static charges on bodies. On the other, there were all the phenomena associated with the processes of charging and discharging of bodies, whether simple conductors or (from 1746) Leyden jars.[1]

A perusal of Franklin's writings on electricity[2] reveals that throughout his electrical investigations he was interested primarily in phenomena of the second kind. His first significant letter on the subject (dated 11 July 1747) began with a discussion of 'the wonderful effect of pointed bodies, both in drawing off and throwing off the electrical fire', after which he went on to describe the circumstances in which a man could receive a charge, and then pass it on to another.[3] In succeeding letters spread over the next few years he was concerned mainly with establishing the identity of lightning and the electrical discharge, with the use of pointed conductors in protecting buildings from lightning, with the mechanism by which clouds might become electrified, and, above all, with the behaviour of the Leyden jar, that extraordinary electrical device whose ability to deliver powerful shocks had been discovered shortly before he began his researches. In these early papers electrical attractions and repulsions were mentioned only incidentally, and then simply as experimental facts; no significant attempt was made to account for them. For example, the repulsion from a charged body of a cork ball suspended on an insulating thread was used at one point as a convenient test of the electrification of either object; the return of the ball when a pointed bodkin was held near either of them was taken as evidence that the point had discharged the body concerned.[4] A 'counterfeit spider' was arranged so that it would leap back and forth between the wire of a charged Leyden jar and an earthed wire nearby.[5] A suspended cork ball was found to be first attracted and then repelled by the wire of a charged jar, and while in the latter condition, it was found to be attracted to the outer coating of the jar.[6]

Franklin set out the basic ideas upon which his theory of electricity was founded in his first communication on the subject, the letter of July 1747. A body showed signs of electricity, he argued, when it contained either more or less than its normal quantity of electrical fluid. Accordingly, he distinguished the two possible states of electrification as 'positive' or 'plus', and 'negative' or 'minus' respectively. In his next letter he argued

[1] This is of course a modern distinction. As we shall see, not all of Franklin's followers would have accepted it.

[2] Except where otherwise stated, all references to Franklin's electrical writings are to his *Experiments and observations on electricity, made at Philadelphia in America* in the edition prepared by I. Bernard Cohen and entitled *Benjamin Franklin's experiments* (Cambridge, Mass., 1941). Hereafter this work will be cited as *Experiments*.

[3] *Experiments*, Letter II, pp. 171–8.

[4] Ibid., p. 172.

[5] Ibid., p. 177.

[6] Ibid., Letter III, p. 182.

that when a Leyden jar was charged in the normal manner, its wire and inner surface became charged 'plus', and its outer surface 'minus'. He was able to conclude from the experiments described above that the cork ball, when charged 'plus', would be repelled by a positively charged body, but attracted by one which was neutral or negatively charged.[7]

What I wish to emphasize here is the order in which Franklin's ideas developed. When he set out to provide a workable theory of electricity, Franklin did not begin by trying to account for the phenomena of attraction and repulsion; on the contrary, his notion of what it was to be electrified arose out of a consideration of his experiments on the sharing of charges between electrified and unelectrified bodies, and it was only after he had formulated this notion that he tried to reconcile the observed attractions and repulsions with it.[8] Hence the generalization set out at the end of the previous paragraph should not be interpreted as a statement of one of the facts to be used in setting up the theory, nor as a statement of something which was an immediate consequence of the theory. It was simply an experimental report, made after the basic theory had been formulated, of another fact with which the theory had yet to be reconciled. Exactly the same view must be taken of a statement to be found in the succeeding letter, Letter IV, in which Franklin recognized the awkward fact that two negatively charged bodies repelled each other just as two positively charged ones did.[9] Similarly, when, as a preliminary to his discussion in Letter V of the origins of atmospheric electricity, Franklin stated that 'every particle of matter electrified is repelled by every other particle equally electrified', he was once again giving only a general summary of the observed behaviour of electrified particles of matter (as distinct from the later theoretical account of the behaviour of particles of electrical matter).[10]

To this stage, then, Franklin had yet to give a detailed account of the nature and properties of his electrical fluid. From his experiments on

[7] Ibid., p. 185.

[8] In this his procedure was in sharp contrast with that of most of his predecessors; indeed, a strong case could probably be made in support of the claim that it was precisely at this point that Franklin took the crucial step towards his novel theoretical views. It might be added that one reason why Franklin paid so little attention in his early work to attractions and repulsions was that generally he did not have recourse to them to test whether or not a body was electrified at any particular stage of an investigation; except in the case mentioned above, he almost invariably tested bodies for electrification by seeing whether he could draw a spark from them.

[9] *Experiments*, p. 199. To be sure, Franklin described this result as one 'which surprises us, and is not hitherto satisfactorily accounted for', which could be taken to imply that he felt he had satisfactorily accounted for the other cases of attraction and repulsion. However, in view of his total failure up to this point to provide any theoretical discussion of these other cases, I prefer to interpret his remark as meaning only that whereas it was reasonable to suppose that an accumulation of fluid might give rise to such forces, it was surprising that an absence of fluid (that is, an absence of a cause) could do so.

[10] Ibid., p. 202. Franklin himself emphasized the point I am making, for in support of his assertion he added:

> 'Thus the stream of a fountain, naturally dense and continual, when electrified, will separate and spread in the form of a brush, every drop endeavouring to recede from every other drop. But on taking out the electrical fire they close again.'

the charging and discharging of bodies he had come to ascribe electrical behaviour to either an excess or a deficiency of the fluid, and, in order to explain the ability of the Leyden jar to accumulate opposite charges on its two surfaces, he had been forced to assert that glass was absolutely impermeable to it. He had made no attempt to account for attractions and repulsions, although he had discovered for himself the empirical laws governing their occurrence. It was only in his paper 'Opinions and conjectures, concerning the properties and effects of the electrical matter . . .', the last major paper included in his original published pamphlet of 1751, that Franklin set out to describe for the first time the properties of the fluid he had invoked.[11] 'The electrical matter', he wrote, 'consists of particles extremely subtile, since it can permeate common matter, even the densest metals, with such ease and freedom as not to receive any perceptible resistance'. Furthermore,

> Electrical matter differs from common matter in this, that the parts of the latter mutually attract, those of the former mutually repel each other . . . But though the particles of electrical matter do repel each other, they are strongly attracted by all other matter.[12]

Historians have often taken it for granted that these propositions were framed specifically for the purpose of explaining the well-known macroscopic electrical attractions and repulsions. Franklin's own subsequent procedure suggests, however, that they should be given a different interpretation, for instead of applying them immediately to these phenomena he used them to explain why the electrical fluid would diffuse over ordinary matter with which it came in contact, as was demanded by his account of the various phenomena associated with the charging and discharging of bodies.[13] Furthermore, although Franklin did not set out in detail anywhere in the paper his understanding of how the laws governing macroscopic attractions and repulsions followed from his propositions, he did say enough about his notion of electrical atmospheres to reveal that he did not see the laws as immediate consequences of those propositions. For reasons difficult for us to grasp today he regarded the atmospheres as a vital intermediate link in the explanatory chain: attractions and repulsions between physical objects at the macroscopic level were to be explained in terms of interactions of some kind between

[11] Ibid., pp. 213–36. The covering letter which accompanied the paper was dated 29 July 1750.

[12] Ibid., p. 213.

[13] He did so in the following terms:
'Thus common matter is a kind of spunge to the electrical fluid. And as a spunge would receive no water if the parts of water were not smaller than the pores of the spunge; and even then but slowly, if there were not a mutual attraction between those parts and the parts of the spunge; and would still imbibe it faster, if the mutual attraction among the parts of the water did not impede, some force being required to separate them; and fastest, if, instead of attraction, there were a mutual repulsion among those parts, which would act in conjunction with the attraction of the spunge. So is the case between the electrical and common matter'. (Ibid.)

an electrical atmosphere and an uncharged (or negatively charged) object or between two atmospheres, as the case might be, and these interactions in turn were to be explained by attractions and repulsions at the microscopic level between the various particles involved.[14]

What were these atmospheres? How did their interactions with matter or with other atmospheres explain the macroscopic attractions and repulsions? How were these interactions to be explained in terms of the forces he postulated between the various particles? From the start Franklin was quite specific about the first of these questions:

> In common matter there is (generally) as much of the electrical as it will contain within its substance. If more is added, it lies without upon the surface, and forms what we call an electrical atmosphere; and then the body is said to be electrified.[15]

That is, an electrical atmosphere was simply a layer of electrical fluid surrounding a charged body; a close analogy could be drawn between such an atmosphere and the common atmosphere surrounding the Earth.[16] A little later Franklin added that 'the form of the electrical atmosphere is that of the body it surrounds'.[17] Concerning the other questions, however, he was much less definite.

In one of his early letters Franklin spoke of a 'sphere of attraction' around an electrified ball, but did so without linking this explicitly with his notion of an electrical atmosphere surrounding the ball.[18] From his discussion of the well-known experiment in which a feather enclosed in a sealed glass vessel was affected by a nearby electrified tube, however, one gains a distinct impression that what was supposed to bring about the

[14] At one stage Franklin did make a very cursory remark which might be taken to be somewhat equivocal on this point, for immediately after stating that the particles of electrical matter mutually repelled each other, he added: 'Hence the appearing divergency in a stream of electrified effluvia'. (Ibid.) If this was a reference to his earlier experiment with a stream of water emerging from an electrified fountain (see note 8), it must stand as an instance in which he did not feel it necessary to invoke electrical atmospheres to explain a macroscopic repulsion. On the other hand, it is more likely that Franklin was referring to the apparent divergence (visible in a darkened room) of the discharge from a pointed conductor, for in his original manuscript he had written of the divergency in a stream of 'electrical' rather than 'electrified' effluvia (*The papers of Benjamin Franklin* [New Haven: Yale University Press, 1959—in progress], iv. 10); furthermore, he had included at this point the footnote referring to John Ellicott's paper in the *Philosophical transactions of the Royal Society* which was transferred in the published version to the end of the next section, and Ellicott had argued at one point in his paper that 'the Existence of these *Effluvia* is proved by all those Experiments in which a Stream of Light is seen to issue from the electrified Body; particularly those Streams which are seen to issue in diverging Rays from the End of the original Conductor, when made of Metal, and reduced to a Point' (*Philosophical transactions*, xlv [1748], 196). In this case it would have been the actual repulsion between the particles which was being observed, and no recourse to atmospheres would have been expected anyway. Even if it was the fountain experiment to which Franklin was referring, however, the emphasis on atmospheres in the rest of the paper, and in succeeding papers as well, is overwhelming.

[15] *Experiments*, p. 214.

[16] The analogy was closer in the eighteenth century than it would be today, for, following Newton, Franklin and his contemporaries regarded air too as being made up of mutually repelling particles.

[17] *Experiments*, p. 216.

[18] Ibid., p. 207.

attraction was an actual contact between a mass of electrical fluid and the feather:

> When an additional quantity of the electrical fluid is applied to the side of the vessel by the atmosphere of the tube, a quantity is repelled and driven out of the inner surface of that side into the vessel, and there affects the feather, returning again into its pores, when the tube with its atmosphere is withdrawn.[19]

In the case of repulsions between two positively charged bodies (forces due to negatively charged bodies were not considered at all), only once did Franklin give any indication of his thinking. He described an experiment in which two balls, A and B, were suspended from insulating threads, with A being positively charged; 'bring A into contact with B', he wrote, 'and half the electrical fluid is communicated, so that each has now an electrical atmosphere, *and therefore they repel each other*.'[20]

Difficulties in the theory

Presumably Franklin did not spell out his ideas in any more detail than has been described because he thought that they were obvious and straightforward. I shall argue in this section, however, that they were neither obvious nor straightforward; and we shall see that, either because they did not understand what he had in mind or, if they did, because they felt it to be unsatisfactory, the majority of his supporters tried to explain the macroscopic attractions and repulsions in rather different ways. Eventually, in 1755, Franklin tried to rectify the situation, but by then it was too late, and in any case his attempt amounted to little more than a re-statement of some of the problems that beset his theory. Many of his most important followers continued to prefer the accounts they had developed themselves.

Consider once more the basis of Franklin's theory: electrically neutral matter contains within it as much electrical fluid as is required to saturate it, and the surplus it acquires when it becomes positively charged gathers around it in the form of an atmosphere. A question immediately arises: why does the atmosphere remain around the body? Franklin gave two separate answers to this question in his paper 'Opinions and conjectures . . .'. The first was that it was held in place by the attraction between the particles of the body and those of the electrical fluid constituting the atmosphere, just as the air is held in place around the Earth by gravitational attraction:

> The form of the electrical atmosphere is that of the body it surrounds . . . And this form it takes, because it is attracted by all parts of the surface of the body . . . Without this attraction, it would not remain round the body, but dissipate in the air.[21]

[19] Ibid., p. 233.
[20] Ibid., p. 216. The emphasis is mine.
[21] Ibid.

Later, in discussing the rubbing of an evacuated glass tube, he invoked the insulating properties of the surrounding air:

> *In vacuo* the electrical fire will fly freely from the inner surface . . . but air resists its motion; for being itself an electric *per se* [i.e. an insulator], it does not attract it, having already its quantity.[22]

Now these two explanations are completely contradictory; if the atmosphere is held to the body by an attractive force, it should not leak off freely *in vacuo*. Furthermore, in the former explanation it is asserted unequivocally that, were it not for the attraction, the atmosphere would 'dissipate in the air'; but in the latter we are told that the air resists such dissipation because of its insulating characteristics. Other problems arise as well, because according to the theory any ordinary body contains a large amount of electrical fluid and this will necessarily exert a repulsive force on any nearby fluid. Unless, therefore, the attraction between particles of ordinary matter and particles of the fluid is consistently greater than the mutual repulsion between particles of the fluid, there will be no resultant attraction to retain an atmosphere in its place; nor will there be any resultant attraction of the kind Franklin wanted to invoke in order to explain the sharing of charge between a charged conductor and an uncharged one when they come in contact, or the attraction between an uncharged body and a positively charged one. Franklin's characterization of the force between the particles of ordinary matter and those of the fluid as 'strong',[23] a term he did not use when discussing the repulsion between particles of fluid, may possibly imply a recognition of this point; but, if so, he certainly did not stress it. What is more, if he did recognize the point, how are we to understand his assertion, when discussing the evacuated tube, that the atmosphere will 'fly freely' from the body *in vacuo*? This surely implies that there is no resultant attractive force retaining the atmosphere in place, and hence that the attractive force due to the ordinary matter of the body is exactly balanced by the repulsive force due to the fluid within it. But, as has just been pointed out, Franklin's explanations of other effects besides the retention of atmospheres on bodies depended on these forces being unequal. Franklin offers us no escape from this dilemma. To allow him the maximum possible consistency in his theory, we should perhaps ignore his statement about the evacuated tube and presume that, for him, the attractive force was always greater than the repulsive one. If we do so, however, we must also remember that, with the possible exception of the single occasion already mentioned, he said nothing which might be taken to indicate his recognition of the need for this additional premise in the theory. His readers in England and on the Continent were

[22] Ibid., p. 233.
[23] Ibid., p. 213.

left to do the best they could to understand what he meant, and it is hardly surprising that they failed to follow him completely.[24]

Consider too the problems associated with the explanation of the observed repulsion between two positively charged bodies that was given by Franklin's theory. It will be recalled that in his first group of published papers Franklin had done no more than hint that when the atmospheres surrounding two bodies came into contact, they repelled each other. But, by analogy with air, it was much more reasonable to suppose that the two atmospheres would simply merge into a single atmosphere, and symmetry would suggest that then the two lumps of ordinary matter would if anything migrate to the centre of the common atmosphere; that is, they would appear to attract each other. This was one of the problems that Franklin tried to solve in 1755, but the best he could do even then was to make a bald assertion that mixing did not occur:

> Electric atmospheres, that flow round non-electric bodies, being brought near each other, do not readily mix and unite into one atmosphere, but remain separate, and repel each other.[25]

Similarly, to explain the facts of electrostatic induction, he had to assume against all analogy that when an atmosphere came into contact with the fluid within a neutral body, the two bodies of fluid remained distinct and did not merge into a single mass:

> An electric atmosphere not only repels another electric atmosphere, but will also repel the electric matter contained in the substance of a body approaching it; and, without joining or mixing with it, force it to other parts of the body that contained it.[26]

As we shall see, it was the kind of inconsistency exemplified here between the behaviour that had to be attributed to the electrical fluid in Franklin's theory, and the normal laws of fluid dynamics, that Franklin's supporters sought particularly to avoid in their own elaborations of the theory.

[24] There are certain dangers in choosing to ignore Franklin's statement about the evacuated tube, for in a later discussion of the relation between an electrical atmosphere and air he made an identical remark:

> 'Dry air assists in confining the electrical atmosphere to the body it surrounds, and prevents its dissipating: for in vacuo it quits easily, and points operate stronger . . . Yet an electric atmosphere and air do not seem to exclude each other, for we breathe freely in such an atmosphere, and dry air will blow through it without displacing or driving it away . . . An electrical atmosphere raised round a thick wire, inserted in a phial of air, drives out none of the air, nor on withdrawing that atmosphere will any air rush in.' (Ibid., p. 248.)

Similar remarks recur throughout the later letters. Perhaps the safest conclusion to draw is that Franklin himself did not fully appreciate the dangers to his whole concept of atmospheres inherent in such remarks.

[25] Ibid., p. 302.

[26] Ibid. Since mixing would only become a problem once contact between the two masses of fluid had occurred, Franklin's concern here to rule it out as a possibility confirms that it was indeed actual contact between the two masses of fluid that he had in mind as the cause of their mutual repulsion.

Finally, nowhere did Franklin attempt an explanation of the observed repulsion between two negatively charged bodies: even in 1755, all he could do was to state the facts of the matter.[27] Some years later his friend and fellow-American, Ebenezer Kinnersley, suggested to him that this and all other observed cases of repulsion could be reduced to differential attractions between the bodies concerned and the surrounding air,[28] but Franklin did not regard the suggestion as a satisfactory solution to the problem:

> You know I have always looked upon and mentioned the equal repulsion in cases of positive and of negative electricity, as a phaenomenon difficult to be explained. I have sometimes, too, been inclined with you, to resolve all into attraction; but besides that attraction seems in itself as unintelligible as repulsion, there are some appearances of repulsion that I cannot so easily explain by attraction.[29]

As a matter of fact, at the time Franklin wrote these lines there had been in print for over two years an account of the place of attractions and repulsions within the theory which was far more satisfactory than the one he had put forward. In his *Tentamen theoriae electricitatis et magnetismi*[30] Franz Aepinus had simply done away with the atmospheres, asserting that the surplus of fluid on a positively charged body was confined within the body itself. He regarded the observed macroscopic attractions and repulsions as immediate consequences of the attractive and repulsive forces between particles of fluid and those of ordinary matter (inserting, in order to render the whole account consistent with observation, the additional hypothesis that, like the particles of fluid, the particles of ordinary matter also mutually repelled one another). He saw no need to invoke, as Franklin had done, an intermediate step in the explanatory chain, and so he was able to avoid the problems Franklin had encountered at this point.[31]

Unfortunately Aepinus's book did not receive the attention it deserved, and many people continued to adhere to the theory of atmospheres for a number of years afterwards. In most cases, however, in their discussions they did not interpret these atmospheres in the same way as Franklin

[27] Ibid.

[28] Ibid., p. 349.

[29] Ibid., p. 365. Does Franklin's admission that attraction was 'unintelligible' mean that he was by then (1762) aware of the problems his theory confronted even in accounting for attractions, or is it simply a recognition that all the forces he had invoked lacked a 'mechanical' explanation of the Cartesian kind?

[30] St Petersburg, 1759.

[31] In a very real sense, then, the action-at-a-distance theory of electricity was due to Aepinus and not to Franklin (a point that has been made already by E. T. Whittaker in his classic *History of the theories of aether and electricity* [London, 1910], p. 48). On the other hand, Franklin's forces between particles were probably not confined in their action to microscopic distances, since his atmospheres extended to considerable distances around charged bodies, and the particles in their outermost layers were apparently still held in place by the attraction of the particles of ordinary matter at the centre. It therefore becomes even less clear why Franklin thought it necessary to invoke atmospheres at all, but invoke them he certainly did.

had done, and I have suggested already that this was due, at least in part, to Franklin's early failure to elaborate on his ideas sufficiently, and probably also in part to a recognition that Franklin's ideas were inadequate. It is to these variants of the theory of atmospheres that we must now turn.

Reactions to Franklin's ideas

Franklin's theory of attractions and repulsions involved, as has already been pointed out, *static* interactions between electrical atmospheres, or between an atmosphere and lumps of ordinary matter. By contrast, in the variant theories about to be discussed, the interactions were *dynamic* ones, involving a transfer of fluid as the observed motions occurred. They therefore bore a superficial resemblance to the earlier excitation theory of electricity, since this too invoked streams of electrical fluid as the cause of the motions. As we shall see, however, the resemblance was no more than superficial, since in the excitation theory the motions were supposed to be a direct mechanical consequence of the impact of the streams of fluid on the bodies moved, whereas this was not the case in the theories now to be considered. Nevertheless, it seems plausible to suppose that the relationship was more than accidental: as they struggled to understand what Franklin was suggesting, his European contemporaries could not help but be influenced by what they knew of the earlier theory.

Beyond this, however, the dynamical approach was simply a natural development of the ideas Franklin had described in his early letters. All ordinary matter, Franklin had suggested, was under normal conditions saturated with its 'natural' quantity of electrical fluid; electrification amounted to an alteration in the distribution of fluid amongst the various samples of matter involved. Since the fluid was also assumed to be highly elastic, such a state of affairs would always be extremely unstable. What was more natural, then, than that Franklin's supporters should have understood him to be saying that all electrical phenomena—including attractions and repulsions—occurred as the equilibrium of the fluid was restored after its initial disruption during the process of electrification? In this regard, the widespread preference among his supporters for the verbs 'condense' and 'rarify', rather than Franklin's own 'accumulate' and 'subtract',[32] when explaining what was involved in charging bodies 'plus' and 'minus', is surely significant: Franklin's readers simply took

[32] See, for example, the report in the *Gentleman's magazine* of the dispute in Paris between Nollet and the Franklinist Jean Baptiste LeRoy (*Gentleman's magazine*, xxix [1759], 319), as also LeRoy's own definition of his terms, as set out in the title of his initial contribution to the debate: 'Mémoire sur l'électricité, où l'on montre par une suite d'expériences, qu'il y a deux espèces d'électricités, l'une produite par la condensation du fluide électrique, & l'autre par sa raréfaction; & qu'elles ont chacune des phénomènes particuliers qui les caractérisent parfaitement' (*Histoire de l'Académie Royale des Sciences* [1753], Mémoires, p. 447). Professor Russell McCormmach has drawn attention, in a somewhat different context, to the tendency of Franklin's supporters to talk in terms of densities, in his unpublished Ph.D. dissertation 'The electrical researches of Henry Cavendish' (Case Western Reserve University, 1967).

it for granted that what counted in electricity, as in all other systems of fluid mechanics, were not so much the absolute quantities of fluid involved, as their densities.

Similarly, and much to Franklin's chagrin, his followers tended to modify the basic law of electrical attractions to read that any two unequally electrified bodies, rather than any two bodies carrying unlike charges, would attract each other.[33] That this might do violence to the facts was forgotten; what mattered was that electrical repulsions then became a very special case indeed (occurring, so it was said, only between two *equally* charged bodies), and the attractions that were supposed to occur in all other cases could be subsumed happily under the single standard 'hydrodynamical' rubric of 'restoring the equilibrium'.

The confusion engendered by Franklin's early failure to discuss attractions and repulsions is evident even in the earliest discussion of his ideas, that given by William Watson early in 1748 (New Style). In a paper he read to the Royal Society on 21 January of that year Watson inserted a long extract from Franklin's first letter on electricity, commenting at the end that it was the remarkable similarity between Franklin's ideas and his own which had prompted him to do so.[34] Yet when we compare the developed theories of the two men,[35] we find that they differed radically in the explanations they gave of attractions and repulsions. Rather than static interactions after the Franklinian model, Watson had invoked a mechanism almost identical to that given earlier by Nollet; he concluded that the bodies moved in response to the impact of two opposing streams of fluid, one converging to the electrified body, the other diverging from it![36] Obviously the source of Watson's 'misunderstanding' of Franklin to the point where he believed they shared the same theory lay in the latter's failure to discuss attractions and repulsions at all, except for the enigmatical remark—retracted in the published version of the letter—that before receiving Watson's paper he had already 'discovered and demonstrated [the] afflux . . ., as well as [the] efflux'

33 The best-known instance of this is Ebenezer Kinnersley's remark in a letter to Franklin (*Experiments*, p. 350), when referring to the electrical fluid, that 'it [is] one of the established laws of this fluid, that quantities of different densities shall mutually attract each other, in order to restore the equilibrium'. Other cases are described later in this paper. Kinnersley's 'law' evoked from Franklin the response that 'what you give as an established law of the electric fluid . . . is, I think, not well founded, or else not well expressed'. (Ibid., p. 367.) That Franklin, by then living in London, was aware of how his ideas had been re-interpreted by some of his followers, is indicated by his comment (following a re-statement of his own views): 'But this you jokingly call "electrical orthodoxy". It is so with some at present, but not with all; and, perhaps, it may not always be orthodoxy with any body'. (Ibid., p. 368.)

34 William Watson, *Philosophical transactions*, xlv (1748), 100.

35 I. Bernard Cohen has pointed out in his *Franklin and Newton* (Philadelphia: American Philosophical Society, 1956), p. 443, that we have no record of any account by Watson which would fit the date of 'very early last spring' that Watson gave for the elaboration of his own views. It seems unlikely, however, that any such account would have differed significantly from that which he had given in October 1746 (*Philosophical transactions*, xliv [1746–7], 704–49), and so I have based my analysis on the latter.

36 Watson, op. cit. (35), pp. 738–9. Cohen has argued (op. cit. [35], pp. 445–9) that the theories differed in other respects as well.

of the electrical fluid.[37] Only as the later letters reached England could it gradually have become clear to Watson and his colleagues that the doctrine of afflux and efflux played no role in Franklin's theory. Perhaps it was his recognition of this point that prompted Watson to remark at the end of an otherwise favourable review (presented to the Royal Society in June 1751) of the first published collection of Franklin's letters, that 'there are in this work some few opinions, in which I cannot perfectly agree with him'.[38] The sympathetic attitude that Watson continued to adopt towards Nollet's theoretical opinions certainly supports such a view.[39]

The doctrine of a continual afflux and efflux of electrical matter in the region around an electrified body is not only clearly inconsistent with Franklin's notion of an electrical atmosphere as a static condition, it also seemed incompatible with the distinction he had drawn between the two possible modes of electrification, 'plus' and 'minus'. Those who accepted Franklin's account found themselves forced to abandon it as a means of explaining the attractions and repulsions. Yet, as we have seen, Franklin had not provided them with an alternative account. A hint of one solution which soon came to be widely adopted is to be found in an important paper presented to the Royal Society by John Canton on 6 December 1753.[40] Canton had found that if two cork balls were suspended by conducting threads and an electrified glass tube was brought towards them, they would separate while the tube was still quite some distance away. On the other hand, if the suspending threads were insulators, the tube had to be brought much closer before this would happen. Canton attempted to account for these observations as follows:

> As the balls in the first experiment are not insulated, they cannot properly be said to be electrified: but when they hang within the atmosphere of the excited tube, they may attract and condense the electrical fluid round about them, and be separated by the repulsion of its particles. It is conjectured also, that the balls at this time contain less than their common share of the electrical fluid, on account of the repelling power of that which surrounds them; though some, perhaps is continually entering and passing through the threads. And if that be the case, the reason is plain why the balls hung by silk, in the second experiment, must be in a much more dense part of the atmosphere of the tube, before they will repel each other.[41]

It is difficult to grasp Canton's precise meaning here, though it may be seen at once that he relied just as much as Franklin did on the

37 *Experiments*, p. 174.

38 Watson, *Philosophical transactions*, xlvii (1751–2), 210.

39 See Watson's review of Nollet's *Lettres sur l'électricité* (Paris, 1753) in *Philosophical transactions*, xlviii (1753–4), 201–26.

40 Canton's paper was published initially in *Philosophical transactions*, xlviii (1753–4), 350–8. It was subsequently reprinted in *Experiments*, pp. 293–9, and my citations are to the latter source.

41 *Experiments*, p. 294.

real existence of an atmosphere of electricity surrounding a charged body such as his glass tube. Towards the end of the passage just quoted, however, he seems to be suggesting also that part of this atmosphere was continually entering the balls immersed in it and passing away through the supporting threads, and that it was the inhibiting of this process that reduced the effect when insulating threads were used. This impression is immediately confirmed by what follows, when Canton turns to a consideration of what happens when a negatively charged body (in this case an electrified wax rod) is substituted for the positively charged glass tube:

> At the approach of an excited stick of wax to the balls, in the first experiment, the electrical fire is supposed to come through the threads into the balls, and be condensed there, in its passage towards the wax; for, according to Mr. *Franklin*, excited glass *emits* the electrical fluid, but excited wax *receives* it.[42]

In another experiment Canton suspended a metal tube from an insulating thread and observed that when it was charged positively, a pair of cork balls hanging from one end repelled each other. If a rubbed glass tube was then brought near the balls, the repulsion between them decreased. Canton's explanation of this experiment reinforces the dynamical overtones of the earlier passages, for the mutual repulsion of the balls is said to be due to fluid 'running off' through them; if this discharge is decreased, so is the repulsion:

> The bringing the excited glass to the end, or edge of the tin-tube . . . is supposed to electrify it positively, or to add to the electrical fire it before contained; and therefore some will be running off through the balls, and they will repel each other. But at the approach of excited glass, which likewise *emits* the electrical fluid, the discharge of it from the balls will be diminished; or part will be driven back, by a force acting in a contrary direction; and they will come nearer together.[43]

Similar passages recur later in the paper. Apparently for Canton electrical repulsions involved a dynamic and not a static interaction. What is more, it seems that he took this to be what Franklin was proposing.

From the passages just quoted, it is not altogether clear whether Canton supposed that fluid streamed constantly out from positively electrified bodies and in towards negatively electrified ones, or that a flow only took place as attractions or repulsions actually occurred. Certain passages in his later writings suggest that it was the former alternative that he had in mind. In a paper he read to the Royal Society in November 1754, for example, he described how, when a polished glass tube is rubbed in the dark, it 'will at every stroke appear to throw out a great number of diverging pencils of electric fire', and in conjunction with this he remarked:

[42] Ibid.
[43] Ibid., p. 295.

> The resinous and vitreous electricity of Mr. Du Fay . . . received no light till the publication of the second part of Mr. Franklin's experiments; wherein it appears, that the one kind of bodies electrify positively, and the other negatively; that excited glass throws out the electric fire, and excited sulphur drinks it in.[44]

The suggestion seems to be that in the experiment he described, the 'electric fire' being thrown out by the excited glass was being observed directly; that is, in exciting the tube positively a surplus of fluid was collected on it, and once the collecting process ceased, this surplus began to flow outwards. Again, some years later, Canton used very similar terminology to describe the remarkable ability of the tourmaline crystal to acquire opposite charges on its two faces when it was heated:

> The curious in these [i.e. electrical] matters, are now acquainted with a substance, that will, without friction, both emit and absorb the electrical fluid, only by the increase, or diminution of its heat.[45]

Despite these affirmations, however, Canton ultimately abandoned the notion that electrified bodies were surrounded by atmospheres made up of streaming electrical effluvia, for in 1767 Priestley was able to report:

> It is *now* also Mr. Canton's opinion, that electric atmospheres are not made of effluvia from excited or electrified bodies, but that they are only an alteration of the state of the electric fluid contained in, or belonging to the air surrounding them, to a certain distance.[46]

What is more, the context in which Priestley made this statement is significant. It occurs near the end of his discussion of the work which had originated with Canton's experiments of 1753. Earlier in the discussion, having remarked that the principle underlying these experiments 'had not been attended to before; nor indeed do I apprehend it was fully understood, till it was explained in all its extent by Mr. Wilke [*sic*] and Æpinus', Priestley had described the experiments themselves and had reproduced Canton's attempts at explaining them—the very passages quoted above—almost verbatim. After a brief summary of Franklin's contributions to the subject he had then given, together with a cursory reference to Beccaria,[47] a long account of the work of Wilcke and Aepinus, drawn very largely from the latter's *Tentamen.* Immediately afterwards came the statement about Canton's more recent views. Priestley's implica-

[44] Canton, *Philosophical transactions*, xlviii (1753–4), 780–2.

[45] Canton, *Philosophical transactions*, li (1759–60), 403. It needs to be emphasized that Canton was talking at this point about the charges developed on the stone, not the process whereby those charges were developed.

[46] Joseph Priestley, *The history and present state of electricity* (London, 1767), p. 263; the emphasis is mine. Priestley's 'now' provides confirmation, of course, that Canton *had* thought earlier that the atmospheres were composed of 'effluvia'. Furthermore, his characterization of Canton's earlier ideas in terms of 'effluvia *from* excited . . . bodies' shows that in representing those ideas as fundamentally dynamical we have understood Canton's earlier discussions correctly.

[47] See below, note 59.

tion seems to be that Canton's change of heart derived from a knowledge of what Wilcke and particularly Aepinus had done.[48]

During the 1750s French electrical investigators were just as unclear as were their English counterparts as to how Franklin proposed to explain the electrical motions. LeRoy, Franklin's most vociferous French supporter, was one of those who interpreted the 'plus' and 'minus' terminology in a rather different way from that which Franklin had intended, as the dynamically more relevant 'condensation' and 'raréfaction'. On the other side of the debate, Nollet clearly did not realize either how radically new a view of attractions and repulsions Franklin had in mind. As we have seen was the case with Canton, Nollet too thought that Franklin, in his theory of 'plus' and 'minus', was merely rejecting the concept of a double stream of electrical matter around electrified bodies in favour of a stream in one direction only; this is clear from the attack he levelled at what he took to be Franklin's account of the motion of a feather inside a hermetically sealed glass container when a rubbed glass tube was held near the outside of the container:

> Je vous demande pourquoi la plume n'est pas constamment repoussée pendant tout le tems qu'on tient le tube à une petite distance du vaisseau? pourquoi éprouve-t-elle des attractions aussi-bien que des répulsions? Je vois bien que le feu électrique peut l'emporter & l'éloigner, de la parois d'où il sort; mais par quelle raison la plume y revient-elle, avant qu'on puisse supposer le retour de ce feu? & dans vos principes on ne le peut pas, tant que le tube approché extérieurement continue de porter de nouveau feu, dans le demi-épaisseur du verre.[49]

Evidently Nollet supposed that, according to Franklin's theory, a stream of fluid flowed constantly away from positively charged bodies (and presumably towards those which were negatively charged). Furthermore, he simply took it for granted that this flow of fluid would be supposed to act in a straightforward mechanical fashion in bringing about the observed attractions and repulsions.[50]

In contrast with the views of Canton and Nollet, a rather different 'hydrodynamical' version of Franklin's theory, and one which enjoyed considerable prestige for a time, was put forward almost simultaneously by the Italian physicist Giambatista Beccaria.[51] Since Beccaria's account

[48] Priestley, op. cit. (46), pp. 246–63. Canton certainly knew of Aepinus's work, and Priestley was aware of this, since the latter, in preparing his *History*, used a copy of the *Tentamen* which he had borrowed through Canton; indeed, Canton may even have recommended the book to him in the first place. I discuss these and some related points in my paper 'Aepinus and the British electricians: the dissemination of a scientific theory', *Isis*, lxiii (1972), 190–204.

[49] Nollet, *Lettres sur l'électricité* (Paris, 1753), p. 63.

[50] Canton at least seems not to have fallen into this latter trap; see the passage from *Experiments*, p. 294, quoted above, p. 142.

[51] Giambatista Beccaria, *Dell' elettricismo artificiale e naturale* (Turin, 1753). Beccaria's ideas became available to a rather wider audience when a long summary of his views was published in French a few months later under the title *Lettre sur l'électricité, par le R. P. J. B. Beccaria . . .* (Paris, 1754).

soon became perhaps the most authoritative statement of the Franklinian theory as a whole,[52] it will repay some attention.

Beccaria, the professor of experimental physics at the University of Turin, began his treatise by defining what he called 'the signs of electricity' characterizing an electrified body; these included the giving of sparks, the sensation of the electric wind, *and the attraction of light objects*. Having done so, he set out what he took to be the fundamental principle of electricity, which was: '*All* the signs of electricity are due to the vapour expanding itself from a body in which it is in greater quantity into one in which there is less.'[53]

For Beccaria, as for Franklin, electrification was essentially a static condition in which a body contained either more or less than its usual complement of electrical fluid. He differed from Franklin, however, by adopting a totally dynamical outlook when it came to explaining the effects of electrification; that is, he supposed that whenever another body was attracted to the one that was electrified, a flow of fluid occurred from the body which had the greater quantity to that which had less, just as it did in giving rise to all the other electrical appearances. Obviously the stream of fluid which Beccaria envisaged could not provide a simple mechanical explanation of the attraction; the cause of the motion remained mysterious and unknown. Beccaria ascribed it to 'the force of the electrical vapour which expands itself from the body in which there is more into the one in which there is less until equilibrium is reached',[54] but when faced with the necessity of explaining how such a force could give rise to the observed motions, he declined to do so, at least for the time being. In support of his stand, he cited Newton's well-known positivistic views on methodology.[55]

Beccaria's 'law' related only to attractions. To account for the repulsions that were also observed, he had to reduce them to differential attractions. The equilibrium could be restored, he said, by a slow flow of fluid (and hence, by his law, an attraction) between each body and the external air. The separation of the bodies followed, he argued, because

[52] Franklin himself was very favourably impressed by Beccaria's book, and upon its appearance he abandoned his own plans for composing a formal reply to Nollet's onslaught on his theory; apparently he regarded Beccaria's efforts as definitive (*Experiments*, p. 307; see also Antonio Pace, *Benjamin Franklin and Italy* [Philadelphia: American Philosophical Society, 1958], pp. 49–53). Similarly, DeLor, who prepared the translation and arranged for the publication of Beccaria's *Lettre*, remarked in his 'Avertissement' to that work that he had been preparing a defence of Franklin's views but had abandoned his plans when he had found that Beccaria had already done so 'd'une maniére bien supérieure à ce que j'aurois pu faire, & tout-à-fait conforme aux principes que j'avois embrassés'. He had therefore contented himself with making the latter's views available to French readers (*Lettre*, pp. vi–vii). One cannot help wondering whether Franklin's acquiescence (at this stage at least) in Beccaria's dynamical theory was intended—or had he simply not paid sufficient attention to what Beccaria was saying?

[53] Beccaria, *Dell' elettricismo*, p. 17; the emphasis is mine. The original Italian reads: 'ogni segno elettrico avvenga pel vapore, che da un corpo, in cui è in quantità maggiore si espande nell' altro, in cui è in minore quantità'.

[54] Ibid., p. 40.

[55] Ibid.

only the external air was involved; the air between the bodies was replete with the fluid constituting the atmospheres of the two bodies.[56]

Beccaria's *Dell' elettricismo artificiale e naturale* was only the first of a number of important works that he was to publish on electricity. In 1758 his next major production appeared,[57] and in it he expounded the same general principles that he had put forward five years earlier. Once again his chief concern was to establish a workable system of electrical fluid dynamics that would explain the various electrical appearances. Hence he argued at the outset that, through its 'natural diffusive force', the electrical 'vapour' would make continual efforts to expand. If the quantity of vapour in one of the bodies in a material system was altered in a different proportion from that of the other, the diffusive force would strive to restore the equilibrium; it was the resulting motion of fluid across the intervening air from one body to another which gave rise to all those effects included under the general rubric 'electricity'.[58] The air offered a considerable resistance to this passage of fluid, however, with the result that an excess of fluid could form temporarily in and around a suitably insulated body (the fluid collected around the body forming the usual Franklinian electrical atmosphere, of course).[59]

Just as in 1753, Beccaria included the attraction of light objects among the phenomena brought about by the flow of electrical fluid. Now, however, he no longer retreated into Newtonian-style agnosticism when it came to explaining how such a flow could cause two bodies to approach each other. Instead he described a series of experiments which demonstrated, he said, that the air too played a crucial role. More specifically, he suggested that what happened was that, in flowing from one body to the other, the electrical fluid drove out the intervening air, upon which the pressure of the external air drove the bodies inwards towards each other, into the resulting central low-pressure region:

[56] Ibid., pp. 26–8.

[57] *Dell' elettricismo. Lettere di Giambattista Beccaria . . . dirette al chiarissimo sig. Giacomo Bartolomeo Beccari . . .* (Bologna, 1758).

[58] Ibid., pp. 4–5.

[59] Ibid., p. 12. Some years afterwards both Priestley (op. cit. [46], p. 262) and Beccaria himself (*Elettricismo artificiale* [Turin, 1772], p. 173) claimed priority for the latter, over Aepinus, as the first person to abandon belief in Franklinian-style atmospheres—though Priestley at least refused to commit himself absolutely at this point and inserted the word 'probably' in his account. In support, both men referred to 'Elettricismo artificiale, p. 54', which can only refer to Book I of Beccaria's *Dell' elettricismo artificiale e naturale* of 1753. The relevant passage is somewhat obscure but does not really support the claim; and neither does the explanation (noted above) of the mutual repulsion of two equally charged bodies, which had been given only a few pages earlier. Quite apart from this, however, Beccaria's description in the passage now being considered of what happens when the prime conductor of an electrical machine draws its charge from the globe being rubbed proves beyond doubt that even in 1758 he was still expounding very orthodox views on the subject of atmospheres; the charge developed on the globe, he assured his readers, was constantly discharged into the chain (i.e. the prime conductor), 'e così e in essa, e intorno ad essa si accumulerà un eccesso di vapore, il quale per la sua forza diffusiva si spanderà alquanto nell' aria, ma dalla resistenza, e inazione di lei sarà intorno alla catena ritenuto, e quasi contro di essa ripercosso'.

> If the vapour diffusing itself from A, which has more of it, into B, which has less, displaces the air between A and B, the air which presses upon A and B from the outside will prevail; whence A and B, impelled by this excess of pressure, will come together with an equal motion proportional to the quantity of vapour which passes between the bodies.[60]

Beccaria's account was a most influential one for a time.[61] For many of those whose area of specialization lay outside electricity, his works became the standard references on the subject. Boscovich, for example, in trying to reconcile electrical phenomena with his revolutionary general theory of matter, characterized the relevant part of Franklin's theory, 'greatly embellished & confirmed . . . by Fr. Beccaria', in the following terms:

> If these bodies approach one another to within that distance, for which the particles surrounding the bodies, & adhering to them like atmospheres, can act upon one another; then from the body that is electric by excess this fluid will immediately flow towards the one that is electric by defect, until equality is reached. During this flow, the substances which respectively yield & receive the fluid will simultaneously approach one another.[62]

Similarly, in their discussions as to whether or not the nervous fluid they had invoked in their researches into muscle physiology was identical with the electrical fluid, Albrecht von Haller and his correspondents Felice Fontana and Marc Antoine Caldani cited Beccaria as their authority in electrical matters, and they assumed with him that only when a flow of fluid occurred would an attraction take place.[63]

Ultimately, however, Beccaria himself—very probably under the influence of Aepinus—quietly discarded many of the central tenets of the theory he had expounded. He did so in his third major treatise on electricity, published in 1772.[64] Two of the changes he introduced are of particular relevance to us here. First, Beccaria, like Aepinus before him,

[60] Beccaria, op. cit. (57), pp. 42–3. The original Italian is: 'Se il vapore diffondendosi da A, che ne ha più, in B, che ne ha meno, discaccia l'aria, ch'è tra A, e B, prevalerà l'aria, che al di fuori preme A, e B; onde A, e B urtati da quest' eccesso di pressione, si avvicineranno con moto eguale, e proporzionale alla quantità del vapore, che tra essi trascorrerà'. Beccaria at first presented his views somewhat hesitantly, but in a letter to Franklin soon afterwards he spoke with rather more confidence (*Philosophical transactions*, li [1759–60], 514–26; English translation in Franklin's *Papers*, vii, 300–14). Beccaria's ideas have been described more fully in Mario Gliozzi, 'Beccaria nella storia dell' elettricità', *Archeion*, xvii (1935), 15–47; they bear a strong resemblance to those of the earlier electrical theorists, particularly Hauksbee (see, for example, my article 'Francis Hauksbee's theory of electricity', *Archive for history of exact sciences*, iii [1967], 203–17).

[61] See above, note 52, for its impact on Franklin himself and on his disciple DeLor.

[62] Roger Boscovich, *A theory of natural philosophy* (Cambridge, Mass.: M.I.T. Press, 1966; English translation of the edition of 1763), p. 181.

[63] See, for example, Albrecht von Haller, *Mémoires sur la nature sensible et irritable, des parties du corps animal* (Lausanne, 1756–60), iii. 207. See also my paper 'Electricity and the nervous fluid', *Journal of the history of biology*, iii (1970), 235–51, for a discussion of this point.

[64] *Elettricismo artificiale*, cited in note 59. I have also used the English translation of this which appeared a few years later with the title *A treatise upon artificial electricity* (London, 1776). Beccaria had certainly read Aepinus by the time he committed his new views to print, since he cited the latter's *Tentamen* explicitly at one point (*Elettricismo artificiale* p. 175). See also note 65 below.

now explicitly denied the existence of electrical atmospheres around charged bodies.[65] Secondly, he completely abandoned the 'hydrodynamical' account of electrical attractions and repulsions which had occupied so prominent a place in his earlier writings. Instead, he drew the same careful distinction that Aepinus had drawn thirteen years earlier in setting out the foundations of *his* theory, a distinction between those phenomena which arise from an actual transfer of electrical fluid, and those (namely attractions and repulsions) which depend rather on the forces exerted by static accumulations of it![66]

Apart from Aepinus and his student Wilcke, Canton and Beccaria were probably the most important Franklinist electricians in the first twenty years of the theory's existence. We have now seen that for a time both of them gave dynamical accounts of the observed electrical attractions and repulsions, but that they both eventually gave up these attempts to render the Franklinian theory of atmospheres more satisfactory and abandoned the theory instead (and we have also seen that in both cases, Aepinus's influence may have been decisive). Canton and Beccaria seem to have changed their views before most of their contemporaries, however, for we learn from elsewhere that the dynamical view continued to be the commonly accepted one for some time to come. Henry Cavendish, for example, in a draft composed some time between 1767 and 1771, reported views very similar to Canton's as the prevailing norm:

> It has usually been supposed that two bodies, whenever the electricity either runs into or out of both of them, repel each other; but that when it runs into one and out of the other, they attract.[67]

Likewise, there were strong dynamical overtones to Priestley's 1767 report of the standard 'Franklinian' account of the interaction between two bodies carrying opposite charges: this interaction, he tells us, was thought to involve a 'restoration of the equilibrium'.[68] Furthermore, Priestley allowed this passage to remain unaltered even in the third (1775) edition of his book.[69] Only gradually did the static theory replace such views, and when it did so, the atmospheres themselves had also disappeared from the scene.

[65] Beccaria, *Treatise*, p. 179. Significantly, it was precisely to support this assertion that Beccaria cited Aepinus.

[66] Ibid., pp. 1–2. Aepinus's statement of the distinction is to be found in his *Tentamen*, p. 10.

[67] *The electrical researches of the Honourable Henry Cavendish, F.R.S.*, ed. James Clerk Maxwell (Cambridge, 1879), p. 100. For the dating of Cavendish's draft see McCormmach, op. cit. (32), pp. 146–7.

[68] Priestley, op. cit. (46), p. 457. It should be noted that in the case of repulsions between bodies carrying like charges Priestley presented Franklin's own account, and not any of those put forward by his commentators, as the standard version.

[69] Ibid. (3rd edn., London, 1775), ii. 28. Priestley also left unaltered a most interesting comment on the famous 'electrified cup' experiment. In his usual eclectic fashion, besides drawing his well-known inference that the electrical force is inverse-square in character, he added: 'Do not these experiments, likewise, favour the hypothesis of S. Beccaria, that there is no electrical attraction without a communication of electricity?' (Ibid., ii. 374).

Conclusion

I have argued in this paper that even though Franklin's theory of electricity invoked attractions and repulsions between the particles of matter, these were not introduced by him to explain the observed macroscopic electrical attractions and repulsions, nor were they regarded by him or by any of his contemporaries before Aepinus as a sufficient explanation of the macroscopic motions. As an intermediary in the explanatory chain, Franklin invoked his notion of electrical atmospheres, but for a number of years he did not give a sufficiently detailed account of these to reveal to his European readers what he had in mind. When he finally did so, some of his most important supporters found his ideas far from satisfying, since the behaviour he was forced to attribute to the electrical fluid did violence to all the normal laws governing the behaviour of fluids. Not surprisingly, in the absence of any lead from the master, these supporters meanwhile had developed Franklin's initial remarks about atmospheres along lines more in keeping with the known laws of fluid dynamics and with accepted ideas about the causes of the electrical actions. Franklin's subsequent elaborations did nothing to persuade them to abandon this approach. It was not until some years later that such leading Franklinists as Canton and Beccaria discarded their dynamical theories, and when they did so, it was not to adopt Franklin's ideas but to abandon the theory of atmospheres altogether.

The transition from the older effluvial theories of electricity to the classical action-at-a-distance fluid theory constitutes one of the most important 'revolutions' carried through by eighteenth-century science. It also marks an important stage in the spread of the Newtonian scientific tradition generally. From what has been said in this paper, it is clear that in this particular case the revolution did not take place all at once; the theory of atmospheres survived for a period of twenty years or so before it was finally abandoned and the way was opened for a more widespread acceptance of ideas of the kind which had been put forward by Aepinus. More importantly, the time lag in this instance was not just the standard one resulting from the longevity of the upholders of the older traditions and their entrenchment in positions of influence; in this case the delay arose from the fact that the essentials of the theory that was to emerge from the revolution were simply not all settled at once, but were worked out only over a considerable period. In the meantime, a significant element of continuity with the older theories remained in the ideas of even the most vociferous upholders of the new doctrines.

Beyond this, it is clear that, despite the successes of Newtonian gravitational theory, there was an extreme reluctance, even on the part of those eighteenth-century scientists who were steeped in the Newtonian tradition, to resort to unexplained electrical forces acting at a distance on a macroscopic level. And this should not surprise us. Until the invention of the Leyden jar at least, the attraction of light objects had always been the principal electrical phenomenon in need of explanation, and so the onus was very much on the proponents of any theory which simply refused to attempt such an explanation to show that there were good reasons for not providing one. Aepinus, with his background in Newtonian-style mathematics and astronomy, obviously found it easy to take the crucial step, but others, less mathematically inclined, took much longer to be persuaded.

X

SOME MANUSCRIPTS ON ELECTRICAL AND OTHER SUBJECTS ATTRIBUTED TO THOMAS BAYES, F.R.S.

AMONG the Canton Papers in the Library of the Royal Society there are two items, written in a very distinctive and different hand from the other documents in the collection, that have been attributed to the mathematician Thomas Bayes (1702–1761), discoverer of the theorem named in his honour that is so well known to present-day statisticians (1). A notebook in a similar hand is preserved elsewhere in London, in the archives of the Equitable Life Assurance Society, and it too has been attributed to Bayes (2). Parts of these documents are in a shorthand similar in structure to but different in some of its details from the systems most commonly used in England in the first half of the eighteenth century, and the fact that the same slightly idiosyncratic system of shorthand is used throughout reinforces the evidence of the handwriting as to their common authorship. All three documents were, it appears, drawn up by one and the same person.

There is nothing about any of the documents to prove beyond all doubt that that person was Bayes, nor does what is known of their provenance link them definitively with him. The attribution to Bayes of the notebook at the Equitable Life offices is in fact certainly derivative, since it was made merely through a comparison of the handwriting with that used in the documents at the Royal Society (2); but nothing positive is known in regard to these latter documents themselves. Indeed, the grounds upon which they were attributed to Bayes in the first place are uncertain, as is the identity of the person who made the attribution. Probably, however, it was based upon a comparison with the very similar (and similarly distinctive) handwriting in a letter from Bayes to Canton preserved elsewhere in the Royal Society's archives (3).

Certainly what we know in general terms of the fate of Bayes's papers after his death makes it perfectly possible that some of them should have finished up among Canton's papers, and others in the archives of the Equitable Life Society. Following Bayes's death his papers were brought to the attention of Richard Price, a fellow dissenting minister who subsequently became, as Bayes had been, a Fellow of the Royal Society. Price evidently went through the papers with some care, and when he found among them the manuscript of Bayes's 'Essay towards solving a Problem in the Doctrine of Chances', the work in which Bayes developed his famous theorem about probabilities, he passed it on to the Royal Society for publication. More significantly from our point of view, Price chose to transmit the manuscript to the Royal Society through Canton, a close friend who, as has been indicated already, had also had contact with Bayes (4). A year later Price transmitted a supplement to the 'Essay' to the Royal Society via the same route (5). It is reasonable to suppose that if he had come across anything else that he thought might be of interest to the Royal Society, or that dealt with Canton's favourite subject, electricity, he would have passed it on too. And each of the 'Bayes' documents in the Canton Papers fits one or other of these descriptions, for, as we shall see, one of them offers comments on a mathematical question that had been raised in the *Philosophical Transactions* a few years earlier, and the other is concerned with the theory of electricity. Finally, Price became in later years one of the principal actuarial advisers to the early managers of the Equitable Life Assurance Society, and a number of his papers finished up in the Society's archives (6). Had Price earlier come across a notebook while going through Bayes's papers, he might well have chosen to retain it; in this case it could still have been among his own papers at his death, and so could have passed to the assurance society.

Finally, the contents of our documents are entirely compatible with the authorship that has been ascribed to them, for they clearly date from the period when Bayes was most active scientifically, the second third of the eighteenth century. Furthermore, as we would expect if they had been written by Bayes, the interests they reflect are chiefly mathematical and, to a lesser extent, natural-philosophical. All the available evidence—and above all that of the handwriting—confirms, therefore, that, as the inscription in the Canton Papers asserts, Bayes was indeed their author. In what follows, his authorship will simply be assumed.

Much of Bayes's notebook at the Equitable Life is given over to what appear to be notes taken down while reading various mathematical texts of the period. Some mathematical arguments are set out, for example, concerning the motions of the heavenly bodies. Several pages are filled with tables of numbers of various kinds, such as tables of the logarithms of 999, 1001, 9999, 10001, 99999, 100001, etc., and tables, first of even, then of odd numbers that can be broken down into sums of perfect squares. One eight-page segment is headed 'Paris. July. 4. N.S. 1746.', and concerns some electrical experiments performed by the French academician Louis-Guillaume Le Monnier. On closer examination this turns out to be merely a slightly abbreviated transcription (with occasional passages in shorthand) of a letter from the Abbé Turbeville Needham that appeared in the *Philosophical Transactions* for that year (7), and that was devoted in the main to describing Le Monnier's elaborations of the famous 'Leyden experiment' that had burst so spectacularly upon the scientific world only a few months earlier. Elsewhere in the notebook there is a ten-page section on ancient units of length measurement. The last four pages of this are written out almost entirely in shorthand, as are occasional passages earlier on. The piece is headed 'Extract from a translation of a dissertation upon Cubits, written in latin by Sr. Isaac Newton'; it is in fact a very full set of notes extracted from one of Newton's least-known works, his posthumously-published 'Dissertation upon the Sacred Cubit of the Jews . . .' (8). Elsewhere again, the system of shorthand due to Aulay Macaulay (9) is written out in full, but this is not the system that is used in the notes themselves.

At the Royal Society, one of the Bayes items is a brief commentary on a theorem developed by the mathematician Thomas Simpson (1710–1761), and published in the *Philosophical Transactions* a few years before Bayes's death (10). The other is a six-page document on electricity, more than half of it in shorthand. The longhand section at the beginning consists of notes taken from a tract by Benjamin Hoadly and Benjamin Wilson entitled *Observations on a Series of Electrical Experiments* (London, 1756), together with critical comments thereon. The shorthand part that follows turns out when deciphered to be merely further extracts from Hoadly and Wilson's book, but now without any comments interposed.

Some years ago, in the course of a brief account of Bayes's life and times, Mr J. D. Holland drew attention to the existence of the manuscripts just described (11). Subsequently, both in an article on Richard Price and elsewhere (12), he announced the successful decipherment by Mr J. I. Mason of the shorthand used in them. The identification of the works from which our author was quoting now enables us to confirm and complete this decipherment (13).

The system of shorthand used by Bayes was basically that devised a century earlier by Thomas Shelton, but with modifications, some of them due to Elisha Coles (14). Like all known shorthands used prior to the 1760s, it was in the main an alphabetic system rather than a phonetic one; words were written out as spelled but using shortened symbols to represent the consonants and omitting vowels for the most part. The values of the latter were specified instead by the relative positioning of the symbols for the consonants that preceded and followed them, according to the following scheme:

a
    e
* i/y
    o
u

Vowels occurring at the ends of words could be indicated by dots in the appropriate positions, while special symbols were used for vowels at the beginnings of words. Including the latter, the symbols used to represent the various letters of the alphabet were:

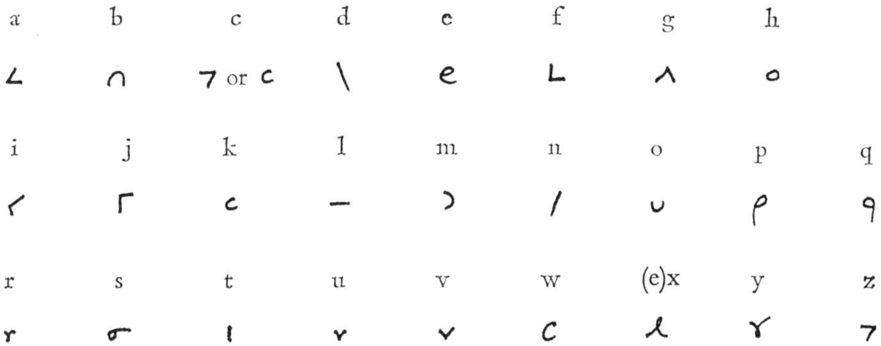

The following examples illustrate the system in operation (15):

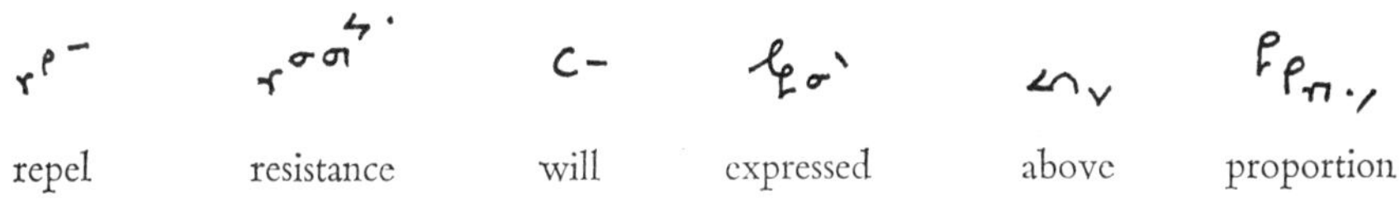

In addition, special symbols were introduced for certain commonly-occurring words and groups of letters, for example:

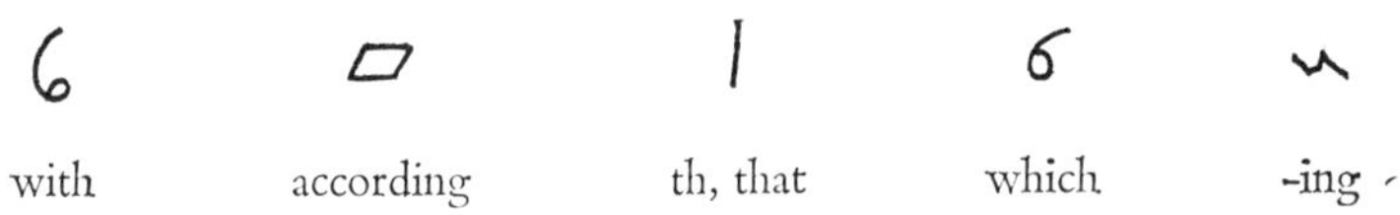

So far as the contents of our documents are concerned, Bayes's comments on the ideas of Hoadly and Wilson, published as an appendix to this note, are the most interesting (though the comments on Simpson's work on the theory of errors also shows considerable insight). They touch, in fact, on some of the central difficulties in electrical theory at that time, and concern the role and indeed the actual existence of the 'atmospheres' of electrical matter that were supposed by the leading theorists of the age to surround electrical bodies. Hence they warrant a little further attention.

The book by Hoadly and Wilson with which Bayes was concerned was but one of a number of works put out by Wilson at about this time with the aim of establishing a theory of electricity that he had developed based on the all-pervading aether whose existence had been postulated some forty years earlier by Sir Isaac Newton in the second English edition of his *Opticks* (16). On the other hand, the notion developed by Wilson in these works of electrical atmospheres as accumulations of electrical matter surrounding electrified bodies was not due to Wilson himself but came rather from Benjamin Franklin, having been set out by the latter in the famous series of letters that he sent from Philadelphia to the Royal Society over a period of years beginning in 1747. It was in these same letters, too, that Franklin had elaborated his important distinction between the two possible modes of electrification, *plus* and *minus* (17). Wilson, in his writings from 1750 on, tried to incorporate these new ideas of Franklin's into his general aether-based account. Naturally he proceeded at once to identify the specifically electrical matter that Franklin had invoked with his own Newtonian-style aether, but Franklin's other ideas gave him rather more difficulty.

In the *Observations* that he and Hoadly published in 1756, Wilson accepted the Franklinian notion that a body could become electrified by either gaining or losing some of its natural share of electrical fluid, or aether. But, he argued, all bodies oppose a certain resistance to either gaining or losing this fluid; most importantly, air offers a powerful resistance to the admission of any excess fluid. In the case of positively-charged bodies, these ideas led Wilson to conclusions similar to those that had been reached by Franklin himself. When we electrify a body positively, he said, not all the surplus fluid involved can be absorbed into the body. The excess, unable to enter the surrounding air either, is confined to the region immediately around the body; that is to say, an atmosphere is formed. Turning to negatively-charged bodies, Wilson went considerably beyond anything that Franklin had said. (The latter had in fact paid insufficient attention to the special problems which emerged in this case.) Here too, Wilson suggested, an electrical atmosphere was formed, this time composed of the fluid that had been driven from the body in the process of electrifying it negatively. When whatever was responsible for the electrification was removed, such an atmosphere would flow back to its original home, but, Wilson insisted, it would do so only gradually, and in the meantime the body would remain negatively charged. Positively-charged and negatively-charged bodies should and do display, according to Wilson, basically the same effects. Where differences appear (as they do particularly in cases of interactions between two or more charged bodies), these are due to the different dispositions of the electrical fluid composing the atmospheres in the two cases: with positively-charged bodies, the fluid has a tendency to flow outwards since the body itself is already replete with fluid, whereas with negatively-charged bodies the fluid tends to flow inwards.

Clearly, if Wilson's account of the formation of an atmosphere when a body became charged negatively had any plausibility at all, it had it only in cases where an insulated body was being charged by induction; as was pointed out almost immediately by Franklin, it was not at all clear how what Wilson had put forward could be applied to any other case (18). But equally grave difficulties arose in trying to reconcile with the theory the observed fact that when equal positive and negative charges came into contact, they neutralized each other. Franklin hastened to point these out: 'How does it appear that the Atmospheres surrounding Bodies electrified plus and minus are similar? If they are both plus Atmospheres, how do they destroy each other on being brought together?' Or again, 'One of the two Atmospheres being suppos'd to enter the Ball electrified minus, what is suppos'd to become of the other?' (19).

Bayes, too, was dissatisfied with the proposed 'negative atmospheres', for, like Franklin, he could not see how they differed in any significant way from the more familiar atmospheres around positively-charged bodies. In response to Wilson's attempted explanation of the repulsion between two positively-charged bodies in terms of the two atmospheres, 'each of them exerting their endeavours to expand into the air, want[ing] more room to do it in', he pointed out that from the principles Wilson had stated earlier, it would appear that nothing in this account would be altered if the word 'plus' was replaced at any point by 'minus'; hence it should follow that, contrary to ordinary experience, *any* two electrified objects should repel each other (20).

Bayes's objections went much further than this, however, and the position he adopted with respect to the whole question of atmospheres was a quite radical one. Franklin was at least prepared to concede that an account in terms of atmospheres could if properly framed provide an acceptable basis for explaining electrical attractions and repulsions: indeed, an assumption that this was so was deeply imbedded in Franklin's own theory, and the explanation he had given for the observed repulsion between two positively-charged bodies was very similar to that now proposed by Wilson. But Bayes would not even concede this much: simply to talk about atmospheres was no substitute, he insisted, for providing a detailed explanation of the forces that had been found to act in these cases, and such an explanation had not been given. As a result, he found himself 'really at a loss to know what it is which according to our A. must naturally drive the balls asunder'. What Wilson had said could, he thought, be interpreted in two different ways, 'but neither of these inform me in what manner or by what force the two atmospheres drive the balls asunder'. (Nor, he might well have added, did the account that Franklin had given.) Thus dissatisfied, Bayes may even have been prepared to deny the existence of atmospheres altogether; at least, this is a possible interpretation of his otherwise obscure argument that 'because bodys Electrifyed plus or minus give the same signs, therefore they are not surrounded with similar Atmospheres ready to expand themselves into any approaching body which resists less than the air'.

Finally, it should be noted that Bayes was not prepared, either, simply to accept without argument the identification of the electrical fluid with Newton's aether. As he pointed out, Hoadly and Wilson nowhere demonstrated or even attempted to demonstrate this identity, they simply assumed it. More than this, Bayes clearly felt that further argument was needed to establish the existence of the aether itself; unlike many British scientists of the period, and Hoadly and Wilson in particular, he recognized that Newton had not proved but had merely postulated this.

I have argued elsewhere (21) that many of Franklin's contemporaries found what he had to say about electrical atmospheres difficult to follow, and his explanations of the observed electrical attractions and repulsions in terms of them unsatisfactory. The attempt by Hoadly and Wilson was but one of many that were made in the 1750s and '60s to improve upon what Franklin had done in this regard. In the end, however, their account was no more successful than Franklin's had been, and the criticisms levelled at it in the document we have been considering were fully justified. In fact Bayes, in pointing to the inadequacy of what had been proposed, was spelling out in a particularly lucid manner the central difficulty in electrical theory in the immediate post-Franklin era. Ultimately it came to be seen that at the heart of the difficulty lay the notion of an atmosphere itself; no evidence could be found of the real existence of such atmospheres (indeed, the invention in 1757 of the so-called 'air condenser' provided something more akin to positive evidence of their non-existence), and also electrical theorizing turned out to be considerably easier without them. And so the notion was gradually abandoned. As we have seen, it is possible that Bayes wanted to call the existence of such atmospheres into question. If so, he was in fact pointing the way ahead. But in any case his comments make plain the nature of the questions that most troubled the electrical theorists of his day. And at the same time they reveal him as a clearer thinker in matters electrical than many of his contemporaries.

APPENDIX

Comments by Thomas Bayes on Hoadly and Wilson's *Observations on a Series of Electrical Experiments* (1756)

Pag. 1. Takes it for granted that according to S^{r}. Isaacc N. there is an Æther which is dispersed thro' all space & which is the cause of gravity

Obs. S^{r}. Is. in his Opt. 3^{d}. Edit^{n}. in order to shew that he did not take gravity for an Essential property of Bodies added a question concerning its cause. And what S^{r}. Is. makes a mere query, in support of which he owns he had no experim^{ts}. is here represented as a point decided by him. Whereas it wou'd be very unfair on the account of what he has said, to assert it was S^{r}. Isaac's opinion that an Ætherial fluid was the cause of gravity, for according to him it is only a possible way of accounting for the Gravity of bodys but as yet supported by no experim^{ts}.

Pag. 3. It being taken for granted that the Phænomena of Elasticity arise from the force & action of a certain fluid, it is proposed to be inquired into whether this Electrical fluid & the Æther be not one & the same fluid.

Obs. From hence one wou'd fancy that it was the principal design of the Treatise to shew this. But I can scarcely see any thing like an attempt to prove this.

P. 17. W^{never} a body is electrifyed either plus or minus, & remains so after ye experim^{t} is over there are similar Atmospheres of ye electrical fluid surrounding them that are ready to expand themselves into any body that approaches that resist less than the Air & this is the reason why bodys give very nearly the same signs, when they are Electrifyed either plus or minus

Observ. On the contrary it seems rather to follow that because bodys Electrifyed plus or minus give the same signs, therefore they are not surrounded with similar Atmospheres ready to expand themselves into any approaching body which resists less than the air.

P. 19. When two balls Electrifyed plus & suspended by two silk strings are brought near to each other; they repel each other & stand for some time at a distance from each other; because ye two Atmospheres, each of them exerting their indeavors to expand into the Air, want more room to do it in; when ye weight of ye balls is not sufficient to prevent it must naturally drive them asunder; till these Atmospheres are dissipated.

Obs. From this paragraph I am really at a loss to know what it is which according to our A. must naturally drive the balls asunder. Will it be said y^{t}. ye two Atmospheres drive ye balls asunder that they may have room to exert their indeavors to expand into ye air? Or do they do this that they may actually expand into ye air? One or other of these must be the meaning of the parag. I think if it has any meaning at all. But neither of these inform me in what manner or by what force the two atmospheres drive the balls asunder. And it is to be observed that according to our author whether a ball be Electrifyed plus or minus it is surrounded with a similar Atmosphere exerting its indeavor to expand into the air. & consequently reading this paragr. without the word plus the reasoning in it will appear full as good & plain as it do's now, & will prove as well that any two Electrifyed balls must repel each other as w^n. both are Electrifyed plus.

P. 19. Par. 2. W^n. two bodys are both Electrifyed minus, suspended by silk strings & brought near one another, they repel each & stand f. some time at a distance, because the condensed Electrical fluid in ye air in order to force it self in at ye surfaces of the balls between their two centers, crouds in, & forces them asunder; till their Atmospheres get all into the balls.

[Observation so heavily cancelled as to be illegible]

P. 19. Par. 3. When two balls electrifyed one plus & ye other minus are brought near one another, they will gradually come together and unelectrify each other. Because the atmosphere of ye ball electrifyed plus is indeavoring to dissipate it self from the center of the ball outwards; & ye atmosphere of ye ball Electrifyed minus is indeavoring to dilate it self from the air inwards to the center of ye ball. The common atmospheres therefore of the two balls thus brought near together exert their forces in one & ye same direction between them; the flow of the Electrical fluid into ye ball Electrified minus is facilitated by ye indeavor of ye Electrical fluid to get out of ye ball electrified plus; & vice versâ: & so the two balls & ye air between them very readily return to their natural states.

[Observation again cancelled, illegible]

[There follow long quotations, without comment, from Hoadly and Wilson, pp. 20–24, on a method proposed for distinguishing between *plus* and *minus* states of electrification, and brief extracts from pp. 69–71 denying the materiality of fire, again without comment.]

Notes

(1) Royal Society, Canton Papers, II, 32. An extended extract from the longer of these items is included, by permission of the Society, as an appendix to this paper. I wish to record my gratitude to Mr N. H. Robinson, the Society's Librarian, for his assistance during the preparation of this paper.

(2) The identification was made by Mr M. E. Ogborn, formerly General Manager and

Actuary of the Equitable Life Assurance Society. I am grateful to Mr Ogborn for allowing me to consult materials from the archives of the Society, and more generally to both him and Mr R. H. Storr-Best, the Staff Manager at the Society, for helping me in my researches.

(3) Royal Society, Miscellaneous Manuscripts, I, 17; this letter was published in *Phil. Trans.*, **53**, 269–71 (1763). The most likely person to have made the attribution is James Yates, F.R.S., who put the Canton Papers in order in 1870. He made no mention, however, of having done so.

(4) The paper appeared in *Phil. Trans.*, **53**, 376–418 (1763), accompanied by a covering letter from Price describing the circumstances of its discovery (pp. 370–375).

(5) *Phil. Trans.*, **54**, 296–325 (1764).

(6) Price's connections with the Society are described in M. E. Ogborn, *Equitable Assurances: The Story of Life Assurance in the Experience of the Equitable Life Assurance Society 1762–1962*, London: Allen & Unwin, 1962, chap. VII.

(7) *Phil. Trans.*, **44**, 247–263 (1746).

(8) 'A Dissertation upon the Sacred Cubit of the Jews and the Cubits of the several Nations. . . . Translated from the Latin of Sir Isaac Newton, not yet published', in John Greaves, *Miscellaneous Tracts, Letters, Poems, &c.*, London, 1736, vol. 2, 405–433.

(9) Aulay Macaulay, *Polygraphy, or Shorthand made easy to the meanest capacity*, Manchester, 1747.

(10) Simpson, *Phil. Trans.*, **49**, 82–93 (1756).

(11) J. D. Holland, 'The Rev. Thomas Bayes, F.R.S., (1702–61)', *J. R. Statist. Soc., Ser. A*, **125**, 451–461 (1962).

(12) J. D. Holland, 'An Eighteenth-Century Pioneer: Richard Price, D.D., F.R.S., (1723–1791), *Notes and Records Roy. Soc. Lond.*, **23**, 43–64 (1968); and also on the 'Letters to the Editor' page of *Electronics and Power*, July, 1965 (p. 238).

(13) Mr Mason has kindly made available to me a copy of a report on his work that he lodged at the Royal Society. His decipherment is reasonably complete, though some constructions defeated him.

(14) Shelton published two different shorthands, the first in 1641 and the other in 1650. It was upon the latter of these that our author based his system. Both are described, along with that proposed by Coles in 1674 and many others, in J. E. Rockwell, *Shorthand Instruction and Practice*, Washington, 1893.

(15) The first two examples are due to Mr Mason.

(16) Other writings by Wilson on the subject included his *Essay towards an Explication of the Phaenomena of Electricity, deduced from the Aether of Sir Isaac Newton*, London, 1746, and his *Treatise on Electricity*, London, 1750; 2nd ed. 1752.

(17) I. B. Cohen, ed., *Benjamin Franklin's Experiments*, Cambridge, Mass.: Harvard U.P., 1941. See also my article 'Franklin's Electrical Atmospheres', *Br. J. Hist. Sci.*, **6**, 131–151 (1972), for an analysis of the place of 'atmospheres' in eighteenth-century electrical theory more generally.

(18) Franklin's extended criticisms of Hoadly and Wilson's tract are reproduced in *The Papers of Benjamin Franklin*, New Haven: Yale U.P., 1959–, vol. 8, pp. 239–263.

(19) Franklin, *loc. cit.*, pp. 244, 246.

(20) See appendix.

(21) In my article 'Franklin's Electrical Atmospheres', cited above (n. 17).

XI

Electricity and the Nervous Fluid

Throughout its history, one of the more pressing questions which physiology has sought to answer has concerned the manner in which muscular contractions are brought about. Until fairly recently the answer to this question has almost always involved recourse to "animal spirits" of one kind or another: following Galen, these spirits were supposed to flow from the brain through the nerves to the muscles, where they stimulated the contraction. Few of the theoretical modifications introduced from time to time disturbed this basic pattern of explanation; such modifications usually concerned instead either the nature of the spirits themselves, or the way in which they induced the contractions, or both.

Within the traditional pattern of explanation just outlined, there remained two very different possibilities: one could regard the "animal spirits" as immaterial influences of some kind, or one could regard them as a material agency. The latter view, always widely accepted, became increasingly popular during the seventeenth and eighteenth centuries as part of the widespread acceptance of the Cartesian demand that all the phenomena of nature be explained solely in terms of motions and impacts among the particles of matter. As time passed, however, it became clear that the simple kinds of mechanism envisaged by Descartes were inadequate, and during the eighteenth century numerous branches of science compromised by introducing for explanatory purposes unexplained forces and subtle fluids of one kind or another. In the case of muscle physiology, the work of Albrecht von Haller constituted an important stage in this development. It is my purpose in this paper to investigate the rationale of Haller's discussions on the nature of the nervous fluid, and in particular to try to reach some understanding of

Reprinted by permission of Kluwer Academic Publishers.

why he dismissed out of hand the suggestion that the nervous fluid might be the same fluid as that held responsible for electrical action. Incidentally to this, I shall try to indicate the extent to which Galvani had to meet Haller's objections when he postulated his "animal electricity" a few years later.[1]

HALLER'S DISCUSSION OF MUSCULAR CONTRACTIONS AND THE NERVOUS FLUID

Much of the early discussion of muscular contraction was vitiated by the presumption that a muscle can contract only in response to an impulse transmitted to it by the nerve. Haller realized that muscular fibre possesses the ability to contract of its own accord, without the intervention of any nervous impulse. Accordingly, he drew a careful distinction between the properties of irritability and sensibility in muscular tissue: any part where a contraction could be observed upon stimulation he called irritable, whereas he reserved the term sensible for those parts whose stimulation caused signs of unrest (in animals) or was consciously noticed (if a human being was involved).[2] From a long series of experiments, he concluded that the only parts which possess sensibility are those which are supplied with nerves, whereas irritability he found to be a property of the muscular fibres themselves.[3]

His distinction enabled Haller to separate the traditional problem into two parts. He could consider firstly the manner in which the nerve transmits a stimulus to the muscle (the stimulus being brought about either by an exercise of the will or by an irritation of the brain or nerve fibre), and he could then treat as a quite

1. I thus reject at the outset the critical attitude adopted by Philip C. Ritterbush in his *Overtures to Biology: The Speculations of Eighteenth-Century Naturalists* (New Haven: Yale University Press, 1964), toward those eighteenth-century physiologists who sought to identify the nervous fluid with the electrical one. In the context of the time, and given the prevailing theories of nervous action and electricity, I regard the possibility of such an identification as a perfectly proper subject of *scientific* inquiry; the two fluids had been ascribed properties which were almost identical, and experiment had shown unequivocally that they were at least related, for electricity had been found to be remarkably effective in stimulating muscular contractions.

2. Albrecht von Haller, *A Dissertation on the Sensible and Irritable Parts of Animals* (Baltimore: Johns Hopkins Press, 1936; a reprint of the English edition of 1755), pp. 8–9. The term "irritability" had been introduced into physiology by Francis Glisson in his *Tractatus de natura substantiae energetica* of 1672, during a discussion of the means whereby bile is discharged into the intestines.

3. Haller, *A Dissertation*, p. 24.

separate problem the way in which the muscle responds to stimulation, whether the stimulus be transmitted *via* the nerve, or whether it be due to an irritant applied directly to the muscle. In this paper I shall be concerned with Haller's answer to the first of these questions. Let us note in passing, however, his answer to the second of them.

Haller rejected the traditional explanation of muscular contraction in terms of elasticity, on the grounds that (i) elasticity is a property of dry fibres, and dried muscular fibres are deprived of all their irritability; (ii) "elasticity is a property of hard bodies, and irritability of the softest"; and (iii) the fibres of old subjects are more elastic than those of young ones, but they are less irritable. In addition, he repeated two of Robert Whytt's arguments: "the motion of the heart ceases and is renewed spontaneously, which is not observed in any elastic fibre, and upon pricking steel with a needle you produce no irritation in it."[4] He rejected Whytt's conclusion, however, that the soul acts to contract the fibres that are touched to prevent their being injured, his rejection being based on his distinction between irritability and sensibility; the soul he held to be active only in the case of the latter. His ultimate conclusion was that "irritability is a property of the animal *gluten*, the same as we acknowledge attraction and gravity to be properties of matter in general, without being able to determine the cause of them."[5] In other words, he concluded that contractility is a power innate to muscle fibre, a power to which he later gave the name "vis insita," a power "found in no other part of the human body."[6]

Beyond this,

> concerning the vis insita, we do not indeed inquire; as this seems to be a more brisk attraction of the elementary parts of the fibre by which they mutually approach each other, and produce as it were little knots in the middle of the fibre. A stimulus excites and augments this attractive force, which is

4. *Ibid.*, p. 40. Whytt had presented his arguments in his *Essay on the Vital and Other Involuntary Motions of Animals* (Edinburgh, 1751).

5. Haller, *A Dissertation*, p. 42.

6. Haller, *First Lines of Physiology*, 2 vols. in one (New York: Johnson Reprint Corp., 1966; a reprint of William Cullen's translation of the third Latin edition of 1767), I, 233. The concept of irritability did not appear in the editions of 1747 and 1751, and it was not until the third edition that Haller included a reasonable account of this aspect of his work (Lester S. King, Introduction to the 1966 reprint, p. xl).

placed in the very nature of the moving fibre. The other explanations are hypotheses. . .[7]

It was remarked at the outset of this paper that the Cartesian demand for mechanical explanations was widely accepted among eighteenth-century physiologists. At least four different mechanical explanations had been suggested, however, for the transmission of impulses by the nerves, and in Haller's time the issues between them had yet to be resolved. Some of these explanations in fact went beyond the traditional one based on "animal spirits" and accounted for the transmission in terms of vibrations, whether gross vibrations of the nerve as a whole, or microscopic vibrations of the particles of which it was composed,[8] or vibrations of the particles of some aethereal medium supposedly pervading it,[9] or some combination of these.[10] At the same time, however, more traditional kinds of explanation invoking the flow of a fluid through the nerves remained popular, being expounded most notably in the writings of Boerhaave and his pupils—amongst whom, of course, Haller is to be numbered.

In the fourth volume of his *Elementa physiologiae*,[11] and again in his *First Lines of Physiology*, Haller presented a number of arguments against all explanations invoking vibrations. Since the nervous fibres are not tense either in their origin or in their termination, he said, they cannot possibly vibrate. Again, in many cases they cannot vibrate because they are firmly tied to

7. *Ibid.*, I, 236.

8. T. Morgan, *Mechanical Practice of Physick*, (London, 1735), pp. 145ff.

9. Isaac Newton, *Opticks, or a Treatise of the Reflections, Refractions, Inflexions and Colours of Light*, (New York: Dover, 1952; based on the 4th edition, London, 1730), Query 24, p. 353. Consistent with his well-known mid-career abandonment of ether hypotheses, Newton, in Query 12 in the first (1704) edition of the *Opticks*, spoke simply of vibrations "being propagated along the solid fibres" of the nerves, without mentioning the ether at all. The wording of this Query remained unaltered in subsequent editions, although of course an entirely new interpretation had to be put on it following the addition of what became Query 24 in 1717. It would be interesting to know what prompted Newton to extend his ether hypothesis to the action of the nerves in the way he did, but so far as I am aware this is a question which Newton scholars have not yet investigated.

10. David Hartley, for example, in his *Observations on Man* (London, 1749), combined the last two possibilities. Haller, in his *Elementa physiologiae* (IV, 358), listed a number of others besides Morgan, Newton, and Hartley who had denied the existence of a nervous fluid, but I have not checked his references to see which particular variant of the vibration theory each espoused.

11. Lausanne, 1762.

solid substance (here he gave as an example the fact that the nerves of the heart are tied to the great arteries). Furthermore, he argued, the nerves are totally devoid of elasticity, since when a nerve is cut in two, it neither becomes shorter, nor are the divided ends drawn back. In addition, "the extreme softness of the medulla in the brain, with all the phenomena of pain and convulsion, leaves no room to suspect any sort of tension concerned in the effects or operations produced by the nerves." Finally, he noted that muscles situated above the point of irritation are never convulsed, only those below. "This," he argued, "is a consequence altogether disagreeing with elasticity; for an elastic cord propagates its tremors every way, from the point of percussion to both extremities." [12]

All of Haller's arguments except the last impinge only upon the hypothesis that the nerve vibrates like a stretched string; they do not rule out either microscopic vibrations of the particles of the nerve fibre, or ethereal vibrations along the lines suggested by Newton.[13] Far more penetrating arguments against vibratory hypotheses in general had been given already by Whytt in his *Essay on the Vital and other Involuntary Motions of Animals.* Whytt had argued first against both those explanations which invoked an elastic power in the nervous fibres and those which postulated oscillations of an animal spirit, on the grounds that inert matter does not possess any power of generating motion. How then, he asked, could a pinprick or a drop of acid cause the large contractions which are observed? He had also rejected the possibility of ethereal oscillations on the grounds that such oscillations must follow the ordinary laws of vibration of elastic bodies, and are hence incompatible with the observed

12. For convenience the quotations are taken from the *First Lines of Physiology* (I, 219–220). In vol. IV of the *Elementa physiologiae,* the same arguments, sometimes expanded a little, appear on pp. 361–365. It is worth noting as well that several of the arguments had been given earlier by Boerhaave in his renowned *Institutiones medicae* (see, for example, the English translation published by W. Innys [London, 1742–46] under the title *Dr. Boerhaave's Academical Lectures on the Theory of Physic* (II, 310–313)). They were repeated by many other authors throughout the rest of the century.

13. Elsewhere in the *Elementa physiologiae* (IV, 378–379), Haller did argue against the ether hypothesis on the grounds that the ether could not be confined within the nerves. He was there concerned to argue, however, that the nervous fluid, whose existence he was by then taking for granted, was not to be identified with the ether; his argument was not designed to rule out the possibility of ether vibrations being confined in the nerves, and it certainly did not do so.

alternate contractions and relaxations of stimulated muscle, since the latter motion has a period which decays in time.

If we have reservations about the effectiveness of Haller's arguments, he certainly had none; he felt that he had shown any explanation in terms of vibrations to be untenable, and so he adopted the alternative and more traditional hypothesis of a nervous fluid:

> If neither the phenomena of sense nor motion can be explained from the nature of elasticity, the only probable supposition that remains is, that there is a liquor sent through the brain, which, descending from thence through the nerves, flows out to all the extreme parts of the body; the motion of which liquor, quickened by irritation, operates only according to the direction in which it flows through the nerve; so that convulsions cannot thereby ascend upwards, because of the resistance made by the fresh afflux of fluid from the brain. But the same liquid being put in motion in an organ of sense, can carry that sensation upwards to the brain; seeing it is resisted by no sensitive torrent coming from the brain in a contrary direction.[14]

The postulation of a fluid matter acting in the manner indicated raised a number of questions. For instance, where are the channels through which the fluid flows, and what happens to the fluid once it has arrived at the muscle? Microscopic analysis had failed to reveal any channels in the nerves, and the failure of nerve fibres to swell when a ligature was applied to them seemed contrary to expectations on any theory of fluid flow. Haller could not account for these observations; he dismissed them rather as indicating only the weakness of our senses, "but [they] have not any validity against the real existence of a juice or spirit in the nerves."[15] He supposed that much of the fluid sent to the extremities was exhaled through the cutaneous nerves; he added, however, that he suspected that "a part [adheres] to the fibre; and thus it happens, that the muscles grow strong with exercise, and their brawny parts become thicker."[16]

Given the existence of a nervous fluid, the manner in which that fluid actually brought about muscular contractions remained unclear; it was no more unclear, however, than the manner in which any stimulus applied directly to the muscle caused a contraction. In this regard, Haller remarked that

14. *First Lines of Physiology*, I, 220.
15. *Ibid.*, p. 221.
16. *Ibid.*, p. 243.

> the direct manner by which the nerves excite motion in the muscles, is so obscure, that we may almost for ever despair of its discovery . . . [but] you may suppose the nervous liquor to be of a stimulating nature, by which means it forces the elementary particles of the muscular fibre to approach nearer to each other . . . That muscle then is contracted which in a given time receives more of the nervous fluid, whether that be occasioned by the will, or by some irritating cause arising in the brain, or applied to the nerve.[17]

He proceeded to discuss the great wastage of power in the muscles, concluding that

> the action of the nervous or animal fluid is very powerful . . . nor does this seem to be otherwise explicable, than by the incredible celerity by which the influx of this fluid obeys the command of the will. But how, or whence it acquires such a velocity, is not in our power to say; it is sufficient, that we know the laws of its motion are such, that a given action of the will produces a new and determinate celerity in the nervous fluid.[18]

What was the nature of this powerful fluid of whose existence he was convinced? Haller remarked that many held it to be "incompressible and watery, but of a lymphatic or albuminous nature."[19] This opinion he rejected, largely on account of the extreme motility the fluid had to have if it was to explain nervous action satisfactorily. "Moreover, it is very thin and invisible, and destitute of all taste and smell; yet reparable from the aliments." In particular, "it is carefully to be distinguished from that visible, viscid liquor exhaling into the intervals of the nervous chords."[20] Haller's fluid, it eventuated, was not at all like ordinary gross fluid matter. This presented no difficulty, however; the science of his day was filled with subtle fluids ranging from the phlogiston of the chemists, through the fluid of fire, responsible in one manifestation for heat and in another for light, to the pair of fluids responsible for those mysterious phenomena, electricity and magnetism. It seemed clear to Haller that the nervous fluid resembled these—that is, it was a *subtle* fluid. The only question remaining was whether it was identical with one of the other fluids, or whether it was instead a new and different fluid, a subtle fluid *sui generis*.

17. *Ibid.*, pp. 236–237.
18. *Ibid.*, p. 240.
19. *Ibid.*, p. 221.
20. *Ibid.*, p. 222.

HALLER'S REJECTION OF THE ELECTRICAL FLUID

In 1745, at about the same time as Haller began his researches on irritability,[21] the advent of the Leyden jar had made it painfully obvious to the recipients of electrical discharges that electricity was remarkably effective as a stimulant of muscular contractions. Since the "electricians" of the time accounted for the effects they produced in terms of the excitation and transmission of a subtle electrical fluid (thought by many to be closely related to, if not identical with, Boerhaave's celebrated fluid, Fire), it was only natural that Haller should entertain the possibility that the nervous fluid was just another manifestation of the electrical one.[22] Suggestions linking the two had appeared in print even before the sensational announcement of the Leyden experiment,[23] and from what has been said in the previous section of this paper, it is clear that the electrical fluid could well have been the ideal substance for Haller's purposes: it was a subtle fluid; it had been shown by Watson[24] to move with

21. This dating of Haller's first researches on irritability is due to Lester S. King, "Introduction" to *First Lines of Physiology*.

22. In view of the persistence in the literature of statements to the contrary, it perhaps needs to be emphasized that Benjamin Franklin was *not* the first to introduce the notion of a subtle electrical fluid into electrical theory, for the earlier effluvialists such as Nollet and Watson had also invoked such a peculiar fluid in their attempt to account for the phenomena. Franklin's innovation was that he regarded an electrified body as one containing either more or less than its normal quota of fluid, whereas his predecessors had described the electrified state in terms of an excitation of the fluid already present.

23. Thus Motte, in his authorized translation of Newton's *Principia* (London, 1726; reprinted London, 1968, p. 393), had described as an "electric and elastic spirit" the ether which Newton had invoked to explain, amongst other things, how "all sensation is excited, and the members of animal bodies move at the command of the will." A few years later, Hales (*Haemastaticks* [London, 1733; reprinted New York, 1964], p. 59), cited some of Stephen Gray's experiments in support of the electrical theory. Then in the very paper in which he announced Musschenbroek's famous discovery to the world (*Hist. l'Acad. Roy. Sci.*, 1746, Mém. p. 16), the *abbé* Nollet remarked that he considered a person who drew a spark from an electrified body "comme remplie on pénétrée d'un fluide subtil, dont la répercussion se fait sentir plus ou moins fort, plus ou moins profondément, à proportion de la grandeur du choc qu'il a reçu . . . Cette commotion n'est qu'un mouvement de pression imprimé à un fluide fort élastique." Nollet's explanation of the appearance of sparks in electrical discharges was framed in the same terms; it involved a "répercussion" of two opposing streams of electrical fluid. Haller cited Hausen, Boissier, and Deshais as other early proponents of the theory (*Elementa physiologiae*, IV, 378); Rothschuh has given a more complete listing in his article "Von der Idee bis zum Nachweis der tierischen Elektrizität" (*Sudhoffs Archiv*, 44 [1960], 25–44).

24. William Watson, *Phil. Trans. Roy. Soc.*, 45 (1748), 76–85.

extreme rapidity; and it had the desired effect on muscular tissue. Yet Haller rejected the electrical fluid, claiming that the nervous fluid was something quite distinct. Why?

In his *First Lines of Physiology* Haller mentioned only very briefly the possibility that the nervous fluid was electrical. "Many of the moderns," he remarked, "will have it to be extremely elastic, of an etherial or of an electrical matter . . . An electrical matter is, indeed, very powerful, and fit for motion."[25] To counter the suggestion he raised two objections, but he did not elaborate on them. One was that the electrical fluid "is not confinable within the nerves." Secondly, "a ligature on a nerve takes away sense and motion, but cannot stop the motion of a torrent of electrical matter."[26] These arguments were sufficient, he felt, to justify the bald assertion that "the animal spirits are not of the nature of an electric torrent."[27]

As usual, Haller set out his arguments in more detail in his *Elementa physiologiae*.[28] In this case, however, the arguments were given even more fully in papers by Felice Fontana and Marc Antoine Caldani (both written in 1757) which Haller had included in the third volume of his *Mémoires sur la nature sensible et irritable, des parties du corps animal*.[29]

Fontana and Caldani had directed their arguments particularly at a M. Laghi, who had dared to make the controversial identification, and who in doing so had apparently tried to explain why, despite the electrical conductivity of the surrounding tissues, the electrical fluid could nevertheless be confined within the nerves.[30] His theory was characterized by Caldani as follows:

> M. Laghi conjecture . . . que les esprits sont de nature électrique, que la matiere de ce nom repandue partout l'univers, est determinée par le suc nerveux à courir par les nerfs; & que les esprits électriques, envelopés par le énacité

25. *First Lines* . . . I, 221.
26. *Ibid.*
27. *Ibid.*, p. 237.
28. IV, 378–380.
29. Lausanne, 1760. Henceforth referred to as *Mémoires*.
30. Tommaso Laghi (1709–1764) was professor of medicine at Bologna at the time. I have not been able to consult any of his works, but since my concern is with the arguments *against* the electrical hypothesis rather than those for it, the omission should not be significant. From remarks made elsewhere by Caldani (*Haller-Caldani Briefwechsel*, ed. E. Hintzsche, [Bern, 1966], p. 16), it seems likely that Laghi's defense of the electrical hypothesis appeared in his *Epistola responsoria ad Caesarium Pozzi* (Bologna, 1756).

> du suc nerveux, ne sont arretés dans les fibres musculaires, qu'autant que celles-ci sont chaudes & flexibles.[31]

If Caldani's characterization is correct, Laghi's "suc nerveux" was supposed to have a rather peculiar ability to retain the electrical fluid despite its being surrounded by tissues which are electrical conductors. Caldani took Laghi to mean that the "suc nerveux" was an electrical insulator: in that case, he argued, it would simply stop the transmission of the signal along the nerve, without in any way preventing the electrical fluid from spreading into all the neighboring tissues. Fontana raised a different but equally powerful objection when he asked:

> Comment se peut-il qu'un muscle seul se meuve au gré de la volonté, & qu'un grand nombre de muscles ne se contractent pas en même temps? Comment empeche-t-on que le vapeur électrique ne s'épande dans tous les muscles, qu'aborde ce nerf, plus voisins même du cerveau?[32]

Much of Laghi's argument for the identity of the fluids seems to have been based on analogy, and Fontana in particular spent much time in heaping ridicule on the analogies drawn. However, both he and Caldani also produced other arguments of more lasting interest against the identification suggested. Caldani, for example, wrote:

> Je ne sai pas d'ailleurs, si on pourra accorder au suc nerveux la sagesse nécessaire, pour mener la matiere électrique dans un nerf préférablement à un autre: ou le pouvoir de dispenser cette matiere des loix de l'équilibre, pour lui faire enfiler un seul des nerfs, qui sortent du cerveau.[33]

Caldani here raised a serious difficulty. He was rather unfair, however, in using it to reject just the electrical theory. The same problem could equally well have been used to refute any other theory of nervous action. The question as to how the brain initiated impulses in a single nerve was unanswerable on *any* theory currently available: equally *ad hoc* modifications would have had to be made to Haller's fluid theory, for example, as were required in the electrical theory, before it could have provided an answer.

In the passage just quoted, Caldani referred to the "laws of equilibrium" of the electrical fluid. To understand his reference, and also to understand an argument offered by Fontana which

31. *Mémoires*, p. 461.
32. *Ibid.*, p. 207.
33. *Ibid.*, p. 463.

will be outlined in a moment, an appreciation of the state of electrical theory at the time is necessary.

Until 1749, all theories of electricity had sought above all to explain the phenomena of electrical attraction and repulsion. They had postulated the emission of an electrical matter from electrified bodies: the most sophisticated of them, that due to Nollet, posited a double stream of electrical matter, one stream moving outward from the electrified body, the other moving inward toward it. In 1749, however, Benjamin Franklin suggested a very different type of theory:[34] all matter has mixed with it a natural quantity of an electrical fluid, and bodies become electrified if they have either an excess or a deficiency of the fluid, acquiring "positive" and "negative" charges respectively in the two cases. Franklin's theory was directed particularly at explaining the phenomena associated with the Leyden jar, and also the general phenomena of conduction; it was not very adequate for explaining the traditional problems of attraction and repulsion. At no stage in his early papers, indeed, did Franklin spell out in detail his views concerning these problems, and it was not until 1755 that he tried to provide an answer in terms of static interactions between the atmospheres of electrical fluid he envisaged surrounding positively charged bodies and other atmospheres or lumps of uncharged or negatively charged matter.[35] His attempt was not very successful, and in any case it came too late to prevent others, nominally adherents of his views, from attempting to explain the appearances in a semi-effluvial fashion quite at variance with what he had had in mind.[36] Of these the most influential was the Italian, Giambatista Beccaria, and it was upon Beccaria's exposition of the theory that Haller and his correspondents relied as they attacked Laghi's assertion that the nervous and electrical fluids were one and the same thing.[37]

34. Franklin's theory was expounded in a series of letters he wrote to his friends in London between 1747 and 1755 (the first detailed exposition being given in 1749). The letters were published under the general title *Experiments and Observations on Electricity, Made at Philadelphia in America*, the first group appearing in print in 1751. A new edition was prepared by I. Bernard Cohen, *Benjamin Franklin's Experiments* (Cambridge, Mass.: Harvard University Press, 1941); this is the edition I have used.

35. *Benjamin Franklin's Experiments*, p. 302.

36. I am currently preparing a detailed study of these aspects of Franklin's work.

37. For example, Fontana (*Mémoires* p. 207) began his critique of Laghi as follows: "Si M. Laghi en admet l'identité, voici ce que j'aurois à lui objecter, *d'après le P. Beccaria.*" (my emphasis).

Beccaria began by stating his adherence to Franklin's doctrine that a body is electrified when it contains either more or less than its natural quota of the electrical fluid, and he accepted wholeheartedly the explanations Franklin had given for the behavior of the Leyden jar, and for a large number of experiments involving the sharing of charges between bodies.[38] However, after carefully defining "the signs of electricity" characterizing an electrified body so as to include the attraction of light objects as well as the giving of sparks and the sensation of the electric wind, he asserted that the fundamental principle of electricity was that "*All* the signs of electricity are due to the vapor expanding itself from a body in which it is in a greater quantity into one in which there is less."[39] That is, for Beccaria, electrical attractions were due to a flow of electrical fluid.[40] That this was the theory that Haller and his correspondents accepted is clear from the phraseology they chose in describing it. Fontana, for example, stated the basic idea as follows:

> La matiere électrique est une espece de vapeur, qui sort d'un corps, où elle est en abondance, pour entrer dans un autre, où il y en a moins, & elle donne des marques de ce passage, dont la force est dans la raison de la différence de la quantité de ce fluide dans ces deux corps.[41]

It was clearly unimportant to them which of the above attitudes one took toward the role of the electrical fluid in electrical phenomena. All that was important to them was that an electrical fluid was postulated: they were then concerned only to argue that it was not the active principle responsible for the transmission of nervous impulses. Which theory one accepted did make a difference, however, as to the sorts of arguments which could be presented to counter suggestions such as that made by Laghi. Because he accepted Beccaria's theory, Fontana was able to develop the following argument in addition to those already cited:

38. Giambatista Beccaria, *Dell' elettricismo artificiale e naturale* (Turin, 1753).

39. *Ibid.*, p. 17 (my emphasis): "ogni segno elettrico avvenga pel vapore, che da un corpo, in cui è in quantità maggiore si espande nell' altro, in cui è in minore quantità."

40. Beccaria was not simply reverting to the traditional effluvial theory, however: as he pointed out himself, the impact of an outward-flowing stream of matter hardly provided a straightforward mechanical explanation for the inward motion of the bodies in its path.

41. *Mémoires*, p. 207.

> La matiere électrique doit être dans les nerfs & dans les muscles dans une proportion inégale, puisqu'aucun mouvement n'en resulteroit, si la quantité étoit égale: elle paroit pourtant l'être, puisque les nerfs & les muscles sont électrifables par communication [i.e. they are conductors]. On sait que le frottement des corps de cette classe, n'ébranle pas l'équilibre de la matiere électrique, ne cause aucun torrent, & *par conséquent aucun mouvement.*[42]

The last phrase in this passage provides the clue, I believe. Evidently, Laghi had interpreted muscular contractions as ordinary electrical attractions caused by a flow of electrical fluid from the nerve to the muscle. Fontana is arguing that in a system of conductors in contact, no flow of fluid can occur of a kind which can give rise to such electrical attractions; according to Beccaria's theory, attractions only occured if there was a flow of fluid across a dielectric (usually air, of course).

By 1760, then, there were in print a number of powerful arguments against any identification of the electrical and nervous fluids. Let us review briefly those which have been discussed in this paper. Firstly, there is the argument that because the nerves are surrounded by electrical conductors, any electric current flowing in them will be dissipated into the surrounding tissues. Haller's argument based on ligatures of the nerves remains unanswered. There is the question of how an electrical impulse originating in the brain can be confined to a single nerve. There is Fontana's argument, just given, that an attraction giving rise to contractions can only occur if inequalities in the distribution of the fluid can be established, and they cannot be since all the tissues involved are conductors. Finally, there is the argument (implicit in Fontana's ridicule) that the analogy drawn between the electrical and nervous fluids is simply a bad analogy. These are weighty arguments. Yet in 1791 Galvani found no difficulty in explaining his experimental results in terms of an "animal electricity" flowing in the nerves. It will be our task in the next section to review the effect of developments in the intervening period on the validity of the above arguments, so as to determine the extent to which Galvani's position was a tenable one.[43]

42. *Ibid.* (my emphasis).

43. Haller's statement cited above (p. 241) that the nervous fluid is "reparable from the aliments" suggests another important difference between it and the electrical fluid. Ordinary matter and the electrical fluid were held to be independent species of being; the total quantity of each

"ANIMAL ELECTRICITY"

In the years between 1760 and 1791, there were two developments in particular, one in electrical theory and one in animal physiology, which are relevant to our present discussion. The change in electrical theory is the least striking of the two. Indeed, the fact that there was a change has not generally been appreciated by historians, since insufficient attention has been paid to Beccaria's work; but it was nevertheless significant, because it imposed still tighter constraints on the role which the electrical fluid could play in physiological theory. In 1756 Franz Aepinus and Johan Carl Wilcke, working together, developed an air condenser; that is, a device which was essentially a Leyden jar, but one in which the insulating layer was air rather than glass.[44] As it became more and more widely known, their discovery sounded the death knell for all electrical theories which involved the emission of a fluid (including Beccaria's), and at the same time it led to the rejection of Franklin's own theory of static interactions between electrical atmospheres. The only remaining alternative, and the one which was gradually if reluctantly accepted by physicists following its formulation by Aepinus,[45] was to suppose that masses of the electrical fluid exert forces on each other and on ordinary bodies at a distance—that is, forces were invoked which were no more explained than was the force of gravity.

This change in the theory in no way supported Laghi's arguments—indeed, it undermined one of them even more than Fontana had done already—but it did lead to a clarification of the issues involved. Laghi, it will be recalled, had ascribed the muscular contractions to a simple "electrostatic" attraction

in the universe was conserved, and one could not be converted into the other by any means. His statement that part of an animal's food intake is converted into nervous fluid suggests that Haller, like Boerhaave before him, took the latter to be simply an extremely subtle form of ordinary matter, and not an independent species of being. Elsewhere, in fact (*Elementa physiologiae*, IV, 381), Haller stated explicitly that while the nervous fluid was certainly extremely subtle, it was yet denser than fire or ether or electric or magnetic matter—since it could be contained in the as-yet-unobserved channels in the nerves! Possibly he was reluctant to raise arguments such as these against the electrical theory directly for fear of being accused of being too devoted to his own particular physiological "system."

44. Aepinus, *Mém. Acad. Berlin* (1756), pp. 119–121. For Wilcke's part in the discovery, see C. W. Oseen, *Johan Carl Wilcke, Experimental-Physiker* (Uppsala, 1939), pp. 50–51.

45. Aepinus, *Tentamen theoriae electricitatis et magnetismi* (St. Petersburg, 1759).

brought about by the flow of electrical fluid from the nerve to the muscle. According to Aepinus's theory, however, such a flow of fluid would not produce an attraction. As in Beccaria's theory, an insulating gap was required between the end of the nerve and the muscle if an electrical attraction was to occur, for only then could electricity accumulate on the end of the nerve to exert its attractive force across the intervening space. If anything, then, the new theory made it even more evident than before that the contractions were not due to electrical attractions. Henceforth, those physiologists who wished to identify the electrical with the nervous fluid had to explain the contractions in terms of a *stimulation* of the muscle by a current of electrical fluid flowing from the nerve; how the fluid effected this stimulation remained unexplained, as did the generation of the nerve-currents in the brain despite the fact that all the tissues involved were electrical conductors. Presumably pressure gradients were developed in the electrical fluid in the brain, but it was not at all clear how this could happen.

The changes in electrical theory thus had some influence on the dispute we are considering in this paper. Much the more striking of the two developments mentioned above, however, was the discovery in the 1770's that the shock administered by the fish known as the torpedo, well known since antiquity, was just an electric shock.[46] More specifically, it was shown that the fish could accumulate at will quite large quantities of electricity of opposite sign at opposite ends of the peculiar cellular structures with which they are endowed. Furthermore, Hunter's anatomical investigations showed that these structures were extraordinarily well supplied with nerves. For the electrically minded physiologist the conclusion was obvious: the cellular structures functioned as a Leyden jar, drawing charge from the nerves (which were thus seen as being analogous to the "conductor" of the electrical machine from which the more common type of Leyden jar was charged). The whole argument was no more than an analogy, of course, but it was an analogy of a particularly convincing kind, for it rested on the clear demonstration that the fish *could* generate electrical charges. It could not be dismissed as lightly as Fontana had dismissed Laghi's crude analogies twenty years earlier.[47]

46. A useful summary of the work which led to this conclusion has been given by W. Cameron Walker in his article, "Animal Electricity before Galvani," *Annals of Science*, 2 (1937), 84–113.

47. It was in fact sufficiently strong to convert Fontana himself to the electrical theory.

The phenomena associated with the torpedo had other consequences for the arguments we are considering. Firstly, the torpedo provided conclusive evidence that gradients could be set up in the electrical fluid even though all the tissues involved seemed to be conductors. If this could happen in the torpedo's cells, it could also happen in the brain, and so could give rise to electrical currents in the nerves. In addition, the fact that the electrical apparatus of the torpedo was confined to a particular tissue was relevant to the argument which claimed that any electrical impulse in the nerves would be dissipated into the surrounding tissue. If the fluid could be confined to one particular tissue in the torpedo, it could be asked, why could it not also be confined to a single tissue, the nerve fibre, in other circumstances?

One thing which would certainly prevent the dissipation of electrical impulses from the nerves would be an insulating sheath around them. There was, of course, no anatomical evidence for the existence of such sheaths, and it was presumably for this reason that Haller and his correspondents, Fontana and Caldani, ignored the possibility in constructing the arguments discussed earlier; Haller, however, could not find any channels to conduct his fluid, either, and this did not prevent him from assuming their existence. At any rate, Galvani, convinced already by his experiments with frogs that muscular contractions were brought about by a flow of "animal electricity," did not hesitate to postulate the existence of insulating "oily" sheaths around the nerves, and, unable to find them, he offered as a "proof" of their existence the results of a chemical distillation of nerve matter which yielded a large amount of oil.[48] With this postulated, however, it became awkward to explain how contact with the outside of the nerve fibre could stimulate the muscle. To answer this, Galvani suggested that the insulating layer was very thin, so that its insulating properties could break down under appropriate circumstances. Since one such set of circumstances was the application of mechanical pressure, Galvani could have easily overcome Haller's objections based on his experiments with ligatures, though he did not in fact do so.[49]

48. Luigi Galvani, *Commentary on the Effects of Electricity on Muscular Motion,* trans. Margarget Glover Foley; notes and introd. by I. Bernard Cohen (Norwalk, Connecticut: Burndy Library, 1953), p. 76.

49. It should be remarked that Galvani's views on the function of the nerves were rather unorthodox in ways quite unrelated to those discussed in this paper. He did not adopt the traditional view of the nervous fluid as the medium responsible for the transmission of impulses from the brain

CONCLUSION

It may be seen, then, that if one was prepared to accept the existence of insulating sheaths on the nerves, all the arguments raised against the proposed identification of the nervous and electrical fluids, except one, could be answered satisfactorily. The single exception involved the question of how an electrical disturbance in the brain could be confined to a single nerve, and, as was indicated earlier, it was scarcely fair to hold this sort of objection against the electrical theory alone. In that case, there remained no convincing argument to show why one should not accept the identification of the two fluids. On the other hand, of course, it remained an open question as to whether there was any convincing argument to show why one *should* accept the identification either. Galvani thought that his experiments provided just such an argument.

---

to the muscles. Rather, he regarded it as something whose circulation from the inside of a muscle to the outside (or perhaps vice versa) *via* a nerve resulted in the muscle contracting (*Commentary*, p. 82). This peculiar feature of his theory does not affect the preceding argument, however, since for Galvani as for everybody else, muscular contractions were still supposed in the end to be due to the flow of a subtle fluid through the nerves, and it was this fluid which he held to be electrical in character.

XII

ELECTRICITY IN FRANCE IN THE POST-FRANKLIN ERA.

Students of the mid-eighteenth century revolution in electrical theory associated with the name of Benjamin Franklin - particularly those writing in English - have usually assumed that Franklin's ideas displaced those of his great French rival, the Abbé Nollet, almost immediately upon their initial publication in the early 1750s. Nollet himself, it is recognized, maintained his opposition to the new theory until his death in 1770, as did one or two of his supporters; but these efforts have generally been seen as a rearguard action opposing in vain what had quickly become a *fait accompli.* Little evidence has, however, been tendered in support of this interpretation, the chief item referred to being Franklin's own later testimony in his autobiography.(1) Yet he was scarcely a disinterested observer. The picture presented is in my opinion defective and indeed actually misleading on several counts. In the present paper I shall focus on just one of its inadequacies, by challenging the commonly-accepted view of the way in which the French scientists of the period reacted to Franklin's work.

To begin with, it is essential, I believe, to separate Franklin's contributions to electrical theory from his even more famous work on lightning discharges. Despite the almost universal scepticism that persisted well into the 1770s concerning the effectiveness and even the desirability of fitting buildings with elevated conducting rods as Franklin had recommended, it nevertheless remains the case that the success of the lightning experiments of 1752 made Franklin's name a household word throughout the learned world. But - and this is the central point of my paper - the merits of Franklin's electrical theory were by no means so universally recognized at this early period, and were in fact not so recognized for a number of years afterwards. And this was especially the case in France.

So far as electrical theory proper is concerned, I would assert that the most important innovation in Franklin's account was his notion that electrification involved the generation of static "charges" on bodies rather than the excitation of matter already present in those bodies. But that this was so was not at all clear to Franklin's contemporaries. To them the chief distinguishing mark of Franklinian electrical theory was its postulation of two contrary modes of electrification in nature, called by Franklin "plus" and "minus". Such a notion was completely incompatible with Nollet's theory of afflux and efflux, and as a result it quickly became one of the central points at issue following the publication in Paris early in 1752 of a French translation of Franklin's first papers on electricity. In particular, Franklin's notion gave rise to a series of fierce debates in the Paris Academy of Sciences in the mid-1750s between Nollet and the Franklinist Jean Baptiste Le Roy.(2)

It is the outcome of these debates that interests me here. And the evidence relating to this is surely enough to make any whiggishly-inclined historian turn in his grave, because it forces us to conclude that, contrary to what has so often been taken for granted, the eighteenth century's most prestigious scientific institution, and indeed the French scientific world in general, opted, after carefully weighing all the evidence, for the "wrong" side in the dispute. Nollet, in other words, emerged victorious. Far from ending his days as leader of a defeated, dispirited and declining sect, he died as he had lived, the acknowledged master among French electrical investigators. Franklin's scientific reputation in France soon came to rest almost entirely on the success of the lightning experiments, and hardly at all on his contributions to electrical theory!

I can do no more here than summarize the evidence that leads me to these rather startling conclusions. Perhaps the most striking items are two editorial notes appended by the Abbé Rozier to articles on electricity in early numbers of his journal, *Observations sur la physique, sur l'histoire naturelle et sur les arts*. When a précis of an article by the ardent Italian Franklinist Giambatista Beccaria appeared in the number for February 1773, Rozier felt it necessary to include for the benefit of his French audience an introductory footnote explaining in the most elementary of terms what Beccaria meant by his frequent references to electricity *en plus* and *en moins*. Otherwise, Rozier feared, these might distract many readers and cause them to lose the thread of what was being said!(3) Fourteen months later, i.e. in April 1774, twenty-two years after Franklin's book first appeared in French, Rozier made an even more revealing remark as he introduced a report on a recent memoir dealing with the Leyden experiment: "This phenomenon", he wrote, "has been explained by Messrs. Nollet and Franklin in different ways. If the majority of physicists has adopted the notion of the *former*, there are many who, with reason, have declared themselves for that of the latter".(4)

There seems no reason to doubt Rozier's testimony or his editorial judgement in these matters. If confirmation is needed, however, that Franklin still had few supporters in France in the early 1770s, and that his central ideas had sunk almost into oblivion, it is to be found in a paper read by Le Roy to the Academy of Sciences at its Easter meeting in 1772. Le Roy too found it necessary to remind his audience what was meant by electricity *en plus* and *en moins*. "It is with much regret", he said, "that I take up again a subject that I treated so long ago..., but I am forced to do so. For I have had only too much occasion to realize from objections that have been made afresh to me quite recently that even today in this country there is only a very small number of people who are really informed as to what constitutes the difference between these electricities, even though Mr Franklin announced them more than twenty years ago, and though they have since been adopted by the most celebrated physicists in Europe".(5)

Nollet's victory in the mid-1750s may be traced in the sequence of electrical articles in Diderot's famous *Encyclopédie*. The early articles "Conducteur (Physique)" and "Coup foudroyant" were written by Le Roy from an outspokenly Franklinist point of view. Le Roy's numerous forward references make it clear that he expected to write the later articles as well, but in the event he did not do so. Instead, the major entries "Electricité" and "Feu électrique" (published in Vols. IV [Nov. 1755] and V [Oct. 1756] respectively) came from the pen of Louis-Guillaume Le Monnier, who adopted a scrupulously non-committal approach to the theory of his subject. It seems clear that Le Roy was replaced as a matter of deliberate editorial policy.

In 1760, Nollet's second volume of *Lettres sur l'électricité* appeared to the general acclaim of the French press.(6) So too did Du Tour's important Nollet-orientated work, *Recherches sur les différens mouvemens de la matière électrique*. Of the numerous other books on electricity published in France during the 1760s, I know of none that espoused Franklin's ideas prior to Sigaud de la Fond's work at the very end of the decade. Small wonder that Dalibard, one of Franklin's few remaining French supporters, wrote to him in despair in 1762, saying that the level of opposition to the Philadelphian's principles that he had had to combat had driven him to "renounce electrical experiments almost entirely".(7) Small wonder, either, that the French saw Joseph Priestley's Franklin-orientated *History of Electricity* (1767) as a piece of polemical writing rather than objective history.(8) Given the nature of the French literature on the subject during the preceding twenty years, it was!

When did the French finally adopt Franklin's theory, then? It is impossible to give an unequivocal answer to a question framed in these terms but, speaking generally, the changeover seems to have occurred in the early-to-mid-1770s. In the last years of the previous decade, Sigaud de la Fond began presenting regular courses of lectures in Paris in which he approached electricity from a Franklinist point of view. In the textbook that parallelled his lectures, Sigaud presented Franklin's theory, we are told, "with all the clarity and precision that it had to be given to counterbalance the authority that the Abbé Nollet had acquired among both physicists and amateurs".(9)

Franklin's own visits to Paris in 1767 and 1769 doubtless gave his remaining adherents new heart and perhaps won him fresh support, and Nollet's death in 1770 removed his principle antagonist from the scene.

But the most influential factors would have been the publication of a French edition of Franklin's *Works* by his friend Barbeu Dubourg in 1773, and Franklin's own return to France in December 1776 as the representative of the American colonies during their war with England. Aldridge has pointed out elsewhere that Franklin's arrival led to a remarkable upsurge of interest in France in his other major published work, *The Way to Wealth*,(10) and undoubtedly the same thing happened with respect to his electrical writings. By the end of the decade, the new generation of French scientists - men such as Coulomb, Lavoisier and Laplace - were able to take the theory of the two contrary electricities for granted. Even in the early 1780s, however, Le Roy still adopted a somewhat defensive tone when speaking of "plus" and "minus" electricity,(11) while on the other side of the debate, Brisson continued to uphold energetically the merits of Nollet's ideas.(12) All in all, the Abbé's essentially Cartesian approach to electrical theory enjoyed a remarkable longevity in France, even if it did not do so to anything like the same extent in learned circles elsewhere.

Notes

1. Franklin, *Autobiography*, ed. H.W. Schneider (New York, 1952), p.153.
2 *Hist. de l'Acad.*, 1753, Mém. pp. 429-514; 1755, Mém. pp. 264-83.
3. *Observations sur la physique...*, I, 1773, 89.
4. *Ibid.*, III, 1774, 253 (my emphasis).
5. *Hist. de l'Acad.*, 1772, Mém. pp. 502-3.
6. See for example *L'annee littéraire*, 1761, Pt. IV, pp.120-38; or *J. de Trévoux*, LXI, 1761, 87-102.
7. Franklin, *Papers*, X, 61.
8. Schofield, intro. to reprint ed., (N.Y. 1966), p. xxiii.
9. *Observations sur la physique...*, II, 1773, 204.
10. Aldridge, *Franklin and his French Contemporaries* (N.Y., 1957), p.42.
11. *Hist. de L'Acad.*, 1783, Mém. p. 615.
12. Brisson, *Dictionnaire raisonné de physique* (Paris, 1781), art. "Electricité".

XIII

ON TWO SUPPOSED WORKS BY LEONHARD EULER ON ELECTRICITY

On 9/20 September 1755, G. F. Müller, the Secretary of the Academic Conference of the St. Petersburg Academy of Sciences, wrote to Leonhard Euler in Berlin to tell him the result of the Academy's recently judged prize competition on the theory of electricity[1]. The successful essay was one that Euler, apparently acting in his capacity as the Academy's chief link at this period with German scientific circles, had sent to St. Petersburg in the previous December[2]. At the time, however, Euler had not given any indication of the paper's authorship. Now, in response to a request from Müller, he revealed that it had been written by his son, Johann Albrecht Euler, and that he himself had also played a rôle in its preparation:

> Die Eigenschaften der elektrischen Kraft und insonderheit die Verwandtschaft derselben mit den Würkungen des Donners hatten mich auf diese Gedanken gebracht, da ich aber nicht wußte, ob es mir erlaubt wäre, für den Preiß zu arbeiten, so habe ich dieselben meinem Sohn *Joh[ann] Albrecht* übergeben, und ihn die überschickte Schrifft ausarbeiten lassen.[3]

On the basis of this 'confession,' a number of authors in recent years have asserted that the essay, published by the St. Petersburg Academy in 1756 together with two of the other entries submitted for the competition[4], should be regarded as the work not of the son, but of his distinguished father[5]. Indeed, one writer, L. S. Minchenko[6], has gone so far as to assert the father's authorship not only of the prize-winning essay, but also of a second paper on electricity published in the son's name some years afterwards[7]. He has pointed out that the ideas put forward in this latter work were a logical development of what had been proposed in the earlier essay (he takes it as established that this was the work of the older man).

[1] A. P. Yushkevich *et al.*, *Die Berliner und die Petersburger Akademie der Wissenschaften im Briefwechsel Leonhard Eulers* (Berlin: Akademie-Verlag, 1959–61), I, 91–2.

[2] Euler to Müller, 20/31 December 1754; *ibid.*, 72.

[3] Euler to Müller, 26 September/7 October 1755; *ibid.*, 92.

[4] *J. A. Euleri Disquisitio de causa physica electricitatis . . . una cum aliis duabus dissertationibus de eodem argumento* (St. Petersburg, 1756).

[5] S. Ya. Lur'ye, in *Leonard Eiler 1707–1783: sbornik stateĭ i materialov k 150-letiyu so dnya smerti* (Moscow/Leningrad: Izd'vo Akad. Nauk, 1935), 155–6; Ya. G. Dorfman, in his edition of F. U. T. Aepinus, *Teoriya elektrichestva i magnetizma* (Leningrad: Izd'vo Akad. Nauk, 1951), 480; V. P. Zubov, in E. Winter *et al.*, *Die deutsch-russische Begegnung und Leonhard Euler* (Berlin: Akademie-Verlag, 1958), 42.

[6] L. S. Minchenko, 'Ob odnoĭ dissertatsii Leonarda Eĭlera po elektrichestvu,' *Trudy Institut Istorii Estestvoznaniya i Tekhniki*, XXVIII (1959), 188–200.

[7] J. A. Euler, 'Recherches sur la cause physique de l'électricité,' *Hist. Acad. Roy. Sci. (Berlin)*, XIII (1757; publ. 1759), 125–59.

And he has drawn attention as well to the existence in the archives of the Soviet Academy of Sciences of a manuscript in Leonhard Euler's handwriting entitled 'Experimenta de electricitate'[8]. Minchenko has argued that the points set out in this document, published in Russian translation as an appendix to his article, are the very ones underlying the modifications introduced into the second of the published pieces. A common authorship can therefore, he says, be assumed.

Now these arguments, however plausible they may be in isolation, are by no means conclusive. Furthermore, there exists, as it happens, additional evidence which in my view is sufficient to undermine the general conclusions that have been reached, and cause us to conclude instead both that Leonhard Euler was not as dishonest a man as has been implied, and that the works in question are in fact properly ascribed to Johann Albrecht Euler. Before introducing that evidence, however, I would like to draw attention to a major incongruity that is forced upon us if we accept the arguments outlined above, an incongruity which is sufficient in itself, I believe, to give rise to considerable scepticism about the arguments themselves.

Consider the older Euler's 'confession' concerning the prize winning essay. This has evidently been taken to mean that Euler actually wrote the piece, and that his son merely provided the final editorial 'polish.' But if this is correct, Euler must stand convicted out of his own mouth of a deliberate attempt, either to defraud the St. Petersburg Academy of the 100 ducat prize it had offered for the competition, or to advance his son's career by fraudulent means. Moreover, Müller stands doubly convicted, first for arranging to have the prize money transmitted to Johann Albrecht Euler even though he knew he had not written the winning essay, and secondly for having the essay printed over the younger man's name rather than that of the father. If Minchenko is right, Euler too was guilty a second time when (presumably to conceal the previous crime) he had the Berlin Academy print the later essay in his son's name rather than his own. And this despite the fact that Johann Albrecht Euler was also a fully-fledged member of the Academy in Berlin at the time! And why were all these deceits perpetrated, we might ask? If they were to enable Leonhard Euler to get his hands on the 100 ducats from St. Petersburg, they were unnecessary, because Euler himself, despite his own doubts on the matter, was, as it turned out, eligible to enter the competition. Müller knew this – indeed, he informed Euler of it in replying to the letter of 'confession'[9] – and so the latter could have been publicly revealed as the author of the essay (and hence the winner of the prize) even at that late stage in the proceedings. On the other hand, if the intention were to advance young Johann Albrecht's career, to resort to fraud was both extremely risky and most unnecessary; the younger Euler was already firmly established as a member of the Berlin Academy, and he was also showing enough promise to give his father every confidence in his future anyway. Finally, if, for whatever reason, the older Euler's

[8] Arkhiv Akad. Nauk S. S. S. R., Leningrad: fond 136, op. 1, no 134, ll. 412–5. I have not seen this document.

[9] Müller to Euler, 31 October/11 November 1755; Yushkevich *et al.*, *op. cit.*, 93–4.

intention were to deceive, why ever did he mention to Müller the rôle he himself had played in the writing of the prize essay? It all seems most unlikely.

Fortunately, the interpretation of Euler's letter that has brought us to this point is not the only one that can be placed on it. It is equally possible – indeed, this would seem to be a more faithful reading of what Euler wrote – to understand him as saying no more than that the initial ideas upon which the paper was based were his, and that it was merely these, not a developed dissertation, that he had passed on to his son. In this case, of course, the credit for the dissertation itself would remain with the latter, and all suggestion of deceit would disappear. And if we suppose that this is how Müller in St. Petersburg read Euler's letter, we can also much more easily understand why he should have responded to it as he did with some very flattering remarks about young Johann Albrecht[10].

Plausibility considerations alone, then, suggest that this is what Euler meant. So too do two further pieces of evidence to which I wish to draw attention in this note. Both of these derive from sources extremely well placed to know who actually wrote the works in question; both Johan Carl Wilcke, the source of one of them, and Franz Aepinus, the source of the other, spent two years in Berlin in the mid-1750's, and during this time they became close friends of the two Eulers as members of a cheerful *Tischgesellschaft* of enthusiastic young men who met regularly at the Eulers' house for meals and scientific conversation[11]. What is more, both Wilcke and Aepinus were themselves engaged during this period in electrical researches, perhaps the most advanced researches in that field then being carried out anywhere in the world. Any shortcomings in the younger Euler's knowledge of electricity would therefore have been transparently obvious to them.

Wilcke's testimony consists of the formal Dedication of the doctoral thesis, *De electricitatibus contrariis*, that he successfully defended before the Philosophical Faculty of Rostock University in October 1757[12]. In this thesis, Wilcke set out in detail the results of his two years of electrical researching in Berlin. Most strikingly, from our point of view here, he elected to dedicate the work to none other than Johann Albrecht Euler! In doing so, moreover, he referred explicitly to Euler's 'lumieres dans cette partie de la Physique, couronnés d'une illustre académie'; these, he said, were 'au dessus de l'ordinaire.' Had he entertained the slightest doubt as to his friend's authorship of the prize-winning essay, he would surely never have acted in this way, since to do so would have meant needlessly risking his own reputation at the very outset of his career.

10 *Ibid.*

11 That there was such a group gathered around the older Euler at this period is not well known. Its existence and membership have been established largely from the correspondence of Johann Albrecht Euler, now preserved in the Academy of Sciences archives in Leningrad (especially fond 1, op. 3, no. 46). Besides Aepinus, Wilcke, and the Euler children, the group included the Swiss mathematician Louis Bertrand and the future St. Petersburg academicians Stepan Rumovskii and Semyon Kotel'nikov, and also perhaps the young Berlin academicians Johann Bernard Merian and Johann Jakob Huber. The mineralogist Johann Gottlob Lehmann may also have been a member of the group; on this, see C. W. Oseen, *Johan Carl Wilcke, Experimental-Fysiker* (Uppsala: Almqvist & Wiksell, 1939), 35–6.

12 J. C. Wilcke, *De electricitatibus contrariis* (Rostock, 1757).

Aepinus's evidence bears directly not on the prize-winning essay but on the later publication in the Berlin *Mémoires*. As Minchenko has indicated, however, the contents of the two documents are intimately related, and there can be little doubt that both were written by the same person. What is significant for us here, therefore, is that when Aepinus (by now installed as a member of the Academy of Sciences in St. Petersburg) wrote a long and friendly letter to Berlin early in 1758 commenting in detail on some of the points raised in the later paper, he addressed his comments to Johann Albrecht Euler, not to the father[13]. And the tone of the whole letter makes it abundantly clear that Aepinus had no doubts at all that the younger man had written the paper, and also that he regarded him as a serious contributor to electrical researches.

The testimony of two such well-placed contemporaries as Wilcke and Aepinus is, I believe, more than sufficient to establish J. A. Euler's claim to the writings in question, and hence also to establish the alternative interpretation of Leonhard Euler's 'confession' that I have suggested as the correct one. What remains to be done to clinch the argument is to account for the further document to which Minchenko has drawn attention, the 'Experimenta de electricitate' in the older Euler's handwriting. But this is easily done. Certainly the points set out by Euler in this document are exactly those that underlay the theoretical modifications introduced in the paper in the Berlin *Mémoires*, but they are also the very points with which Wilcke dealt at length in his thesis, and which he and (as we learn from the Berlin *Mémoires* paper) Aepinus had investigated so thoroughly during their time together in Berlin. More than this, they were the crucial points at issue in the wider controversy over electrical theory that raged throughout the learned world during the 1750's and beyond, between the supporters and opponents of Benjamin Franklin's theory of *plus* and *minus*. Small wonder, then, that Euler *père* should have chosen to draw up for his own use a summary of the experimental evidence bearing on this debate. Small wonder, either, that he should have accepted the distinction between *plus* and *minus* electricity advocated so powerfully by Aepinus and Wilcke (though not the further conclusions they wanted to draw from this) when he came to write his *Letters to a German Princess* a few years later[14]. But it does not follow that his son could not have been similarly but independently persuaded. It does not follow that because the older Euler drew up his summary, he need have been in any way responsible for the inclusion of similar material in the paper in the Berlin *Mémoires*. This material had a common currency at the time, nowhere more so than over meals at Euler's house in Berlin. It was certainly not in Leonhard Euler's possession alone.

There remains, then, nothing to prevent our rejecting the interpretation that has been put on Euler's 'confession' in recent years. Neither the prize-winning essay nor the later paper in the Berlin *Mémoires* was written by Leonhard Euler. Both came, as indicated on the works themselves, from the pen of his son Johann

[13] Aepinus to J. A. Euler, 28 February 1758; Arkhiv Akad. Nauk S. S. S. R., Leningrad: fond 1, op. 3, no. 46, l. 53. Aepinus's choice of phrase suggests that he may have been commenting not on the full paper but on a précis of this that Euler had sent to him.

[14] English translation by Henry Hunter (London, 1802), II, 99.

Albrecht. All that the father claimed to have contributed were some initial ideas. And neither of the Eulers, nor Müller either for that matter, need have his reputation sullied any longer by imputations of deliberate deceit and literary fraud in this matter.

XIV

Aepinus, the Tourmaline Crystal, and the Theory of Electricity and Magnetism

"THE TOURMALINE IS ONE of the most interesting species in the mineral kingdom"—so *The Popular Encyclopedia* assured its readers almost a century ago.[1] As with Trembley's polyp, however, the tourmaline enjoyed its greatest renown during the eighteenth rather than the nineteenth century, and by the time the above remark was published, it had long since ceased to occupy the center of the scientific stage in the way it once had done. In this paper I intend to look briefly at some of the circumstances surrounding the initial upsurge of interest in the stone in the middle years of the eighteenth century. As we shall see, important developments in the theories of electrical and especially magnetic action flowed from this first detailed investigation of its properties. I shall not extend the discussion, though, to the later years of the century, when the general recognition that the electricity acquired by a heated tourmaline was constantly associated with a particular axial direction in the crystals gave the stone a renewed importance, this time in the rapidly developing science of crystallography.[2]

The key figure in the early period was the young German physicist and astronomer Franz Aepinus, today the least known of all the major eighteenth-century electrical investigators. What is more, Aepinus' analysis of the electrical properties of the stone played a key role in his own scientific development and led directly to his writing his important essay on the theory of electricity and magnetism *Tentamen theoriae electricitatis et magnetismi* (St. Petersburg, 1759). His part in the story is of considerable interest in its own right, and it will be my principal concern in what follows.

[1] "Tourmaline," in *The Popular Encyclopedia; or, Conversations Lexicon*, 7 vols. (London, 1880).

[2] This aspect of the phenomenon was one of the most obstinate difficulties confronting René Just Haüy as he developed his purely geometrical theory of crystal structure, and it became a significant factor in the eventual overthrow of Haüy's doctrines. John G. Burke, *Origins of the Science of Crystals* (Berkeley/Los Angeles: University of California Press, 1966), pp. 133–138.

Aepinus was born and raised in the ancient Hanseatic port of Rostock, where his father taught philosophy and theology for over fifty years. He entered the local university in 1740, at the age of fifteen, and remained there as a student until 1744, when he transferred to Jena. In Rostock he had concentrated particularly on mathematics, but in Jena he studied medicine as well under the noted teacher G. E. Hamberger. He returned to Rostock early in 1747 and some months later defended a thesis on intermittent fevers.[3] His interest in medicine, however, was but slight—he is reported to have presented a thesis on a medical topic only to make his father happy[4]—and instead of taking out his medical degree he remained attached to the university as a *Privatdozent*, giving regular courses of lectures on mathematical subjects for a number of years. At the same time, on the basis of several reports published in the Rostock *Gelehrte Nachrichten*,[5] he developed a local reputation as an astronomer, and it was this that led to his appointment early in 1755 to the chair of astronomy in Frederick the Great's Academy of Sciences in Berlin.[6] Aepinus remained in Berlin for only two years before departing for St. Petersburg, where he had in the meantime been appointed professor of physics. It was in the Prussian capital, however, that he began the epoch-making series of electrical and magnetic researches upon which his reputation almost entirely rests.

In Berlin Aepinus developed a particularly close friendship with the great mathematician Leonhard Euler, the man who had been primarily responsible for his appointment to the Academy,[7] and with Euler's son Johann Albrecht, a talented young man of about his own age. At the time both the Eulers were actively concerned with the study of electricity, and it was almost certainly they who awakened Aepinus' interest in the subject.[8] Then a few months

[3] Aepinus, *Meditationes de causa et indole febrium intermittentium* (Rostock, 1748). An earlier publication on the paths of falling bodies is often said to have been his doctoral dissertation, but in fact it was a *programma* advertising his first course of lectures following his return to Rostock.

[4] J. C. Koppe, *Jetztlebendes gelehrtes Mecklenburg* (Rostock/Leipzig, 1783–1784), Vol. I, p. 11.

[5] *[Rostock] Gelehrte Nachrichten*, 1753, pp. 170–173, 182–184, 210–213, 284–288, 316–319, 489–492; 1755, pp. 108–109, 499–504, 525–528.

[6] That Aepinus had already acquired a reputation as an astronomer by that time is attested by the comments of the eminent Tobias Mayer upon learning of the appointment: "Der Herr M. Aepinus in Rostock, von welchem ich vernehme, daβ er nächstens nach Berlin gehen, und das dortige Observatorium zum Gebrauch bekommen werde, besitzet alle diejenige Einsicht und Geschicklichkeit, welche ihn des Besitzes dieser Stelle würdig mache." Tobias Mayer to L. Euler, Feb. 23, 1755, in Yu. Kh. Kopelevich, "Perepiska Leonarda Eĭlera i Tobiasa Maĭera," *Istoriko-astronomicheskie issledovaniya*, 1959, 5:422; English translation in Eric G. Forbes, ed., *The Euler-Mayer Correspondence* (London: Macmillan, 1971), p. 97.

[7] A series of letters from Aepinus to Euler concerning the appointment are preserved among the latter's papers in Leningrad (Arkhiv Akademiya Nauk S.S.S.R., Leningrad; fond 136, Op. 2, No. 5).

[8] Shortly before Aepinus arrived in Berlin the younger Euler had submitted an entry to the prize competition on the cause of electricity which had been announced by the St. Petersburg Academy of Sciences. (L. Euler to G. F. Müller, Dec. 20/31, 1754; in A. P. Yushkevich, *et al.*, eds., *Die Berliner und die Petersburger Akademie der Wissenschaften im Briefwechsel Leonhard Eulers*, Berlin: Akademie-Verlag, 1959, Vol. I, p. 72). But we know that the father was also interested in the subject, for when it was announced some months later that his son's paper had won the prize, he confessed that the original ideas upon which the paper was based had been his, but that not being sure of his eligibility to submit an entry to the competition, he had passed

after Aepinus took up his appointment he was joined in Berlin by his friend and one-time student Johan Carl Wilcke, who was likewise fired with enthusiasm for electrical experimentation. During 1756 Wilcke pursued his electrical researches vigorously, and so, apparently, did Aepinus, for by early the following year he was displaying an extremely high level of sophistication in both electrical theory and experimentation.[9] The two men were sharing a house at this period,[10] and no doubt they would have done much of their experimenting together. Wilcke in fact later testified to the help he had received from Aepinus;[11] while for his part, Aepinus, in giving a circumstantial account of his discoveries, made clear his indebtedness to "a friend," who, from the context, was plainly Wilcke.[12]

Nor were Wilcke and Aepinus left to pursue their various electrical researches on their own. The younger Euler continued to work on the subject, and he evidently kept in close touch with what his friends were doing, for he subsequently frankly admitted the powerful impact Aepinus' work had on his thinking.[13] We may be sure that electricity would also have been a favorite topic of conversation for the other members of the group of talented young men who gathered around Leonhard Euler and who met and dined regularly at his house. This group included not only Euler *fils*, Aepinus, and Wilcke, but also the Swiss mathematician Louis Bertrand, two Russian visitors, Stepan Rumovskiĭ and Semyon Kotel'nikov, the young academicians Johann Bernard Merian and Johann Jakob Huber, and (apparently) the mineralogist Johann Gottlob Lehmann.[14]

It was Lehmann who, toward the end of 1756, drew Aepinus' attention to the peculiar ability which a small crystal of tourmaline he possessed had of drawing ashes to itself when placed on a burning coal.[15] This behavior had become well known in the Dutch and German jewel trade in the early years of the eighteenth century, soon after samples of the stone were first imported

---

them on to his son to be worked out in the necessary detail (Euler to Müller, Sept. 26, 1755; *ibid.*, p. 92). On this point, see R. W. Home, "On Two Supposed Works by Leonhard Euler on Electricity," *Archives internationales d'histoire des sciences* (forthcoming).

[9] F. U. T. Aepinus, "Mémoire concernant quelque nouvelles expériences électriques remarquables," *Histoire de l'Académie Royale des Sciences et Belles Lettres à Berlin*, 1756, *12*:101–121.

[10] A. J. D. Aepinus, introduction to J. C. Wilcke's *De electricitatibus contrariis* (Rostock, 1757), p. 10.

[11] Wilcke described (*ibid.*, p. 93) how he reported some of his results to Aepinus, about whom he added "quem in hisce laboribus socium quandoque habuisse, plus semel profuit."

[12] F. U. T. Aepinus, *Tentamen theoriae electricitatis et magnetismi* (St. Petersburg, 1759), p. 75; cf. Wilcke, *De electricitatibus contrariis*, pp. 83–88.

[13] J. A. Euler, "Recherches sur la cause physique de l'électricité," *Hist. Acad. R. Sci. Berlin*, 1757, *13*:125.

[14] That the elder Euler had such a group gathered around him is not well known. Its existence and composition have been determined largely from the correspondence of his son, now preserved in the Academy of Sciences archives in Leningrad (esp. fond 1, Op. 3, No. 46). For Lehmann's possible membership of the group, see Wilcke's report, quoted in C. W. Oseen, *Johan Carl Wilcke, Experimental-Fysiker* (Uppsala: Almqvist & Wiksell, 1939), pp. 35–36.

[15] Aepinus, "Mémoire," p. 107. The dating is provided by Aepinus, *Tentamen*, p. 2; but cf. J. C. Wilcke, "Geschichte des Tourmalin," *Abhandlungen der Königl. Schwedischen Akademie der Wissenschaften*, 1766, p. 99, where a slightly later date is indicated.

from Ceylon (then the only known source of supply) for use in jewelery. Names such as *Aschentrekker* and *Aschenzieher,* which were coined for the stone at this time, reflect its most striking property. Yet, despite the fact that a reasonably accurate account of the phenomenon was published in 1717 by Louis Lémery (though without any mention of the role of heat in bringing about the effect),[16] it was then more or less forgotten in scientific circles until Lehmann brought it to Aepinus' notice.[17] Symptomatic of this neglect is the fact that the stone went by a bewildering variety of different names during the intervening period, and the name "tourmaline," derived from its Singhalese name, *tóramalli,* only became established as a result of Aepinus' introduction of it into the scientific literature.

When Lehmann brought the tourmaline to Aepinus' attention, he may also have suggested that its behavior ought to be investigated electrically. In any event, when Aepinus took up the matter, he found at the first try that the effect was indeed an electrical one.[18] More important, he quickly recognized that he was dealing with something very different from the usual kinds of electrical phenomena. A tourmaline crystal, he discovered, became electrified simply by being warmed; none of the usual rubbing was required. What is more, he found that when it was warmed, the tourmaline did not become electrical in the usual manner. Instead of simply acquiring an overall charge, the crystal acquired opposite electrical charges on two opposite faces.[19]

The electrical properties of the tourmaline crystal are extremely complex, and it is hardly surprising that Aepinus did not altogether succeed in sorting them out. To get as far as he did, however, was an impressive achievement which testifies to his scientific insight as well as to his skill and patience as an experimenter. The key to his success was his analysis of the behavior of the stone in terms of the Franklinian doctrine that there were two contrary modes of electrification possible in nature—called by Franklin "plus" and "minus." Aepinus found that if he immersed his tourmaline in hot water for a time and then withdrew it, it would consistently acquire a positive electrical charge on one particular face and a negative one on the opposite face. He regarded a stone in this condition as being in its "natural" electrical state, for reasons which will soon become apparent. By comparing the orientation of its "poles" with that of another tourmaline which had been cut rather

[16] *Histoire de l'Académie Royale des Sciences* (Paris), 1717, Hist. pp. 7–8.

[17] Linnaeus gave a brief description in the preface to his *Flora Zeylanica* (Stockholm, 1747), p. 8, and even coined the name *lapis electricus* for the stone. However, this seems to have been nothing more than an inspired guess.

[18] Aepinus nowhere made any mention of Lehmann's having suggested that electricity might be involved, and indeed by his choice of phrase he indicated quite strongly that the idea was his own: "La premiere fois que j'entendis parler d'une singularité aussi remarquable, je formai aussi-tôt la conjecture qu'elle devoit son origine à l'Electricité" ("Mémoire," p. 107). The suggestion that Lehmann might have been responsible for introducing electricity into the discussion is due to Wilcke: "Herr Lehmann [hatte] berichtet: er besitze einen kleinen Stein, der, auf Kohlen gelegt, mit der Asche spielte, und verdiente durch die Elektricität untersucht zu werden. Herr Aepin . . . nahm ihn zu sich, und fand bey der ersten Probe die Sache richtig" ("Geschichte des Tourmalin," pp. 99–100).

[19] Aepinus, "Mémoire," pp. 110 ff.

differently, he showed that the location of the poles in no way depended on the external form of the sample and hence had to be a function of its inner structure. If a crystal was heated by placing it on a hot surface or by using a lens to focus sunlight on it, it would acquire a pair of charges opposite to those it displayed when in its "natural" state; but subsequently it would of its own accord revert to the "natural" condition. (Its tendency to do so was, of course, what led Aepinus to regard the latter state as the natural one in the first place.) Aepinus was inclined to attribute the "unnatural" charges in the second series of experiments to the uneven heating which the stone received in these cases, and the restoration of the "natural" condition to the achievement of an equilibrium in the heat distribution, but he was by no means dogmatic on this point. Finally, he recognized that if the stone was rubbed, it behaved just like glass; that is, it acquired a positive charge. But he also succeeded in showing that this kind of electrification was entirely independent of the other, which was, so far as he knew, peculiar to the tourmaline.

Aepinus read a paper embodying his results—the paper from which the above account was drawn—to the Berlin Academy on March 10, 1757, only three weeks before he left to take up his appointment in St. Petersburg.[20] In the paper he included a report of another remarkable experiment in which he had successfully repeated Musschenbroek's famous *expérience de Leyde* using a layer of air as the dielectric instead of the glass that was normally used. Just prior to his departure he sent a *précis* of his paper to Paris, where it was promptly published in F.-V. Toussaint's *Observations périodiques sur la physique, l'histoire naturelle et les arts.*[21] Eighteen months later a summary of his work, based this time on the original paper itself (which had by then appeared in print in the Berlin Academy's *Mémoires* for 1756), was also published in England, in the supplement to the 1758 volume of the *Gentleman's Magazine.*[22]

Once Aepinus' discovery became known to the electrical investigators of western Europe, it aroused great interest, and for good reason. To begin with, Aepinus had hit upon an entirely new means of generating electricity, quite different from the usual frictional methods. Furthermore, since his method involved nothing more than warming his crystal, it promised to shed much light on the difficult question, then of great moment in electrical inquiry, of the true relationship between electricity and "fire."[23]

Aepinus' work also promised, however, to shed considerable light on electrical theory per se. During the mid-1750s the great debate between the Abbé Nollet and his supporters on the one hand and Benjamin Franklin and his followers on the other was at its height. Any new discovery, such as the one Aepinus had made, was bound to be scrutinized closely by both sides to see what it might contribute to their arguments. As we have seen, Aepinus had found

[20] E. Winter, ed., *Die Registres der Berliner Akademie der Wissenschaften 1746–1766* (Berlin: Akademie-Verlag, 1957), p. 231.

[21] *Op. cit.*, 1757, *2*:341–345 (in the number for May 1757): cf. F. U. T. Aepinus, *Recueil de différents mémoires sur la tourmaline* (St. Petersburg, 1762), p. 130.

[22] *Gentleman's Magazine*, 1758, *28*:617–619.

[23] Cf. Wilcke, "Geschichte des Tourmalin," p. 95.

Franklin's conception of two contrary modes of electrification—"plus" and "minus"—extremely helpful in understanding his observations, and his report was couched in thoroughly Franklinian terms. Inevitably, therefore, it was treated at once as an important contribution to the general debate. In England, Franklin, John Canton, and Benjamin Wilson all quickly verified Aepinus' most important results and strongly supported his interpretation of them in terms of the Franklinian theory.[24] In Paris, by contrast, an attempt to understand the behavior of the tourmaline in terms of Nollet's theory of electricity was published soon afterward. This purported to be the work of a Neapolitan nobleman, the Duc de Noya Carafa, but in fact the noted naturalist and friend of Nollet, Michel Adanson, was in large measure responsible for it.[25] Adanson was not nearly so successful as Aepinus had been, however, in reconciling the phenomena with his general theoretical position, so that in most parts of Europe Aepinus' work on the tourmaline considerably strengthened the position of the Franklinists in the general debate, while the position of Nollet and his effluvialist supporters was correspondingly weakened. On the other hand, in France, where Nollet's ideas continued to hold sway until well into the 1770s, Adanson's work, despite its shortcomings, was regularly cited for many years as the standard work on the subject.[26]

Aepinus' work on the tourmaline catapulted him suddenly into the front rank of electrical investigators. Yet even his English supporters did not accept uncritically all his conclusions concerning the behavior of the stone. Canton, for example, introduced an important modification when he recognized that the tourmaline became electrified if it was cooled, just as it did when it was heated, with the difference that the positive and negative charges were now reversed. But, as the young Torbern Bergman pointed out explicitly a few years later when he made the same discovery,[27] this was in fact quite consistent with Aepinus' results, once it was appreciated that in the latter's experiments

[24] Franklin to William Heberden, June 7, 1759; in his *Experiments and Observations on Electricity* (4th ed., London, 1769), pp. 375–378. Canton, *Gentleman's Magazine*, 1759, *29*:424–425 (this report being published over the *nom de plume* "Noncathoni," an obvious anagram for "Iohn Canton"). Benjamin Wilson, *Philosophical Transactions of the Royal Society of London*, 1759, *51*:308–339.

[25] *Lettre du Duc de Noya Carafa sur la tourmaline, à Monsieur de Buffon* (Paris, 1759). In the catalogue of the Bibliothèque Nationale, Paris, it is assumed that "Duc de Noya Carafa" was simply a pseudonym used by Adanson. Yet there really was such a person as the Duc, and what is more, he took an active interest in philosophical matters. Whether he actually wrote much of the book, however, is debatable. The *Lettre* began with an account of how the Duc had first become acquainted with the tourmaline, and during the course of this it was frankly conceded (p. 7) that most of the research to be described was done by Adanson. The latter certainly regarded the book as his own, for in the catalogue of his library he listed it as "Duc Noya Adanson Lettre sur la tourmaline" (George H. M. Lawrence, ed., *Adanson: The Bicentennial of Michel Adanson's "Familles des plantes,"* Pittsburgh: Hunt Botanical Library, 1963–1964, Vol. I, p. 273). Yet this claim does not necessarily conflict with what was said in the *Lettre;* probably what happened was simply that the Duc wrote the introductory section but then left Adanson to do the rest. For the latter's friendship with Nollet, see Lawrence, ed., *Adanson,* Vol. I, p. 31.

[26] I discuss Nollet's long-continued dominance of French electricity in my paper "Electricity in France in the Post-Franklin Era," *Proceedings of the XIVth International Congress of the History of Science, Tokyo-Kyoto, 1974* (Tokyo, 1975), Vol. II, pp. 269–272.

[27] Wilcke, "Geschichte des Tourmalin," p. 112.

with crystals heated in boiling water, he was in fact observing the charges that developed during the *cooling* phase.

Far more awkward were some of the results announced by Benjamin Wilson, who quickly established himself during this period as the leading English authority on these matters, to the point where his investigations were crowned by the Royal Society with the Copley Medal for 1760.[28] While agreeing wholeheartedly with the general tenor of Aepinus' results, Wilson directly rebutted one of them. Aepinus had stated that the tourmaline when heated invariably acquired opposite charges on two opposite faces. Wilson reported on the contrary that when the particularly large stone he had at his disposal was heated unevenly, it acquired the same charge on both faces.

Aepinus found this result most perplexing, and when he published his *Recueil de différents mémoires sur la tourmaline* in mid-1762 he discussed the point at some length. His initial reaction was to suppose that the trouble stemmed from slight differences in the experimental arrangements, but even when he repeated Wilson's experiments exactly as the Englishman had described them, he still could not reproduce the latter's results. Obviously embarrassed about seeming to contradict "un aussi habile observateur" as Wilson, and yet confident that he had not made a mistake himself, he freely confessed his bewilderment: "Il m'est extrèmement difficile, de supposer, que Mr. Wilson soit tombé ici dans une méprise, néanmoins je suis convaincu, que de ma part, il n'y a assurement, aucune faute. Ainsi je ne sais qu'en juger."[29] Wilson in turn reiterated his faith in the results he had obtained, and in doing so apparently felt free to adopt a rather patronizing tone toward a colleague so far from the scientific centers of western Europe. Aepinus, he somewhat unjustly remarked, had not really described his exact procedures sufficiently clearly, but "I am inclined to believe some material circumstance has been omitted in the method; or that our apparatus's are essentially different; (though you seem to have had a regard to some of the necessary requisites for making the experiment properly:) otherwise, I cannot apprehend why you were not able to succeed: because the experiment always answers with me."[30]

Joseph Priestley, when he came to write his *History of Electricity* a few years later, found the whole affair most confusing[31] but in the end decided that Wilson's apparatus was more accurate than Aepinus', so that his results could be better depended upon.[32] However, Priestley's judgment here cannot be regarded as definitive, since he reached it without ever reading Aepinus' *Recueil,* the only place where the latter described in any detail the apparatus he had used in his experiments.[33] In Sweden, Wilcke and Bergman proceeded more

[28] Royal Society, *Journal Book,* entry for Dec. 1, 1760.

[29] Aepinus, *Recueil,* p. 192.

[30] Benjamin Wilson, *A Letter . . . to Mr. Æpinus . . .* (London, 1764), p. 15; also *Phil. Trans.,* 1763, *53*:446.

[31] Robert E. Schofield, ed., *A Scientific Autobiography of Joseph Priestley (1733–1804)* (Cambridge, Mass.: M.I.T. Press, 1966), pp. 18, 28.

[32] Joseph Priestley, *The History and Present State of Electricity* (2nd ed., London, 1769), p. 301.

[33] *Ibid.,* p. [714].

circumspectly: they sought to reconcile the conflicting results by arguing that though in normal circumstances the two opposite faces of a tourmaline acquired opposite charges, in reality they acted quite independently of each other, so that under appropriate conditions both Aepinus' and Wilson's results were possible.[34] Still, uncertainties remained, and the debate was never satisfactorily resolved.

The behavior of the tourmaline is, as stated earlier, extremely complex. In this instance, the conflicting observations almost certainly arose from the fact that when a crystal is heated unevenly, gross and unpredictable piezoelectric effects become superimposed on the primary pyroelectric-cum-piezoelectric phenomenon, and these depend critically on, among other things, crystal size. Given that Wilson's crystal was a particularly large one, this by itself could explain the result he obtained. In addition, Wilcke and Bergman may well have been correct: Wilson's stone may have been large enough for its face farthest from the source of heat actually to have been in its cooling phase while the opposite face was still being heated. This would explain his results equally well. In any event, both Aepinus' and Wilson's results could have been quite valid. There simply was no definitive answer.

From Aepinus' own point of view, his work on the tourmaline played a crucial role in the development of his thought. On the one hand, his discoveries convinced him, if he still needed convincing, that Franklin's ideas about electricity were basically correct. On the other, having recognized that the electricity of the tourmaline was essentially dipolar in character, Aepinus came to view an electrified tourmaline and a magnetized piece of iron as being analogous, and this in turn led him to develop an important new theory of magnetism in close parallel with the Franklinian electrical model. If the electrical properties of the tourmaline could be understood in terms of a surplus of electric fluid on one face of the crystal and a deficiency on the other, then a magnet could likewise be seen as a body with a surplus of an analogous magnetic fluid on one pole and a deficiency on the other. More generally, Aepinus concluded, magnetic phenomena might be understood in terms of the interactions of this magnetic fluid with ordinary matter in just the same way as electrical phenomena were explained in terms of the interactions between ordinary matter and the electric fluid.

Inspired by the tourmaline analogy, Aepinus began his magnetic researches immediately,[35] but the imminence of his departure for St. Petersburg forced a temporary halt. Once settled in the Russian capital, he resumed his magnetic investigations in earnest, and only a little over a year later he was able to present an initial report on the subject at a public meeting of the Russian Academy before the Empress Elisabeth and the assembled dignitaries of her court.[36] In this he limited himself to demonstrating that each of the principal

[34] Wilcke, "Geschichte des Tourmalin," p. 112.

[35] Aepinus, *Tentamen*, p. 3.

[36] Aepinus, *Sermo academicus de similitudine vis electricae atque magneticae . . .* (St. Petersburg, 1758). A Russian translation was published simultaneously under the title *Rech' o skhodstve elektricheskoĭ sily s magnitnoyu. . . .*

magnetic phenomena had a parallel in the electrical realm, and to outlining very briefly at the end of his address the main points of the new theory he had devised to account for them. Despite its limited scope, his essay evidently excited a great deal of interest, for it quickly appeared in at least three separate German editions besides the Latin and Russian versions which had been published in St. Petersburg originally.[37]

During the winter of 1758–1759 Aepinus suffered a protracted illness which interrupted his work for a time,[38] but nevertheless by the following June the completed manuscript of his great work, *Tentamen theoriae electricitatis et magnetismi,* was ready for presentation to the Academy for publication.[39] A week later, after two of his fellow academicians had reported favorably on it, the manuscript was sent to the printer with a request that the work be expedited as much as possible—the result of the Academy's prize competition on the construction of artificial magnets was due to be announced early in September, and, since Aepinus had treated the same topic at some length in his book, his colleagues were anxious that his account should, if possible, be in print before the announcement of the result of the competition.[40] In the event, the academicians' hopes were not realized; with the prize announcement made and printing still unfinished, in September Aepinus submitted two further short dissertations to be included in his book as appendices,[41] and it was early December before copies of the completed work finally reached the Academy's book shop.[42] Because his work had thus been overtaken by events, Aepinus included in his preface a graceful acknowledgement of the priority of the author of the winning entry in the competition (a M. Antheaulme, *Syndic des tontines* in Paris) in the invention of the improved method of magnetizing steel bars described at length in his book[43]—and this despite the fact that, as is clear from the dates involved, he had actually developed the method quite independently.

But Aepinus' work went far beyond anything that Antheaulme had attempted, both in the practical realm to which Antheaulme had in fact confined his attention, and, more importantly, in the realm of theory. Aepinus' contributions to the theory of magnetism were fundamental—even more so, perhaps, than the reformulation of Franklinian electrical theory that he also undertook in his great work. His book represented a truly remarkable advance on the traditional

[37] The three German editions I know of, all entitled "Akademische Rede von der Aehnlichkeit der elektrischen und magnetischen Kraft," appeared in the *Hamburgisches Magazin,* 1759, *22*:227–272, in the Leipzig *Allgemeines Magazin der Natur, Kunst und Wissenschaften,* 1761, *11*:90–131, and as a separate publication (Leipzig, 1760). The original Latin edition was reviewed in the *Göttingische Anzeigen von gelehrten Sachen* early in 1759 (pp. 116–119).

[38] Aepinus, *Tentamen,* p. [xiii].

[39] *Protokoly zasedaniĭ konferentsii Imperatorskoĭ Akademii Nauk s 1725 po 1803 goda* (St. Petersburg, 1897–1911), Vol. II, p. 428.

[40] *Ibid.*

[41] *Ibid.,* p. 436.

[42] Personal communication from Mr. D. E. Bertel's, Archives Department, Academy of Sciences of the U.S.S.R. A number of recent authors, apparently following the lead given by Poggendorff in his *Biographisch-Literarisches Handwörterbuch zur Geschichte der exacten Wissenschaften,* Vol. I (Leipzig, 1863), have stated that Aepinus' book was published in Rostock. This is certainly wrong.

[43] Aepinus, *Tentamen,* p. [xix].

Cartesian-style vortex theories that had been generally accepted until then. Just as he had done in his public lecture a year earlier, but now much more rigorously and in much greater detail, Aepinus systematically transferred his improved and semi-quantitative version of the Franklinian theory of electricity to the magnetic realm. In almost every case, the phenomena quickly yielded to his analysis, and as a result the theory of magnetism was for the first time placed on a reasonably satisfactory footing.

So far as the study of the tourmaline itself was concerned, little further progress was made for many years. Aepinus, in his *Recueil de différents mémoires sur la tourmaline* of 1762, reprinted his initial paper on the subject, and also provided a great deal more detail concerning his experimental procedures. In addition, he republished the papers of Wilson and the Duc de Noya Carafa, together with his comments on these. Despite the book's controversial character, it failed to generate much discussion. Attempts were made both to classify the stone mineralogically and to analyze it chemically,[44] but the results obtained were inconclusive. Wilcke held high hopes that the tourmaline would provide the key to a grand unification of the theories of heat, electricity, and magnetism,[45] but these, too, came to nothing. Benjamin Wilson observed at one point that his tourmalines looked darker when viewed along the line joining their electrical poles than they did when viewed in a direction perpendicular to this,[46] but even such a promising hint led nowhere at the time. In fact, the only other definite advances that were made before the 1790s were two further discoveries of Wilson's. First, he found that the behavior Aepinus had described was not peculiar to tourmaline but occurred with a number of other semi-precious stones as well. Also, some of the stones he used in these experiments were uncut, and this enabled him to make the important observation that the electrical charges that developed when the stones were heated were constantly aligned with the crystal grain.[47] Thirty years later, Haüy was to seize on this as the starting point for his crystallographically oriented reinvestigation of the entire problem.[48]

So far as the early period was concerned, however, the potentialities of the tourmaline had been exhausted. Nevertheless, science had already gained lasting benefits from its study. Franklin's theory of electricity had gained significant new support. Above all, the theory of magnetism had been transformed, and here Aepinus' work was of fundamental importance. His ideas represented a completely new departure, and led directly to the work of Coulomb and Poisson in the years that followed. And they stemmed directly, as we have now seen, from his initial study of the electrical properties of that most interesting of crystals, the tourmaline.

[44] E.g., S. Rinman, *Abhandl. Königl. Schwed. Akad. Wiss.*, 1766, pp. 46 ff., and K. A. Gerhard, *Nouveaux mémoires de l'Académie Royale des Sciences et Belles-Lettres à Berlin*, 1777, pp. 14–24.

[45] Wilcke, "Geschichte des Tourmalin," p. 95.

[46] Wilson, *A Letter . . . to Mr. Æpinus . . .*, p. 17; also *Phil. Trans.*, 1763, *53*:448.

[47] Wilson, "Observations upon some Gems Similar to the Tourmalin," *Phil. Trans.*, 1761–1762, *52*:443–447.

[48] R. J. Haüy, "Extrait des observations sur la vertu électrique que plusieurs minéraux acquièrent à l'aide de la chaleur," *Journal d'histoire naturelle (Lamarck)*, 1792, 1:449-61.

XV
Science as a Career in Eighteenth-Century Russia: The Case of F. U. T. Aepinus

I

IT is a commonplace that the modes of thought which we characterise today as peculiarly 'scientific' emerged during the 17th century, out of what has come to be called by historians 'the scientific revolution'. However, this was a revolution confined very largely to a few of the countries of western Europe. In Russia, by contrast, even at the very end of the 17th century attitudes towards the natural world remained firmly rooted in traditional learning. Therefore when the new science came at last during the first decades of the next century, it did not do so as the result of any gradual evolution of long-standing Russian ideas. It came instead as something foreign and quite unfamiliar, imposed in large measure from above by Peter the Great as part of his remarkable attempt to transform his realm all at once into a modern European state.

As was natural in such circumstances, it was only very gradually that significant numbers of Russians began to feel any affinity with the ideals of the new science. When Peter founded his Academy of Sciences in St Petersburg in 1725, he had to staff it entirely with foreigners, mostly Germans, and even a hundred years later, half of the academicians were still of foreign birth.[1] What is more, for most of the 18th century the Academy remained the only centre of scientific activity in the country. In the early years in particular, life must have been difficult for its members. Not only did they find themselves in a foreign land where their day-to-day activities and interests were completely incomprehensible to the society around them upon whose support they depended; Peter's premature death also deprived them of their principal political support, just when they needed it most.

The outburst of anti-German nationalist feeling which swept Elisabeth to the throne in 1741 made the position of the essentially German Academy less secure than ever. Several of its original members had returned to their homelands long before; in the political

[1] B. L. Modzalevsky (ed.), *Spisok chlenov Imperatorskoy Akademii nauk 1725–1907*, St Petersburg, 1908.

turmoil of 1741, Leonhard Euler, the most outstanding of all the early academicians, decided to accept the invitation of Frederick II of Prussia to move to Berlin. Not surprisingly, the replacements were in general not men of the same calibre as those who had left; the standards in the Academy went into a general decline.

In the years that followed, the Academy might well have become more of a foreign enclave within the Russian capital than ever, had not a steady increase which had been going on for several years in the number of native Russians in the lower echelons of the academic hierarchy culminated at about the same time in the appointment of the first native Russians to professorships within the Academy. More important still for the subsequent history of the Academy was the character of the most notable of these appointees, Mikhail Vasil'yevich Lomonosov.

The extraordinary nature of Lomonosov's achievements has received wide recognition in recent years. There is thus no need to rehearse here his scientific contributions, nor his work as a poet, as an historian or as a grammarian. My concern here is, rather, his impact on the Academy and on the lives of his fellow academicians.[2]

Lomonosov was a man of violent temper and passionate enthusiasms, an ardent nationalist who bitterly resented both the continuing German dominance within the Academy and the patronising attitude which many of the German professors adopted towards their Russian hosts. He strove constantly to make the Academy more responsive to the needs of Russian society, and presumably it was with this in mind that he had himself appointed in 1757 as one of the Councillors in the Academy's Chancellery, the administrative apparatus which enjoyed under the Academy's charter a considerable measure of control over the academic professors.

The powers granted to the Chancellery, and their ruthless exploitation by the long-time secretary, Johann Schumacher, had already led to much discontent within the Academy proper. Following Lomonosov's appointment, relations between the Chancellery and the academicians became very much worse. Lomonosov's passions and enthusiasms became constant thorns in the flesh of the other professors. He argued with them, he insulted them, he bullied them. In response, the other academicians came to demand more and more insistently in the years that followed that the Academy's structure should be reformed: 'the Academy will then take on quite another appearance,' the astronomer Stepan Rumovsky wrote to his friend Johann Albrecht Euler in Berlin, 'the members of the Academy will

[2] For English readers, the most convenient account of Lomonosov's career remains B. N. Menshutkin, *Russia's Lomonosov*, Princeton, N.J., 1952.

have no chicanery to fear, and the despotic power to which Mr Lomonosov aspires will be reduced to nothing'.[3]

Eventually, in 1765, this wish was granted; the Chancellery not only lost all its powers over the academicians, it was abolished altogether—though by the time this happened, Lomonosov was already dead. With the return to St Petersburg of the Eulers, father and son, a few months later, the Academy entered a rather more tranquil phase in its history. At the same time, the return of the Eulers put the seal on a period in which the scientific standards within the Academy had gradually been restored to their original high level. This process had begun, however, some years earlier.

II

On 12 May 1757, in the midst of all the turbulence of the Lomonosov era, yet another German scholar arrived to take his place in the Academy. Franz Aepinus was thirty-two years old at the time. Born and raised in the ancient Hanseatic port of Rostock, where his father taught philosophy and theology for almost forty years, Aepinus had spent some years at the local university before proceeding to Jena, where he studied medicine and mathematics under the noted teacher G. E. Hamburger. Following his return to Rostock in 1747, he had taught mathematics in the university there for some years, and at the same time had built up a local reputation as an astronomer. On the basis of the latter, he had been appointed early in 1755 to the chair of astronomy in Frederick the Great's Academy of Sciences in Berlin.[4] His appointment less than two years later as professor of physics in the St Petersburg Academy was seen by his friends as a considerable step upwards—clear evidence that despite all its troubles, the Academy was still held in high esteem by the learned world, if it was not yet so regarded within the Russian society which gave it a home.[5]

[3] Rumovsky to J. A. Euler, 22 November/2 December 1764; Arkhiv Akad. nauk SSSR, Leningrad, fond 1, op. 3, no. 46, l. 274–5: 'L'academie prendra alors tout autre face, les membres de l'academie n'auront rien a craindre de la Chicane, et le pouvoir despotique, au quel aspire Mr Lomonosoff, sera reduit a zero.'

[4] Professor John L. Heilbron, who prepared the entry on Aepinus for the *Dictionary of Scientific Biography*, was apparently unaware of Aepinus's early interest in astronomy. This interest is attested, however, by the series of reports by the latter which appeared in the Rostock *Gelehrte Nachrichten* during the early 1750s (1753: pp. 210–13, 489–92; 1755: pp. 108–9, 499–504, 525–8). Furthermore, that these papers attracted attention in high places is revealed by the comments of the eminent astronomer Tobias Mayer upon learning of Aepinus's appointment at Berlin: 'Der Herr M. Aepinus in Rostock, von welchem ich vernehme, dass er nächstens nach Berlin gehen, und das dortige Observatorium zum Gebrauch bekommen werde, besitzet alle diejenige Einsicht und Geschicklichkeit, welche ihn des Besitzes dieser Stelle würdig mache' (Mayer to L. Euler, 23 February 1725; in Yu. Kh. Kopelevich, 'Perepiska Leonarda Eylera i Tobiasa Mayera', *Istoriko-astronomicheskiye issledovaniya*, V, 1959, p. 422).

[5] Upon learning of Aepinus's appointment, Rumovsky wrote to J. A. Euler on 28 October/7 November 1756, as follows: 'Je vous prie de faire mes complimens . . . surtout

It was only during his last months in Berlin that Aepinus had embarked upon the brilliant series of electrical and magnetic researches upon which such fame as he enjoys today almost entirely rests: he in fact presented his first formal paper on the subject at an academic gathering in Berlin just a few days before his departure for St Petersburg.[6] Once established in the Russian capital, he pursued these researches vigorously, but he also continued to work on a number of other topics which had attracted his attention during his time in Berlin, or in some cases even earlier. Among these, the most noteworthy is his long-standing interest in the improvement of optical instruments, and in the resolution of certain problems in higher mathematics. His new post offered Aepinus considerable scope for pursuing his experimental researches in particular, for it gave him access to—indeed, the supervision of—the excellent physics 'cabinet' which had been built up, with Lomonosov's help, by his predecessor in the chair of physics, Georg Wilhelm Richmann. Most notably, Aepinus found himself heir to a first-rate set of electrical apparatus, for Richmann, until his unfortunate death in 1753, had been particularly interested in electricity (indeed, it was this interest which had cost him his life, for he was killed by a lightning bolt which struck him full on the forehead while he was investigating the electricity of thunder-clouds). As well as the collection of apparatus, Aepinus had at his disposal the very comprehensive workshop facilities which the Academy had built up over the years.

In short, in St Petersburg the opportunities for research were first-class. Aepinus, then at the height of his powers, was not slow to take advantage of them, and in the next few years paper after paper flowed from his pen in a seemingly unending stream. Most of these were subsequently published in the *Novi commentarii* of the Academy. A survey of some of their titles shows the breadth of Aepinus's scientific interests at this time: 'A description and explanation of certain new electrical experiments', 'Optical observation of the variation in apparent diameter of narrow apertures brought close to the eye', 'General demonstration of the Newtonian theorem concerning a binomial raised to an indefinite power', 'An improvement of the solar microscope', 'Astronomical dissertation on the effect of parallax in the transit of planets across the sun', 'An examination of the magnetic

---

à Monsieur Aepinus et de le feliciter sur son employe a qui fut il elevé dans si peu de temps' (Arkhiv Akad. nauk SSSR, Leningrad; fond 1, op. 3, no. 46, l. 13). At the time he wrote these lines, Rumovsky had only recently returned to St Petersburg after studying in Berlin under the auspices of Euler *père*. He was therefore well placed to make the necessary comparisons.

[6] E. Winter (ed.), *Die Register der Berliner Akademie der Wissenschaften 1746–1766*, Berlin, 1957, p. 231.

theory proposed by the celebrated Tobias Mayer'.[7] And there were many others.[8]

Nor did Aepinus restrict himself to writing articles for the *Novi commentarii*. In late 1759, his electrical and magnetic researches reached their climax with the publication by the Academy press of his great 390-page *Tentamen theoriae electricitatis et magnetismi*, a book widely acknowledged as one of the most important in the whole history of electricity. Two and a half years later, he published a second major book on electricity, his *Recueil de differents memoires sur la tourmaline*.[9] Furthermore, three times within the space of six years, Aepinus read one of the papers at the Academy's annual public meeting in the presence of the Empress and the assembled dignitaries of her court. Naturally, the Academy always tried to present itself in the most favourable light possible on these occasions, so the fact that his colleagues allowed Aepinus to represent them so regularly bears eloquent testimony to the high opinion they held of his talents. On each occasion, the paper was printed in advance, both in the Latin or German original, and in Russian translation.[10] The first paper in particular evoked considerable interest within the scientific community of the day, for Aepinus took advantage of the occasion to announce his revolutionary new approach to the theory of magnetism. Within three years, the paper had appeared in at least three separate German editions besides the Latin and Russian versions which had been published in St Petersburg originally.[11]

Ceremonial addresses such as those delivered by Aepinus must have served, however imperfectly, to expose those leading members of the Russian upper classes who constituted his audience to the modern scientific attitude towards the natural world. Their publication in Russian translation must have increased their effectiveness in

[7] 'Descriptio ac explicatio novorum quorundam experimentorum electricorum', *Novi commentarii Academiae Scientiarum Imperialis Petropolitanae*, VII, 1758–9, pp. 277–302; 'Observatio optica de mutabilite diametri apparentis foraminis angusti, oculo propinque', *ibid.*, pp. 303–8; 'Demonstratio generalis theorematis Newtoniani de binomio ad potentiam indefinitam elevando,' *ibid.*, VIII, 1760–61, pp. 169–80; 'Emendatio microscopii solaris,' *ibid.*, ix (1762–3), pp. 316–25; 'Dissertatio astronomica de effectu parallaxeos in transitu planetarum sub sole,' *ibid.*, x (1764), pp. 433–53; 'Examen theoriae magneticae a celeberr. Tob. Mayero propositae,' *ibid.*, XII, 1766–7, pp. 325–50.

[8] A full listing is given in *Tableau général . . . des matières contenues dans les publications de l'académie . . . depuis sa fondation. I. Publications en langues étrangères*, St Petersburg, 1872.

[9] St Petersburg, 1762.

[10] (a) *Sermo academicus de similitudine vis electricae atque magneticae . . .*, St Petersburg 1758, and its translation, *Rech' o skhodstve elektricheskoy sily s magnitnoyu . . .*; (b) *Cogitationes de distributione caloris per tellurem . . .* St Petersburg, 1761, and its translation *Razsuzhdeniye o razdelenii teploty po zemnomu sharu . . .* (and in this case the Academy also saw fit to issue a French version, *Reflexions concernant la distribution de la chaleur sur le globe de la terre . . .*); (c) *Abhandlung von den Luft-Erscheinungen . . .*, St Petersburg, 1763, and its translation *Razsuzhdeniye o vozdushnykh yavleniyakh. . . .*

[11] *Akademische Rede von der Aehnlichkeit der elektrischen und magnetischen Kraft . . .*, Leipzig, 1760; and within the pages of two well-known periodicals, the *Hamburgisches Magazin*, XXII, 1759, pp. 227–72) and the Leipzig *Allgemeines Magazin der Natur, Kunst und Wissenschaften*, XI, 1761, pp. 90–131.

this regard immeasurably, in that it made their contents available on a permanent basis to a much wider audience of literate Russians. More effective still in spreading the new ideas was the monthly journal *Yezhemesyachnyya sochineniya k pol'ze i uveseleniyu sluzhashchiya* (*Monthly works of benefit and amusement*) which was brought out regularly by the Academy for a number of years under the editorship of the historian G. F. Müller.[12] Here, articles on scientific subjects were presented in the Russian language at a frankly popular level, and with much success. Aepinus, for example, contributed four excellent pieces relating to his own principal interests. Firstly, the appearance of a spectacular comet late in 1757 prompted a discourse on the nature of comets in general, and on Halley's prediction in particular that the great comet of 1682 should reappear some time in the late 1750s; a quick calculation showed that the comet then in view was not the one predicted by Halley.[13] A little over a year later, Aepinus outlined in two papers published a couple of months apart some practical benefits arising out of his recent researches on magnetism; in one he described some methods he had devised for improving the reliability of the mariner's compass, and in the other he suggested a method whereby the strength of naturally occurring magnets could be increased.[14] Finally, in October 1760, he contributed a piece explaining why it was that the scientific world was preparing with such eagerness to observe the forthcoming transit of Venus across the Sun's disc, and outlining in general terms the considerations which astronomers had to take into account in their calculations concerning this event.[15]

In general, Aepinus succeeded in these articles in striking a nice balance. On the one hand, he combined a precise and uncluttered literary style (something which was, of course, partly to the credit of his Russian translators[16]) with an eminently straightforward manner of exposition, and managed thereby to make what he was saying readily comprehensible; yet on the other hand he usually managed to avoid giving the impression that he was talking down to his

[12] The title of the journal was altered slightly from time to time, as follows: *Yezhemesyachnyya sochineniya k pol'ze i uveseleniyu sluzhashchiya*, 6 vols., 1755–7; *Sochineniya i perevody, k pol'ze i uveseleniyu sluzhashchiya*, 10 vols., 1758–62; *Yezhemesyachnyya sochineniya i izvestiya o uchonykh delakh*, 4 vols., 1763–4.

[13] 'Razmyshleniya o vozvrat ekomet, s kratkim izvestiyem o nyne yavivsheysya komete', *Yezhemesyachnyya sochineniya*, October 1757, pp. 329–48.

[14] 'Razsuzhdeniye o nekotorykh novykh sposobakh, prinadlezhashchikh k popravleniyu magnitnykh strelok i morskago kompasa', *Sochineniya i perevody*, November 1758, pp. 483–506; and 'Kratkoye izvestiye o novoizobretennom sposobe k umnozheniyu sily v natural'-nykh magnitakh', *ibid.*, January 1759, pp. 48–53.

[15] 'Izvestiye o nastupayushchem prokhozhdenii Venery mezhdu Solntsem i zemleyu', *ibid.*, October 1760, pp. 359–71.

[16] In the case of the first article at least, the translator has been identified as Vasily Lebedev (V. V. Bobynin, *Russkaya fizikomatematicheskaya bibliografiya*, Moscow, 1886–1900, I, 3, p. 50).

readers. He was not always successful in the latter regard, however, for his article on the transit of Venus provoked a furious outburst from Lomonosov, who found his approach far too disdainful. Subsequent events provide a good illustration both of the way disputes were conducted within the Academy, and of two radically different views which had emerged within it of the relationship which ought to obtain between science and the community at large; they therefore merit a little closer attention.

III

As will be shown, considerable ill-will had already developed between Lomonosov and Aepinus by the time the latter's article appeared, so that the charges and counter-charges that they levelled at each other in this instance should probably all be taken with a large grain of salt. It is clear, however, that on this particular occasion Lomonosov's resentment was kindled especially by Aepinus's decision to avoid using technical terms like 'ecliptic' and 'horizon' in his account, and to rely on more common phrases instead: it was this that Lomonsov found excessively disdainful; it was, he said, an insult to those for whom the article had been written.[17] Thus offended, Lomonosov's procedure was apparently to spread reports around the city to the effect that Aepinus's article was worthless—at least, he made no attempt to deny having done this when Aepinus accused him of it. Aepinus responded by preparing a defence of what he had written, with the intention of having it included in the next month's issue of the journal. A week after he had lodged this with the Academy, Lononosov read a spirited rejoinder at a session from which Aepinus, in order (so he said) to avoid further disputes, had purposely absented himself.[18] In the end, neither Aepinus's defence nor Lomonosov's rejoinder was published, but Aepinus's original article was re-issued separately as a pamphlet.[19]

Regrettably, the bitter polemics of the argument (together, it might be added, with an unfortunate tendency amongst historians to take everything that Lomonosov ever said at face value) have somewhat obscured the most important point which was at issue. Aepinus

[17] Lomonosov's criticisms are set out in full in *Protokoly zasedaniy konferentsii Imperatorskoy Akademii nauk s 1725 po 1803 goda*, St Petersburg, 1897–1911, hereafter called *Protokoly*, II, pp. 459–60; they are given in Russian translation in P. S. Bilyarsky, *Materialy dlya biografii Lomonosova*, St Petersburg, 1865, pp. 478–80. Both versions are reprinted in Lomonosov's *Polnoye sobraniye sochineniy*, Moscow-Leningrad, 1950–59, IV, pp. 325–31.

[18] Aepinus to Müller, 8 December 1760: 'Ich glaube also, daß es zur Vermeidung alles Wort-Streites (in welchen ich mich einmahlen einzulassen fest entschlossen bin) am besten seyn werde, wenn ich aus der heutigen Conference zurückbleibe' (Arkhiv Akad. nauk SSSR, Leningrad; raz. V, op. 7, no. 28). Cf. *Protokoly*, II, pp. 459–60.

[19] *Svodnyy katalog russkoy knigi grazhdanskoy pechati XVIII veka 1725–1800*, Moscow, 1963–6, p. 439. Unfortunately, no indication is given as to who authorised this publication.

and Lomonosov evidently differed considerably in their attitudes towards the Academy's monthly journal. More significantly, their differences reflect two different viewpoints which were held within the Academy as to the rôle that body as a whole ought to play in Russian society, and, indeed (as I have indicated already) two different attitudes towards the rôle of science in society generally. Lomonosov's claim that Aepinus's article was worthless was based in the end on more solid ground than his resentment at the latter's having thought it necessary to avoid the use of technical terms. He argued as well that the diagram which Aepinus had provided to show how the transit would appear was seriously in error, because no account had been taken in it of the fact that the orbit of Venus was inclined at a small angle to the ecliptic. As a result, Lomonosov charged, the path that the planet would appear to follow across the Sun's disc had been seriously misrepresented.

Aepinus conceded at once that he had ignored the inclination of the orbit of Venus towards the ecliptic in preparing his diagram. But he obviously regarded this as a relatively unimportant factor whose incorporation into his discussion might simply have rendered the whole thing too complicated for the majority of his readers; and he had, after all, introduced his original account by remarking that

> My intention is solely to give society an adequate idea of this phenomenon, and to show the real and incomparable benefit which the whole human race as well as the sciences can expect from a precise observation of it.[20]

Quite clearly, Aepinus had as his aim in writing his article, simply the amusement and edification of his readers; it was meant merely to give them some understanding of what he and his colleagues were doing, and why they were doing it. Equally clearly, this was not at all what Lomonosov wanted. He envisaged the article being used by untrained observers in the remote parts of Siberia as a detailed set of instructions as to what they might expect to see.[21] More generally, he assumed that the article would be used by people who themselves would want to make precise observations of the transit, whereas Aepinus had assumed that his readers would want nothing more than to be given a general idea of what was going on. For Lomonosov, that is, Everyman still had an active and useful rôle to play, even in this most esoteric kind of scientific research. For Aepinus, on the other hand, science had become more specialised; it was for him the prerogative of the scientists alone, theirs to practise and then to explain and perhaps to justify to the general public.

In 1760, Aepinus's attitude was one which was not yet widely held

[20] Aepinus, *Sochineniya i perevody*, October 1760, p. 360.
[21] Bilyarsky, *op. cit.*, p. 479.

within the scientific community. Yet there is certainly no need to see in it, as Lomonosov apparently did, a disdain for Russian society in particular. More likely, it may simply reflect Aepinus's background: he was, after all, trained primarily as a mathematician, and mathematics more than any of the other sciences had become by then a subject open only to the initiated. On the other hand, the continuing intellectual isolation of the St Petersburg Academy within Russian society could only have encouraged the development of such ideas: the differences in outlook between the scientists and the society which supported them were much more marked in St Petersburg than they were in most other centres at this time.

IV

Aepinus's publication record during his first few years in the Russian capital is, all in all, a most impressive one, not only because of the large number of items involved, but also because of their uniformly high quality, and the wide range of subject-matter discussed in them. Here, unquestionably, is the record of a first-rate scientist applying himself diligently and profitably to his work.

And yet, even at this early stage, there were shadows lurking beneath the surface. Even though articles by Aepinus continued to grace the pages of the *Novi commentarii* for most of the 1760s, a study of the *procès-verbaux* of the Academy's weekly meetings reveals a sharp decline in his output as early as 1761, with his last significant contribution being made in August 1763. In October of the following year he presented one final, short, and relatively unimportant paper, and thereafter, nothing. The articles which appeared in the *Novi commentarii* in later years were all in fact read before the Academy quite early in the decade.

Aepinus's decreasing output was due in large measure to his having had a number of extra-academic tasks assigned to him in the meantime. He had also become involved, however, in the bitter personal wrangle with Lomonosov of which we have seen a sample already, and this must surely have diverted him somewhat from his researches as well. The various highlights of this dispute are worth rehearsing simply for the fascinating light they throw on conditions within the Academy at the time, but it is also high time that some much-needed balance was restored to the picture which has traditionally been given, a picture built up almost exclusively from the contributions made by Lomonosov to the proceedings.

For most of the first year after Aepinus's arrival in St Petersburg, there is no evidence of any hostility between him and Lomonosov: indeed, the latter subsequently acknowledged the friendliness which

Aepinus had shown towards him at this period.[22] Not until April 1758 do we find a first hint of discord, when Lomonosov reported to the Academy that Aepinus had communicated to him some criticisms of his paper proposing the development of a telescope which would enable dimly-lit objects to be seen more clearly.[23] Aepinus regarded Lomonosov's plan as impracticable; indeed, on well-established theoretical grounds, as impossible. Lomonosov rejected Aepinus' arguments, and suggested that the perfecting of his scheme should be announced by the Academy as the subject for a competition. The other academicians responded by requesting him first to make Aepinus's criticisms available to them; this Lomonosov promised to do. Since the matter was not thereafter taken any further, it would appear that Aepinus's arguments prevailed.

This, unfortunately, was not the end of the affair. Lomonosov remained unconvinced by Aepinus's arguments, and at a session of the Academy held in June of the following year, he demonstrated a new telescope which his friend Ivan Shuvalov had just received from England, and whose function, Lomonosov asserted, was the same as that of the instrument he had proposed.[24]

Lomonosov's triumph was short-lived. Two weeks later, Aepinus delivered a devastating critique of his claims—and on this occasion, he was not content simply to set out the theoretical difficulties confronting Lomonosov. He went on to report how, when he had put the newly arrived instrument to the test, he had found its performance in dull light to be not even as good as that of an ordinary reflecting telescope. Upon dissection of it, he said, he had found it to be constructed in a 'completely bizarre manner' using twice as many lenses as necessary; it thus betrayed at once, he continued, how it was that its maker had been able to make such fantastic claims for it—he was totally ignorant of even the first principles of optics. Aepinus asked in the most pointed fashion why the English learned journals, usually so eager to defend the priority claims of British inventors, had remained utterly silent about the new device; he provided his own answer in the most insulting of terms: 'Without doubt the unfortunate effort of an ignoramus is seen in England as unworthy of discussion'.[25] Finally, he suggested that, when Lomonosov had been given an opportunity to prepare a reply to his report, both the report and the reply should be sent to the Académie Royale des Sciences in Paris for judgment, with the authors' names suppressed.[26] At this point, how-

22 *Ibid.*, p. 074.

23 *Protokoly*, II, p. 406.

24 *Ibid.*, p. 430.

25 Arkhiv. Akad. nauk SSSR, Leningrad, raz. V, op. 7, no. 11: 'Ohne zweiffel sieht man in Engelland, den unglückl. Versuch eines Ignoranten, für unwürdig an, von ihm zu reden.'

26 *Protokoly*, II, p. 432.

ever, Lomonosov finally seems to have given up the struggle, for there is no record of his having taken up Aepinus's challenge.[27]

Aepinus's whole report was couched in extremely blunt and undiplomatic language, and Lomonosov certainly resented it bitterly: in a memorandum written some time afterwards, he even went so far as to identify it as the starting point of all the subsequent quarrels.[28] On the other hand, a tone of such exasperation pervades Aepinus's document that it is difficult not to see it instead as the culmination of a series of increasingly acrimonious exchanges on the subject between the two men. What is more, other points of disagreement between them had arisen, some stated, some not, by the time Aepinus delivered his report. Less than a month earlier, for example, Aepinus had submitted to the Academy the manuscript of his great work on the theory of electricity and magnetism, his *Tentamen*, requesting approval to have it published.[29] For many years previously, Lomonosov had been keenly interested in electricity. At a public meeting of the Academy in November 1743, he had delivered an address in which he speculated at length on its rôle in various meteorological phenomena.[30] Three years later he had begun but failed to complete a memoir in which he outlined a theory ascribing electricity to the motion of particles of the extremely subtle aethereal matter which, he supposed, filled the pores of all bodies.[31] It must have been extremely galling to him to find that Aepinus had no sympathy whatever for the kind of approach he had adopted, but was happy instead to base his account on unexplained Newtonian-like forces acting between particles at a distance. Not once was either Lomonosov's name or the kind of explanation he had advanced for electrical and magnetic phenomena mentioned in the whole three hundred and ninety pages of Aepinus's book: worse, when Aepinus tendered his manuscript to the Academy, Lomonosov was not even chosen as one of those who were to advise on whether the book merited publica-

[27] S. I. Vavilov has argued that throughout this dispute the main issue lay between Lomonosov's great experimental insight which had enabled him to glimpse a truth not again recognised until only a few years ago, and Aepinus's over-rigid adherence to a theory of optics which we now know to have been incorrect. Unfortunately not enough evidence has survived to enable us to reach a definite conclusion in the matter; most notably, no trace has been found of Aepinus's chief critique of Lomonosov's ideas. It is to be doubted, however, that Lomonosov really carried his experiments far enough to justify Vavilov's argument, for otherwise it would be extremely difficult to understand, in the light of Aepinus's criticisms of it, how Lomonosov could have waxed so enthusiastic over the telescope Shuvalov received from England. Surely he would have been able to recognise its failings himself if he had got very far at all with his own experiments? (A. I Andreyev and B. L. Modzalevsky (eds.) *Lomonosov: Sbornik statey i materialov*, II, Moscow–Leningrad, 1946, pp. 74–80).

[28] Quoted by P. Pekarsky, *Istoriya Imperatorskoy Akademii nauk v Peterburge*, St Petersburg, 1870–3, II, p. 649.

[29] *Protokoly*, II, p. 428.

[30] M. V. Lomonosov, *Oratio de meteoris vi electrica ortis*, St Petersburg, 1753.

[31] *Ibid.*, 'Theoria electricitatis methodo mathematica concinnata', *Polnoye sobraniye sochineniy*, III, pp. 265–313.

tion.[32] Granted that we have no evidence of any ill-feeling over this sequence of events, it would nevertheless be astonishing if a man of Lomonosov's fiery temperament could have resisted feeling considerable bitterness at what had come to pass, or giving vent to his bitterness from time to time.

A further source of friction which developed at this time—and on this occasion, one that is thoroughly documented—arose from Lomonosov's efforts to resuscitate the by then moribund university attached to the Academy. At the very time that Aepinus was preparing his report on Shuvalov's so-called night-vision telescope, he was challenged publicly by Lomonosov at a session of the Academy as to whether he was willing to give lectures in the university. The matter was evidently a very sensitive one, for rather than answering immediately, Aepinus chose to submit a carefully considered reply in writing.[33] He there took some pains to point out that he was not required under the terms of his contract to deliver such lectures; specific provision had been made in the regulations of the Academy for the appointment of a second professor of natural philosophy who should be charged with that task. Furthermore, no provision had been made in his salary for such lecturing duties. Nevertheless, he said, he was prepared to co-operate, but in the circumstances he felt entitled to lay down the conditions under which he would undertake the work. Lomonosov later described these conditions as 'impossible', and he regarded Aepinus's insistence upon them as tantamount to a refusal to undertake lecturing duties;[34] six months later, upon having his proposals for the reorganisation of the university accepted, Lomonosov used the new powers he thus acquired to overrule all of Aepinus's carefully formulated conditions and order him peremptorily, along with a number of other academicians, to begin lecturing at once.[35]

Lomonosov's opinion notwithstanding, the conditions which Aepinus had set out appear, on the surface at least, to be fairly minimal. He wanted it made clear that in agreeing to lecture, he was not committing himself to doing so in perpetuity. He also wanted it spelled out that his academic duties would remain his primary responsibility, and that he would only be expected to lecture so long as his health—then somewhat uncertain—permitted. He demanded freedom to fix the times at which his lectures would be given, and he insisted that the students should attend at his house to hear them. This

[32] *Protokoly*, II, p. 428.

[33] *Ibid.*, pp. 430–1. As early as September 1757, Aepinus had been ordered to give lectures on experimental physics (Lomonosov, *Polnoye sobraniye sochineniye*, IX, p. 471), but apparently he had simply ignored this instruction. It is, therefore, easy to understand why he adopted a very cautious approach when the matter was brought up again.

[34] Bilyarsky, *op. cit.*, p. 470. [35] Pekarsky, *op. cit.*, II, p. 680.

latter cannot have loomed as too great an obstacle for potential students, however, since had Aepinus not set such a condition, he himself would apparently have been expected to walk to the Academy whenever bad weather prevented the use of a carriage.[36] If Aepinus's conditions really were impossible, as Lomonosov charged, then what made them so must undoubtedly have been the remaining one, namely, 'No students must be given to me to instruct other than those of whom it is known for certain, that my trouble with them is not taken in vain'.[37] Here, indeed, is the rub. Throughout its entire history, the basic problem confronting the Academic University had always been its inability to attract suitable students,[38] and it may well have been the case that no students could have been found who could satisfy Aepinus's requirements regarding their competence. If so, his conditions were certainly not as ingenuous as they might have appeared at first. Yet at the same time it is hard not to feel considerable sympathy with him in his unwillingness to waste his time lecturing to a completely uncomprehending audience—particularly if this was not one of the tasks he had contracted to do when he had moved to St Petersburg only two years earlier.

From this point onwards, relations between Aepinus and Lomonosov went from bad to worse. Towards the end of 1760, Lomonosov even went so far as to submit a formal statement of his dissatisfaction with Aepinus's management of the physics apparatus to the President of the Academy, Count Kiril Razumovsky—though on this occasion at least, he was completely ignored.[39]

It was the transit of Venus which led to the worst disputes of all. We have seen already how a storm erupted over Aepinus's discussion of the forthcoming event in the Academy's monthly journal. As the long-awaited day approached, Aepinus found himself charged with the primary responsibility within the Academy for all the observations that were to be made throughout the Russian Empire: Grischow, the professor of astronomy and director of the Academy's observatory, had died in mid-1760, and Aepinus had been placed in charge until a replacement arrived (he had apparently been offered

[36] *Protokoly*, II, p. 431. [37] *Ibid.*

[38] See, for example, the discussion in Alexander Vucinich, *Science in Russian Culture: A History to 1860*, London, 1963, pp. 79–80.

[39] Pekarsky, *op. cit.*, II, p. 701. Lomonosov charged on this occasion that the collection of apparatus had been 'lying thrown about in corners mouldering and rusting without being used at all', to the point where hardly any trace remained of its former excellent condition. Aepinus, he said, had 'disregard[ed] his duty [in this respect] since his very entry into the service of the Academy'. It is quite clear from Aepinus's publication record, however, that during the period in question, he was pursuing an extremely active programme of high-quality experimental investigations in a number of different fields. It is exceedingly difficult to reconcile such objective evidence with Lomonosov's accusations of neglect. Furthermore, it is hard to understand how Razumovsky could have ignored Lomonosov's complaint in the way he did—for the latter was, after all, a high official in the Chancellery of his Academy—unless he was convinced that it was without foundation.

the position himself, but had turned it down[40]—though to no avail, since the observatory remained in his charge for some years). His duties in connection with the transit were two-fold. Firstly, he was charged with instructing Stepan Rumovsky, then an Adjunct in the Academy, in the art of making the necessary observations, and also with outfitting the expeditions of Rumovsky and Nikita Popov to Siberia to observe the transit from two different locations there. Secondly, he was expected to make observations himself in St Petersburg.

It was the way in which Aepinus prepared for the latter task that gave rise to the friction with Lomonosov. In order, so he said, to avoid being disturbed as he made the observations, he refused to admit two of Lomonosov's protégés to the observatory, even in the face of a direct instruction from the latter (using his authority as a councillor of the Chancellery) to do so. An appeal was promptly lodged with the Senate, which ruled against Aepinus. When the latter presented a memoir stating again his reasons for wanting to observe the transit alone, Lomonosov drew up an extraordinarily vituperative point-by-point reply.[41] He insinuated that it was not interruptions that Aepinus feared, but the presence of witnesses to his own clumsiness and ignorance as an observer. And what did it matter anyway, he wanted to know, if those witnesses by their shuffling around should drown out the ticking of the clock? Any competent observer should be able to count off the seconds just as well in his head! He insisted that his two nominees were widely acknowledged to be able astronomers, despite the aspersions Aepinus had cast upon their abilities. Elsewhere he denigrated Aepinus's own prior experience in astronomy, remarking that in Berlin 'there is not a single good instrument, but perhaps a rusty quadrant'—a quite ludicrous assertion in fact.[42]

It is hard to tell at this distance how much Aepinus' objections were based on genuinely-held grounds, and how much on his by then lively antipathy towards Lomonosov and all his works. At one point, the latter alleged that Aepinus, despite his professed need for isolation if he was to make the observation properly, had in the meantime in-

[40] Müller to L. Euler, June 1760; in A. P. Yushkevich *et al.* (eds.), *Die Berliner und die Petersburger Akademie der Wissenschaften im Briefwechsel Leonhard Eulers*, Berlin, 1959, I, p. 151.

[41] Pekarsky, *op. cit.*, II, pp. 730–3.

[42] Bilyarsky, *op. cit.*, p. 493. There exists in Berlin (Archiv der deutschen Akademie der Wissenschaften, Abschnitt I, Abth. xiv, no. 27) an inventory drawn up by Aepinus of the equipment in the observatory there at the time he took charge of it. The apparatus he had at his disposal at that stage amounted to much more than the single rusty quadrant alleged by Lomonosov; the list is actually a very long one indeed. What is more, Aepinus's early reputation as an astronomer had been built largely on his skilful observation of a previous transit, namely that of Mercury across the face of the Sun in 1753! (Aepinus, 'Nachricht vom Durchgang des Merkur', (*Rostock*) *Gelehrte Nachrichten*, 1753, pp. 210–13, 489–92; and also J. C. Koppe, *Jetztlebendes gelehrtes Mecklenburg*, Rostock/Leipzig, 1783–4, I, p. 12).

vited along a number of his non-scientific cronies to keep him company.[43] If this was in fact the case, Aepinus was certainly being less than candid in his protestations. On the other hand, he obviously held very strong feelings on the whole subject, for when the Senate's decision was confirmed, he washed his hands of the transit altogether and left Lomonosov's nominees to observe it unaided. As he had predicted, their efforts turned out to be unsatisfactory.[44]

V

In all Lomonosov's outbursts against Aepinus, two themes in particular provided a constant refrain. Firstly, Lomonosov time and again ascribed both Aepinus's general hostility towards him, and the individual actions which particularly aroused his ire, to the evil machinations of Johann Taubert, Schumacher's nephew and successor as secretary of the Academic Chancellery. This was not a feature peculiar to his indictments of Aepinus, however; Lomonosov seems to have had an almost paranoiac obsession with Taubert, such that whenever he was crossed he tended to see the secretary's influence at work. Besides this fixation, one can also discern in Lomonosov's attacks on Aepinus a passionate resentment against the latter's rapid social advancement during his first years in St Petersburg.

In this regard it is easy to feel considerable sympathy for Lomonosov, for Aepinus's success was extremely rapid indeed, and his fortunes prospered exceedingly—at the expense, as I have noted already, of his scientific pursuits—at the very time that many of Lomonosov's favourite projects were languishing for want of support. Aepinus' rise to imperial favour may well have begun with his public lecture before the Empress Elisabeth and her court on 7 September 1758. Whether it was then or on some subsequent occasion that he first attracted attention, what is certain is that only a few months later, in mid-1759, he was commissioned to prepare an exposition of the system of the world for the edification and amusement of the then Imperial Grand Duchess, soon to become the Empress Catherine the Great. His essay was extremely well received—so well, indeed, that he was appointed there and then by Catherine as her personal tutor in natural philosophy—and from this time onwards, we are told, he was always treated by Catherine with marks of 'the most exceptional favour' (*die ausnehmendste Gnade*).[45]

[43] Pekarsky, *op. cit.*, II, p. 731. [44] *Ibid.*, II, p. 734.

[45] Koppe, *op. cit.*, I, pp. 12–13; see also Rumovsky's letter to J. A. Euler cited above (n. 3), and Aepinus's own dedication to his essay, which was published a few years later under the title *Beschreibung des Welt-Gebäudes*, St Petersburg, 1770. A Russian translation of this, *Razsuzhdeniye o stroyenii mira*, was published simultaneously, and it evidently proved popular, for notwithstanding the fact that 600 copies had been printed initially, a second edition of 1200 copies was required by 1783 (*Svodnyy katalog*, III, pp. 439–40). The

Presumably as a result of his success with Catherine, Aepinus was appointed early in 1761 as Director of Studies at the Imperial Corps of Noble Cadets. His new position evidently carried with it a considerable amount of prestige, since it involved his being advanced one degree in the table of ranks, to the level of Collegial Counsellor (*kollezhskiy sovetnik*).[46] When he was first appointed to the Academy, it was at a salary of 860 roubles per annum.[47] For his post at the Corps of Cadets, there was now added to this an extra 900 roubles p.a.[48] Both the salary differential and the higher rank commanded by his new position testify to the very modest status enjoyed by the Academy in St Petersburg society at the time, for, even though the Corps of Cadets was an extremely highly regarded institution where the scions of the Russian nobility received their education, Aepinus's situation within it was still a somewhat subordinate one.[49] From Aepinus's point of view, the fact that he was drawing a double salary would have placed him in an entirely different income bracket from the one in which he had been until then, that occupied by most of his fellow academicians. If we accept Lomonosov's assertion that he had (at Taubert's instigation, of course!) begun to spend more of his time on frivolous social pursuits than on his scientific duties,[50] we may be sure that he would have found the extra money most welcome. Subsequently, however, Aepinus's close association with Catherine probably made life most uncomfortable for a time, during the brief reign of the latter's husband Peter in 1762. Aepinus's sense of relief at the success of Catherine's coup against her husband in July of that year is manifest in the dedication to the new Empress, dated only a few days after the coup, which he managed to include in his *Recueil de differents memoires sur la tourmaline* just before the book left the printery.

Despite the obligations he had incurred in accepting his new position—for it was certainly no sinecure—Aepinus continued to attend the meetings of the Academy regularly until early 1765, and to play an active rôle in the affairs of that body. Fortunately his dispute with Lomonosov quietened down considerably, and, although much ill-feeling evidently remained, things never again became as heated as they had over the transit of Venus. Aepinus obviously took his

---

entry on Aepinus in the *Russkiy biograficheskiy slovar'* (St Petersburg, 1912), is certainly wrong in dating his winning of Catherine's favour to 1763, and to the public lecture that he gave in July of that year.

[46] Müller to L. Euler, 24 April/5 May 1761; in Yushkevich *et al.*, *op. cit.*, I, p. 173. Lomonosov was particularly enraged at Aepinus' sudden promotion to a rank which he himself had reached only after many years' service. He was even more upset when his efforts to win himself a new promotion were unavailing; typically, he blamed his failure on the machinations of Taubert and Aepinus (Pekarsky, *op. cit.*, II, p. 720).

[47] Arkhiv Akad. nauk SSSR, Leningrad, fond 21, op. 3, no. 3, l. 6.

[48] Lomonosov, *Polnoye sobraniye sochineniy* X, p. 690.

[49] *Adres-kalendar' rossiyskiy na leto ot Rozhdestva Khristova 1765*, St Petersburg, 1765, p. 47.

[50] Pekarsky, *op. cit.*, II, p. 698.

educational responsibilities seriously, and in 1764 he brought out a new and augmented edition of G. W. Krafft's popular textbook on geography: in line with what seems to have by then become standard practice with the Academy's publications, both German and Russian versions of the book were printed.[51] A year later, when the Academy was considering what should be done to round off the education of one of its students, Aepinus submitted a very sensible set of proposals which was eventually adopted by his colleagues.[52] So far as the Academy itself was concerned, Aepinus seems to have devoted himself primarily to the administration of the Observatory, rather than to the pursuit of his physical researches; it was under his direction, for example, that a substantial rebuilding programme was launched in the Observatory in 1763.[53]

Early in 1765, however, all this changed, for now, as a further mark of her favour, Catherine appointed Aepinus tutor to her son, the Tsarevich Paul Petrovich. Since this involved his waiting upon his royal pupil each morning, Aepinus could clearly no longer be expected to attend the regular sessions of the Academy, as they too were normally held in the morning. To cover this situation, Catherine issued a special decree in Aepinus's favour whereby he was to retain all the rights and benefits of his academic position (including a salary which had by then increased to 1000 roubles p.a.) but was at the same time relieved of all his academic duties.[54] The Academy was authorised simultaneously to appoint a new professor of physics to take over those duties. As a result of these arrangements, Aepinus now found himself drawing three salaries, each of 1000 roubles; one from the Academy, one from the Corps of Cadets, and one from the Grand Duke's allowance, presumably to cover the latest appointment.[55] It appears that he relinquished his position at the Corps of Cadets within a few months, [56] but it is unlikely that he would have suffered a reduction in salary thereby, since at about the same time he began to be seconded with increasing frequency to official governmental duties in the College of Foreign Affairs.[57]

During the next few years, Aepinus was called upon occasionally by the Academy for advice—most notably in connection with the

[51] *Hrn. Georg Wolfgang Kraffts Kurze Einleitung zur mathematischen und natürlichen Geographie. . . . Zweite mit Anmerckungen vermehrte Ausgabe von F.U.T. Aepinus*, St Petersburg, 1764; and the Russian translation *Rukovodstvo k matematicheskoy i fizicheskoy geografii. . . .*

[52] *Protokoly*, II, p. 540; for Aepinus's actual proposals, see Arkhiv Akad. nauk SSSR, Leningrad; fond 3, op. 1, no. 287, ll. 27–8.

[53] *Protokoly*, II, p. 502.

[54] Arkhiv Akad. nauk SSSR, Leningrad; fond 3, op. 1, no. 2240, l. 26.

[55] Rumovsky to J. A. Euler (see n. 3).

[56] He was not listed among the staff there in the *Adres-kalendar'* for 1766, whereas he had been listed a year earlier.

[57] S. Poroshin, *Zapiski, sluzhashchiya k istorii Yego Imperatorskago Vysochestva . . . Pavla Petrovicha* (2nd ed.), St Petersburg, 1881, col. 279.

preparations for observing the second transit of Venus in 1769.[58] His academic colleagues evidently still thought highly of him, for when an election was held within the Academy in 1767 to select a representative to send to Catherine's famous Commission on Laws, only G. F. Müller and Jakob von Stählin, the two most senior professors, received more votes than he did.[59] In practical terms, however, his scientific activities virtually ceased at this time, as his other responsibilities made increasingly heavy demands on him. At first, of course, these responsibilities were centred chiefly on his royal pupil, but as the young Tsarevich grew to manhood, they shifted more and more to the College of Foreign Affairs. By 1771 at the latest, Aepinus's appointment there had become a full-time one.[60]

Aepinus's career from this point onwards illustrates one fascinating yet generally unrecognised way in which mathematical talents were put to use in the 18th century—and not only in Russia, for that matter—for the post he occupied within the College of Foreign Affairs was Head of the Cipher Department![61] Nor was he the first academician to be appointed to this sensitive office: from 1742 until his death in December 1764, Christian Goldbach, one of the foundation professors who had been brought to St Petersburg in 1725, had performed the same duties.[62] It is clearly no coincidence that it was within a couple of months of Goldbach's death that Aepinus was first called upon for help. He evidently performed his new duties to the satisfaction of his employers, since in 1773 or 1774 he was promoted again, to the rank of Active Councillor of State (*deystvitel'nyy statskiy sovetnik*),[63] and by 1780 only two officials in the College were ranked ahead of him, namely Panin, the minister, and Osterman, the vice-chancellor.

Very little evidence has survived from this period of Aepinus' life. We know that he joined the important and influential St Petersburg Free Economic Society in 1766, soon after its foundation, and that in 1770 he actually served a term as its president.[64] In the latter year he published an essay in the *Trudy* of the Society on the use of lightning

[58] *Protokoly*, II, pp. 641, 689. [59] *Ibid.*, II, p. 592.

[60] *Adres-kalendar'*, 1771, p. 69. I have not been able to consult a copy of the volume for 1770, but Aepinus's name was not included in the College entry for 1769, so presumably at the time this was printed, his duties there had not yet been established on a regular basis. Aepinus was promoted to the rank of Councillor of State (*statskiy sovetnik*) some time during the first half of 1769 (cf. *Protokoly*, II, p. 689): perhaps this happened in conjunction with the formalisation of his appointment at the College.

[61] P. van Wonzel, *Etat présent de la Russie*, St Petersburg/Leipzig, 1783, p. 85. Aepinus's position in the College was consistently characterised in the *Adres-kalendar'* as 'pri osoblivykh dolzhnostyakh'.

[62] A. P. Yushkevich and E. Winter (eds.), *Leonhard Euler und Christian Goldbach: Briefwechsel 1729–1764*, Berlin, 1965, p. 5.

[63] *Adres-kalendar'*, 1774, p. 66. I have not seen the volume for 1773.

[64] H. Mohrmann, *Studien über russisch-deutsche Begegnungen in der Wirtschaftswissenschaft (1750–1825)*, Berlin, 1959, p. 111.

conductors for the protection of buildings.[65] Furthermore, it seems that he kept up his scientific interests over the years, for in 1778 he took the trouble to have one of the then newly invented achromatic telescopes sent to him from England.[66] This particular event led to a brief flurry of quite important scientific publications, in some of which he presented some very interesting views on the possible origin of the craters on the Moon,[67] while in others he announced his design of an achromatic microscope which was actually a suitably modified version of his telescope.[68] By 1782, however, official duties caught up with him once more, since in that year Catherine asked him to draw up a plan for establishing a school system throughout her Empire; and upon the acceptance of his scheme, he was appointed to the three-man commission set up to implement it.[69] At the same time, he continued to serve the College of Foreign Affairs. Finally, in the 1790s, his mental faculties became seriously impaired.[70] In 1798 his erstwhile pupil, now Tsar Paul, rewarded him for his services to his adopted country with the customary final promotion, and released him from all his duties. He retired to Dorpat in Estonia, and there he died in August 1802.

VI

Aepinus was clearly not altogether typical of the foreign scholars brought to St Petersburg during the 18th century to staff Peter's Academy of Sciences. His purely scientific achievements surpassed those of all but a handful of his fellows; none of the others succeeded in winning the imperial favour so quickly, or so well; none abandoned his scientific career so suddenly, and only Goldbach did so more completely; none made such important extra-scientific contributions to the life of his adopted country. In all, Aepinus achieved enough in his lifetime to make him of considerable interest as an

[65] 'Razsuzhdeniye o sredstvakh sluzhashchikh k otvrashcheniyu povrezhdeniya proizvodimago gromovym udarom i molnieyu domam i drugim stroyeniyam,' *Trudy Vol'nago Ekonomicheskago Obshchestva*, xvi, 1770, pp. 257–85.

[66] Aepinus, 'Ueber den Bau der Mondfläche, und den vulcanischen Ursprung ihrer Ungleichheiten', *Schriften der Berlinischen Gesellschaft naturforschender Freunde*, II, 1781, p. 1.

[67] *Ibid.*; also a Russian translation of the first part of this article, in *Akademicheskiya izvestiya*, no. 8, May–July 1781, pp. 918–33; and a further paper, 'Lettre . . . sur les volcans de la Lune', *Nova acta Academiae Scientiarum Imperialis Petropolitanae*, II, 1784, Hist., pp. 50–52. On the importance of these papers, see my article 'The Origin of the Lunar Craters: An Eighteenth-Century View' (*Journal for the History of Astronomy*, III, Cambridge, 1972, pp. 1–10).

[68] 'Annonce d'un microscope achromatique . . .', *Nova acta Academiae Scientiarum Imperialis Petropolitanae*, ii, 1784, Hist. pp. 41–9; and *Description des nouveaux micrsocopes inventés par M. Aepinus*, St Petersburg, 1784. On Aepinus's microscope, see S. L. Sobol', *Istoriya mikroskopa i mikroskopicheskikh issledovaniy v Rossii v XVIII veke*, Moscow, 1949, pp. 331–9.

[69] D. A. Tolstoy, 'Gorodskiya uchilishcha v tsarstvovaniye imperatritsy Yekateriny II', *Zapiski imperatorskoy Akademii nauk*, LIV, 1887, Appendix 1.

[70] J. M. Wehnert, *Mecklenburgische Gemeinnützige Blätter*, VII, Parchim, 1803, p. 99.

individual. Nevertheless, beyond this, many aspects of his career remain standard for a scientist in 18th-century Russia. Aepinus was a German, and a man with a thoroughly modern outlook upon the natural world, living in an alien society which was only just emerging from its ancient modes of thought. As such, he had to withstand both a general sense of intellectual isolation, and the jingoistic assaults of a Lomonosov—and though he perhaps experienced the latter in a more violent form than most, he was certainly not alone in being subjected to them. The Academy provided excellent facilities for the practice of science, but at the same time it provided a highly volatile environment for its members to work in; within its walls, personal animosities could flourish with an astonishing vigour.

On a more general level, Aepinus's record provides an excellent illustration of the way in which the Academy tried to narrow the gap which separated it from the community it was supposed to serve. Like Aepinus, many of the other academicians contributed popular-level expositions of the new science to Müller's journal, and later to the *Trudy* of the Free Economic Society. The remarkable sales achieved by the Russian translation of the essay Aepinus prepared for Catherine show the success the continuing programme of popularisation was having by then in developing in Russia a taste for the new learning. The existence of Russian-language editions of many of Aepinus's other works also testifies to the Academy's efforts at building up a more sophisticated scientific readership within the Russian community. Aepinus, through his work at the Corps of Cadets, and later on Catherine's Education Commission, made far more important contributions than his fellow academicians to the development in Russia of a modern educational system. Furthermore, certain aspects of his labours—particularly the preparation of textbooks to cater for the growing educational market—once again mirror an important aspect of the Academy's contribution to the transformation which took place in Russian intellectual life during the 18th century. Finally, of course, Aepinus's later career shows how the Academy could occasionally serve even then as a source of expertise which the government would otherwise have had the greatest difficulty in procuring. Like many of the other foreigners recruited into the Academy over the years, Aepinus subsequently served his new homeland faithfully and well: the contribution that he and his fellows made to the life of the nation warrants more attention than it has so far received. It was certainly much greater than we have generally been led to believe.[71]

[71] Research for this paper was supported in part by a grant from the Australian Research Grants Committee. Extracts from the Leningrad Archives of the Soviet Academy of Sciences are quoted by permission.

XVI

Aepinus and the British Electricians: The Dissemination of a Scientific Theory

I

FRANZ ULRICH THEODOSIUS AEPINUS set out his principal contributions to electrical theory in his massive 390-page *Tentamen theoriae electricitatis et magnetismi*, published in St. Petersburg in 1759. As is well known, his version of the one-fluid theory of electricity differed from that which had been put forward ten years earlier by Benjamin Franklin in two crucial ways in particular. First, he denied the existence around electrified bodies of any extended "atmospheres" of electrical matter of the kind which had been postulated by Franklin. Second, where Franklin had invoked two kinds of interparticulate force in his account—a mutual repulsive force between particles of the electric fluid and an attractive force between particles of the fluid and particles of ordinary matter—Aepinus supposed in addition that, Newton's gravitational principle notwithstanding, the particles of ordinary matter mutually repelled each other.

It is equally well known that some years later Henry Cavendish put forward an almost identical version of the theory, in a paper he read to the Royal Society of London in two parts, on December 19, 1771, and January 9, 1772.[1] Furthermore, Cavendish freely acknowledged Aepinus' priority, although at the same time he claimed both that he had arrived at the theory independently (despite Aepinus' account having been published over a decade earlier) and that he had carried his treatment of it much further than Aepinus had done:

> Since I first wrote the following paper, I find that this way of accounting for the phaenomena of electricity is not new. Æpinus, in his *Tentamen Theoriae Electricitatis et Magnetismi*, has made use of the same, or nearly the same hypothesis that I have; and the

* Department of History and Philosophy of Science, University of Melbourne, Parkville, Victoria 3052. I am indebted to Professors Russell McCormmach and Robert E. Schofield for their encouragement and assistance during the preparation of this paper. I must also gratefully acknowledge the cooperation of a number of librarians around the world, particularly Dr. Whitfield J. Bell, Jr. (American Philosophical Society Library), Mr. D. E. Bertels (Archives Department, Academy of Sciences of the U.S.S.R.), Mr. I. Kaye (Royal Society Library), and Mr. T. S. Wragg (Devonshire Collections). The research for this paper was supported by a grant from the Australian Research Grants Committee.

[1] Henry Cavendish, "An Attempt to Explain some of the Principal Phaenomena of Electricity, by means of an Elastic Fluid," *Philosophical Transactions of the Royal Society*, 1771, *61*:584–677; reprinted in *The Electrical Researches of the Honourable Henry Cavendish*, ed. James Clerk Maxwell (Cambridge, 1879), pp. 3–93.

> conclusions he draws from it agree nearly with mine, as far as he goes. However, as I have carried the theory much farther than he has done, and have considered the subject in a different, and, I flatter myself, in a more accurate manner, I hope the Society will not think this paper unworthy of their acceptance.[2]

It seems that Cavendish was by no means the only Britisher to remain unaware for many years of what Aepinus had done, for, some years afterward, John Robison justified his inclusion of a long account of Aepinus' theory in the *Supplement* to the third edition of the *Encyclopaedia Britannica* on the grounds that the latter's *Tentamen*, "unquestionably one of the most ingenious and brilliant performances of this century," was "little known in this country."[3] The parallel neglect of Cavendish's own contributions to electrical theory suggests that the general British neglect of Aepinus' work was at least partly due to certain very general features of British science at the time (such as, perhaps, a widespread inability among experimental scientists to cope with the intricacies of mathematical analysis), but this cannot explain how Cavendish himself could have remained unfamiliar with Aepinus' work for so long. This is particularly so in view of the fact that Joseph Priestley's account of it in his *History and Present State of Electricity*[4] would have revealed Aepinus as an obviously kindred spirit so far as electricity was concerned—and Cavendish had certainly read Priestley, since there is a reference to the first edition of the *History* in the earliest of his surviving manuscripts which is concerned with electricity.[5] Priestley stressed both of the important modifications noted above which Aepinus had made to the Franklinian theory,[6] and as well he emphasized Aepinus' use of mathematics, remarking that "Æpinus has lately given us an excellent specimen of what use MATHEMATICS, and especially algebraical calculations may be of to an electrician."[7]

Unfortunately, Priestley was not content simply to describe what Aepinus had done. He also took it upon himself to dismiss Aepinus' lengthy calculations out of hand in an entirely unjustified fashion:

> He that reads the first chapter, as well as many other parts of [Aepinus'] elaborate treatise . . . , may save a good deal of time and trouble by considering, that the result of many of his reasonings and mathematical calculations cannot be depended upon; because he supposes the repulsion or elasticity [of] the electric fluid to be in proportion to its condensation; which is not true, unless the particles repel one another in the simple reciprocal ratio of their distances, as Sir Isaac Newton has demonstrated, in the second book of his Principia.[8]

---

[2] Cavendish, *Electrical Researches*, p. 3.

[3] Article "Electricity," *Encyclopaedia Britannica*, 3rd ed. *Supplement*, Vol. I (Edinburgh, 1803), p. 538. Robison's article was subsequently reprinted *in toto* in Vol. IV of his posthumous *System of Mechanical Philosophy* (Edinburgh, 1822).

[4] 1st ed., London, 1767.

[5] Cavendish, *Electrical Researches*, p. 102. The reference is certainly to the first edition of Priestley's *History*, since the pagination does not match up in the later editions. The chronology of Cavendish's electrical writings is discussed below. Russell McCormmach has gone so far as to suggest ("Henry Cavendish: A Study of Rational Empiricism in Eighteenth-Century Natural Philosophy," *Isis*, 1969, *60*: 302) that Priestley "may have provided the stimulus for Cavendish to elaborate his ideas on electricity."

[6] Priestley, *History*, pp. 462–463.

[7] *Ibid.*, p. 503.

[8] *Ibid.*, p. 463. Contrary to what Priestley reported, Aepinus did *not* suppose that the repulsion of the electric fluid was proportional to its condensation. Rather, he reiterated over and over in his book that he did not know what the law of force was, and that because of this he was not prepared to make any assumptions about it; all he was prepared to assert was that the force must decrease with increasing distance, and it must be

It was presumably to this passage that Robison was referring when he attributed the subsequent British neglect of Aepinus' work to "the very slight and almost unintelligible account which Dr. Priestley has given of it,"[9] while more recently McCormmach has suggested that Cavendish in particular was discouraged from reading Aepinus' book by Priestley's unfavorable comment.[10]

There can be no doubt that Priestley's remarks would eventually have had considerable influence, since his book came to enjoy a high reputation among electricians. In Cavendish's case, however, we can scarcely regard this as the whole story (as McCormmach himself recognizes[11]). To begin with, Cavendish would have known through his contacts at the Royal Society that Priestley was a relative newcomer to electrical studies at the time the *History* appeared; there is nothing to suggest that Cavendish would automatically have bowed to Priestley's judgment on electrical matters. Then again, except for the single passage quoted above, Priestley's remarks about Aepinus were generally favorable, and in particular (as has been noted already) he reported the latter's crucial theoretical innovations in an admirably clear fashion, which should have left Cavendish in no doubt as to Aepinus' importance to his own work. Finally, assigning the blame to Priestley sheds no light at all on the period prior to 1767. Why had Cavendish not already read Aepinus by the time Priestley's *History* was published? The *Tentamen* had after all been in print for over seven years by then, and, because of the enormous prestige which Aepinus enjoyed as a result of his experiments on the pyroelectric properties of the tourmaline crystal, as well as his successful repetition of the Leyden experiment using a layer of air as dielectric rather than the usual glass,[12] one would have expected it to have been required reading for all aspiring electrical philosophers.

In what follows I shall argue that during the 1760s Aepinus' *Tentamen* was more widely known in the English-speaking world than Cavendish's statement would suggest. First of all, however, I shall argue that Cavendish himself probably acquired a copy of the book as early as mid-1766—that is, well before Priestley's *History* was published and more than five years before Cavendish presented his own paper on electrical theory to the Royal Society. This rather unexpected conclusion raises yet more difficulties for our understanding of Cavendish, and I shall explore some of these before passing on to the more general question of British familiarity with Aepinus' work.

II

Late in 1765 Joseph Priestley, then a teacher at the dissenting academy at Warrington, visited London, seeking support for his first major scientific venture, his projected *History of Electricity*. He succeeded in meeting three of the leading British electricians

---

proportional to the *quantity* of fluid involved. At one point he assumed that the law was inverse square in form, but only in order to illustrate an argument by way of a numerical example (*Tentamen*, p. 304). On another occasion, and for a similar purpose, he supposed (as Priestley accused him of doing throughout) that the force law was one of simple inverse proportionality. But in this case he also explicitly warned his readers that although the actual form of the law was unknown, it was known that it was certainly *not* a simple inverse proportionality (*ibid.*, p. 281).

[9] Robison, "Electricity," p. 538.

[10] R. McCormmach, "The Electrical Researches of Henry Cavendish" (unpublished Ph.D. thesis, Case Western Reserve University, 1967), p. 297.

[11] *Ibid.*, p. 486.

[12] Both sets of experiments were reported by Aepinus in his "Mémoire concernant quelques nouvelles expériences électriques remarquables," *Histoire de l'Académie Royale des Sciences et Belle Lettres à Berlin*, 1756 (publ. 1758), *12*:101–121.

of the day—John Canton, Benjamin Franklin, and William Watson—and also Richard Price, who like the others was a Fellow of the Royal Society. Following Priestley's return to Warrington, his newly found friends assisted him in his project by furnishing him with the books he needed.[13]

On February 14, 1766, soon after his return to Warrington, Priestley addressed an initial courtesy letter to Canton in which he made his first recorded request of one of his new mentors for assistance in obtaining a book. It is perhaps significant that the only author he mentioned specifically on this occasion was Aepinus: "I have desired Mr. Johnson to apply to you for Epinus. If you have any other books that will be of service to my design, I am sure you will not fail to communicate them."[14] We cannot of course be absolutely positive that Priestley was referring here to the *Tentamen*. The indications, however, are that he was.[15] In any case it seems that Priestley had no trouble in obtaining a copy, since only a short time afterward, on April 13, 1766, he was able to report to Franklin, "I greatly admire Æpinus."[16] Furthermore, a note has survived which reveals how Canton subsequently sought the return of the book:

> Mr. Canton presents his Complimts. to Dr. Franklin and begs to be inform'd whether, or not, in Writing to Dr. Priestley, he has mention'd Mr. C's wanting Æpinus's *Tentamen Theoriae Electricitatis & Magnetismi* for Mr. Cavendish; if not, Mr. C would beg the Favour of a Frank.[17]

---

[13] Robert E. Schofield, ed., *A Scientific Autobiography of Joseph Priestley (1733–1804)* (Cambridge, Mass.: M.I.T. Press, 1966), Ch. 2. Throughout this paper I use the term "British" in a broad sense so as to include Franklin and subsequently some of his fellow colonists in North America as well. I do so mainly for ease of reference, but at the same time I justify my usage on the grounds that the North American settlements were still British colonies throughout the period with which I am concerned, and their inhabitants were still properly classified as British citizens.

[14] *Ibid.*, p. 16. Johnson was a well-known bookseller through whom Priestley sometimes worked.

[15] We know from the "Catalogue of Books" which Priestley appended to his *History* that somehow or other he eventually obtained access to a copy of the *Tentamen*. We also know that this was the only separately printed work by Aepinus which he consulted; in particular, though he included it in his catalogue, he did *not* read Aepinus' other major electrical work, his *Recueil de differents memoires sur la tourmaline* (St. Petersburg, 1762). But if the latter was what he had wanted from Canton, he would surely have followed up his initial request, and there is nothing in his subsequent correspondence to suggest that he did so. Thus the only piece other than the *Tentamen* which he could have been asking for is Aepinus' "first account of the tourmaline," published in *Hist. Acad. R. Sci.* (Berlin) for 1756; indeed, Schofield has suggested that this was in fact what he was seeking (*Autobiography*, p. 401). But there is a marked contrast between the confidence of Priestley's first request for "Epinus" and his later plea for assistance in locating the piece on the tourmaline, which was couched in the following terms: "I have deferred writing the history of experiments on the *Tourmalin*, till I can get (which I wish you could assist me in) *Aepinus's first observation* upon it" (Priestley to Canton, March 29, 1766, *Autobiography*, p. 28). The contrast suggests that he was asking for different things in the two cases: in the first he was evidently sure he had gone to the right person for help, in the second he was simply casting around desperately for a copy of what he wanted (having already asked Price for help in locating the piece on the tourmaline some weeks earlier; *Autobiography*, p. 17).

[16] *Autobiography*, p. 30. Once again, Priestley's reference cannot be to the *Receuil de differents memoires sur la tourmaline*, since he did not read it. The "first account of the tourmaline," being only a brief account in the Berlin *Mémoires*, was hardly the kind of thing to warrant such praise, particularly at a time when Priestley was still struggling unsuccessfully to make sense of the differences between Aepinus and Benjamin Wilson over their tourmaline experiments. Hence, by elimination, Priestley must have been talking about the *Tentamen*.

[17] *The Papers of Benjamin Franklin*, ed. Leonard W. Labaree (New Haven, Conn.: Yale Univ. Press, 1959—), Vol. XIII, p. 544. Another possible interpretation of this note is that Canton was seeking on Cavendish's behalf to borrow Priestley's personal copy of the *Tentamen*: in view of the earlier correspondence we may be sure that if

In view of the occurrence of Cavendish's name in Canton's note, it is most unfortunate that the latter is not dated, since if it was, this would enable us to determine when Cavendish became interested in Aepinus. Even more frustrating, at first glance at least, is the incomplete dating on the following note, now held in the Canton Papers at the Royal Society:

> Marlborough street june 23.
>
> Sir,
>
> Since I saw you I have met with Æpini tentamen theoriae electricitatis & magnetismi, which I suppose is the book you spoke of, at Nourses. I am sorry I did not enquire there sooner; but I thought there was so little likelyhood of meeting it there, that I neglected it. I hope however it is not too late to prevent your writing to Dr. Priestly for the book.
>
> Your Obedient humble servant
> H. Cavendish.[18]

Fortunately, however, there is enough evidence available for us to be able to pin down the date of this letter from Cavendish—and hence also that of the one from Canton to Franklin—with a fair degree of probability.

To begin with, the two letters just reproduced are clearly related. Evidently Canton, having earlier lent Priestley a copy of the *Tentamen* (or perhaps having arranged for someone else to lend a copy to him), had now told Cavendish he would retrieve the book so that he in turn could have a look at it. (There is in fact something of a suggestion in Cavendish's note that Canton had recommended the book to him. Could Canton have likewise told Priestley about it earlier? If so, this would explain why the latter had been so confident that Canton could lend him a copy.) Apparently expecting Franklin to be writing to Priestley sooner than he would be doing himself, Canton had asked him to include in his letter a request that the *Tentamen* be returned. Perhaps the book did not then arrive as quickly as Canton had expected, for the first of our two notes is clearly a reminder to Franklin. In the meantime, however, Cavendish, browsing in Nourse's (the well-known eighteenth-century London bookstore), had to his surprise found a copy on their shelves. Presumably, he bought it.[19] The question which concerns us, then, is when did all this happen?

To help us fix the date, we know that Priestley finished his *History* early in 1767, and by April 21 of that year he was able to write to Canton, "Electricity is at present very much out of my thoughts."[20] Priestley was at the time very eager to make a name for himself in the scientific community of the day. Hence it is most unlikely that he would have endangered his good standing with his influential mentors by keeping the books he had borrowed during the preparation of his *History* for very long after he had

---

Priestley had at that earlier stage acquired a copy of the book for his own collection, Canton would have known of it. The tone of Priestley's earlier note to Canton is, however, much more that of a man seeking to borrow a book than one seeking to buy it. Furthermore our (admittedly incomplete) knowledge of the contents of Priestley's library confirms his not having owned a copy (Eric Robinson, "Priestley's Library of Scientific Books: A New List," *Studies in History and Philosophy of Science*, 1970, *1*:145–160).

[18] Royal Society of London Library, Canton Papers, Vol. II, p. 31. This and subsequent extracts from the Society's archives are quoted by permission of the Society.

[19] Mr. T. S. Wragg, Librarian and Keeper of the Devonshire Collections, informs me that the *Tentamen* is listed in a MS catalogue of Cavendish's library which is held at Chatsworth and which he dates to around 1795.

[20] *Autobiography*, pp. 43–44. Priestley's book was advertised in *London Chronicle*, March 31–April 2, 1767 (Franklin, *Papers*, Vol. XIII, p. 421 note).

finished with them—certainly not for long once the book was finished. Surely, then, he would not have carried borrowed books with him when he moved from Warrington to Leeds some months later, in September 1767.[21] We can therefore assume that the copy of the *Tentamen* he had borrowed through Canton would have been returned by that date, and so, unless Priestley subsequently re-borrowed it, in June of any year later than 1767 Canton would not have been seeking its return. Most probably, however, Priestley would have returned the book to Canton well before June 1767. In this case, Cavendish's letter to Canton—and thus presumably his acquisition of his copy of Aepinus' book as well—must be dated June 1766, that is, five and a half years before he read his paper on electrical theory to the Royal Society![22]

III

If Cavendish indeed acquired a copy of the *Tentamen* as early as mid-1766, how are we to understand his remarks at the beginning of the paper he read in December 1771, where he claimed that he had been unaware of Aepinus' ideas until after he had written his own paper? Clearly we can no longer simply attribute his neglect of Aepinus to Priestley's unfavorable report, since our mid-1766 dating *antedates* the appearance of Priestley's *History* by more than seven months. So far as I can see, only three possibilities remain.

First, Cavendish may not have been telling the truth. But to suggest that Cavendish of all people should have thus risked his honor in an obviously futile attempt to tarnish Aepinus' claim to priority is ludicrous. To begin with, there was no need for him to do such a thing: his paper clearly constituted an important advance over anything Aepinus had done, and so it was already assured of a prominent place in any history of electricity. But beyond this, to do such a thing would have been totally out of character with what we know of the man; this was after all the same Cavendish who subsequently was so little interested in establishing his legitimate claims to priority in various other matters that the bulk of his researches—including the first experimental determination of the inverse-square character of the electrical force—lay undiscovered for over half a century after his death.

A second possibility is that Cavendish had already written his paper by mid-1766, but that for reasons unknown he deferred publishing it for five and a half years thereafter. This seems most unlikely, however, for at least two reasons. One of these concerns the overall sequence of Cavendish's electrical researches. Besides his published

[21] It is possible that Priestley had previously obtained Canton's permission to keep the *Tentamen* a while longer. If so, however, it is surprising that there is no mention of this in the surviving correspondence. Furthermore, why should Priestley have wanted to retain Aepinus' book in particular, especially in view of his adverse opinion of the latter's mathematics, which formed so large a part of the work? The only possible explanation I can think of for his having made such an arrangement is that he was already actively planning the *History of Magnetism* volume of his projected multi-volume *History of Experimental Philosophy*. But in fact this does not seem to have been the case. When Priestley finally began work on his series of histories, it was not the *History of Magnetism* but the *History of Vision* with which he began. The earliest reference we have to a *History of Magnetism* dates from February 1771! (*Autobiography*, p. 76.)

[22] If Priestley re-borrowed the book, he would presumably have done so at about the time he gave notice in February 1771 of his intention to prepare a *History of Magnetism*, in which case our letters would have to be dated instead to June 1771. But we have absolutely no evidence that he ever in fact did re-borrow it; our 1766 dating therefore remains the most likely one.

paper, Cavendish also drafted a paper which he headed "Thoughts concerning Electricity,"[23] in which a theory of electricity was developed which differed significantly from the published version. Since the "Thoughts" contain a reference to Priestley's *History*, they must have been set down some time after March 1767. But Maxwell has presented evidence drawn from a detailed comparison of the two theories which suggests that the "Thoughts" were composed *earlier* than the published paper,[24] and McCormmach has pointed more recently to an even stronger reason for accepting this conclusion, namely, that Cavendish assumed in all his subsequent electrical work the theory set forth in the published paper, not that of the "Thoughts."[25] Quite apart from all this, however, if Cavendish had already written the paper he subsequently published by mid-1766 and he then read Aepinus' book, his reaction would surely have been either to publish his paper at once or to decide never to do so at all in that particular form. He would surely not have simply read the paper unchanged, more than five years later.[26]

Only one possible explanation remains, therefore, for Cavendish's statement about Aepinus, and that is that although he may have acquired a copy of the *Tentamen* by mid-1766, he did not in fact read it until much later. It is not unknown even today, of course, for someone to buy a book but not to read it for some time afterward. In the present case, however, at the time he acquired it, Cavendish was clearly looking for a copy of Aepinus' book; he did not simply pick it up on a whim to add to his collection. Furthermore, as we have seen, only a few months after Cavendish made his purchase, Priestley exposed the importance of Aepinus' theoretical innovations in unmistakable fashion—and Cavendish certainly read Priestley.

How then are we to account for Cavendish's continued failure to read Aepinus? I am unable here to do more than offer the most tentative of suggestions: is it possible that despite his having received a typical eighteenth-century classical education, Cavendish did not enjoy reading Latin? Perhaps, though he must have been able to read Latin if necessary, he preferred not to do so, and works written in that language tended to be put to one side until the reading of them could be deferred no longer?[27] If this is in fact what happened, Cavendish must have received an unpleasant surprise when he eventually settled down to read what Aepinus had written.

IV

Because Cavendish only recognized Aepinus' achievement so belatedly, there has been a tendency to assume that the *Tentamen* was very little known among British electricians generally during the 1760s. But, if nothing else, the arguments set out in the

[23] Cavendish, *Electrical Researches*, pp. 94–103.

[24] *Ibid.*, pp. 409–410 (editorial notes).

[25] McCormmach, Ph.D. thesis, p. 147.

[26] McCormmach informs me that there are no earlier drafts of Cavendish's paper to be found among his unpublished MSS.

[27] George Wilson tells us in his *Life of the Honble Henry Cavendish* (London, 1851), p. 16, that Dr. Newcome, Cavendish's teacher at the Hackney school he attended, was "a strict disciplinarian." It seems at least possible that by his disciplinary measures he might have generated a lifelong distaste for the classics in such an extraordinarily shy and retiring individual as Cavendish. To be admitted to Cambridge, Cavendish must have shown a certain facility with the classics, and in fact during his student days he actually published some passable Latin verse (in *Academiae Cantabrigiensis luctus in obitum Frederici celsissimi Walliae principis*, Cambridge, 1751), but still the language may not have come pleasantly for him.

preceding sections of this paper show that Cavendish's remarks should be treated with considerable caution. If we want to know how well known Aepinus' ideas became in the English-speaking world during their first ten years in print, we should look elsewhere for our answer; we should look for evidence instead in the writings of the other leading electricians of the day. When we do so, it emerges that Aepinus' work was by no means as little known as has been supposed.

At the time the most renowned British electrician was Benjamin Franklin. Following his first, five-year sojourn in London as agent for the colony of Pennsylvania, Franklin returned home to Philadelphia in 1762, arriving there on November 1. Evidently he either carried a copy of the *Tentamen* with him, or one was sent to him shortly afterward,[28] because only a few months later, in a letter dated February 26, 1763, he was able to tell his friend Cadwallader Colden about the book:

> Professor Æpinus of Petersburgh, has publish'd a Work on Magnetism and Electricity, in which he endeavours to apply my Theory of the latter to the Explanation of certain Phenomena in the former . . . I have while in England made many Experiments with Magnets, and own myself inclined to Æpinus's Opinion, tho' there are some Difficulties in his Doctrine that as yet I do not see how to solve.[29]

In a letter to Ezra Stiles dated May 29, 1763, he made it clear that it was definitely the *Tentamen* he was describing and not Aepinus' earlier and very much briefer account of his ideas, his *Sermo academicus de similitudine vis electricae atque magneticae* (St. Petersburg, 1758):

> Æpinus, a Member of the Academy of Sciences at Petersburg, has lately published a Latin Work in 4to. entituled, Tentamen Theoriae Electricitatis et Magnetismi; wherein he applies my Principles of Electricity to the Explanation of the various Phenomena of Magnetism, and I think with considerable success.[30]

From a postscript to the same letter we learn that Franklin "enclos'd a little Piece [of] Æpinus's" in the package he sent to Stiles, but this seems to have been the *Cogitationes de distributione caloris per tellurem* (St. Petersburg, 1761) rather than the *Tentamen*.[31] Still, subsequent letters show that both Stiles and John Winthrop read Franklin's copy of the *Tentamen*. Franklin had intended that James Bowdoin should also see it, but apparently in the end he did not do so.[32] Had Franklin carried the book back to London with him when he returned there late in 1764, he would of course have been able to lend it to Priestley when the latter embarked upon his *History* a little over

[28] It now emerges that Franklin received his copy *after* his return to America: see his letter to Aepinus dated June 6, 1766, quoted in N. N. Bolkhovitinov, "V arkhivakh i bibliotekakh S.Sh.A.: Nakhodki, vstrechi, vpechatleniya," *Amerikanskii ezhegodnik*, 1971, p. 331.

[29] Franklin, *Papers*, Vol. X, p. 204. There is some evidence that Franklin read the *Tentamen* much earlier than this even, for there are strong Aepinian overtones to his discussion of the theory of magnetism in a letter to Ebenezer Kinnersley dated Feb. 20, 1762 (*ibid.*, p. 45; also published in Franklin's *Experiments and Observations on Electricity*, ed. I. B. Cohen, Cambridge, Mass.: Harvard Univ. Press, 1941, p. 366). On this, see also I. B. Cohen's *Franklin and Newton* (Philadelphia: American Philosophical Society, 1956), p. 540.

[30] Franklin, *Papers*, Vol. X, p. 266.

[31] *Ibid.*, pp. 267, 351, 389. The *Tentamen* could hardly be described as a "little Piece."

[32] *Ibid.*, p. 351; Vol. XI, pp. 22, 230, 246, 254. Winthrop's opinion of the book is quoted below.

twelve months later. We learn from elsewhere, however, that he had left it behind in Philadelphia.[33]

In all, then, we find that Franklin became well acquainted with the *Tentamen* only two or three years after it was published, and that thereafter he was careful to see that the leading natural philosophers among his fellow colonists were also given the opportunity to read it. We may be sure that he would have continued to recommend the book to his English friends following his return to London. Indeed, he was apparently still recommending it to his learned acquaintances over a decade later, for he seems to have mentioned it during negotiations with his most vociferous Continental supporter, Giambatista Beccaria, over the preparation of an English edition of the latter's definitive *Elettricismo artificiale* (Turin, 1772).[34]

Franklin was not the only British scientist to learn of Aepinus' work very early. Indeed, the attention of all the other leading electricians of the day was drawn to the *Tentamen* as early as the fall of 1762, for on November 4 a copy of the book was presented to the Royal Society. It was subsequently deposited in the Society's library, along with a copy of Aepinus' *Sermo academicus de similitudine vis electricae atque magneticae* and two copies of his *Cogitationes de distributione caloris per tellurem*, which had been presented on the same day. All three works thereupon became available for use by members of the Society, of course.[35]

After Franklin, the three leading electricians in Britain at the time were William Watson, John Canton, and Benjamin Wilson. Watson's standing rested almost entirely, however, on what he had done much earlier; in keeping with his growing professional stature in the medical world, his publications after 1760 dealt almost exclusively with medical and botanical rather than electrical subjects. I know of no evidence which might tell us whether he read Aepinus' book or not. Fortunately, though, we can be much more positive in the cases of Canton and Wilson.

Canton was one of the first British electricians to recognize the importance of Aepinus' pioneering studies on the electrical properties of the tourmaline, following the publication in the supplement to the 1758 volume of the *Gentleman's Magazine* of a summary of the latter's paper in the Berlin *Mémoires*.[36] Canton began his own important researches on the subject at once, and he published a preliminary account of the results he had obtained in the September 1759 number of the same journal over the

[33] Franklin to Barbeu Duborg, Mar. 10, 1773; in *The Writings of Benjamin Franklin*, ed. A. H. Smyth (New York: Macmillan, 1906), Vol. VI, pp. 23–26, and quoted by E. Dvoichenko-Markova, "Benjamin Franklin, the American Philosophical Society, and the Russian Academy of Science," *Proceedings of the American Philosophical Society*, 1947, *91*:252.

[34] A letter from Beccaria to Franklin, published in Antonio Pace, *Benjamin Franklin and Italy* (Philadelphia: American Philosophical Society, 1958), pp. 376–378, begins: "Dopo che mi avete fatta addimandare di Epino, io lo ho riletto anche la seconda volta . . . ." Pace implies that this communication is to be dated 1776, the year in which the English translation of Beccaria's book appeared (*ibid.*, p. 59). To confirm Beccaria's assertion that he had already read Aepinus' work once, and to establish that he had done so before 1772, we may note the occurrence of a reference to the *Tentamen* in the Italian edition of Beccaria's book (p. 175).

[35] Royal Society, *Journal Book*, entry for Nov. 4, 1762. In the early 1760s there was a prohibition on the lending of the Society's books to nonmembers (Royal Society, 1752 Statutes, p. 101; amended in the Society's Council Minutes, Jan. 4, 1768). This would have prevented Priestley's friends from drawing upon the resources of the Society's library on his behalf until after his election to membership on June 12, 1766—by which time, as we have seen, he had already read the *Tentamen*.

[36] *Gentleman's Magazine*, 1758, *28*:617–619.

nom de plume "Noncathoni" (an obvious anagram for "Iohn Canton").[37] He continued his researches on and off for several years, though he showed no particular eagerness to publish his later results, many of which were in fact reported for the first time in Priestley's *History*.[38]

So far as evidence of Canton's early familiarity with the *Tentamen* itself is concerned, we have seen already that it was almost certainly he who was instrumental in making a copy of the book available to Priestley. Presumably, therefore, he had also had ample opportunity to peruse it himself; in fact, I drew attention above to the possibility that he might have been responsible for first arousing Cavendish's interest in what Aepinus had done and that he might also have talked to Priestley about it earlier.

Beyond this, we must turn to Priestley's *History* itself for evidence. One of the most distinctive features setting Aepinus' electrical theory apart from those that had gone before was his denial of the existence of Franklinian atmospheres of electrical matter around charged bodies. Priestley discussed Aepinus' views on the subject at considerable length at one point in the *History* and also (very much more briefly) some similar ideas which he attributed to Beccaria. Because he felt that Aepinus' ideas derived directly from Canton's pioneering experiments on electrostatic induction, Priestley presented them in conjunction with his account of these. He began by describing the experiments themselves, along with Canton's initial, rather unsuccessful attempts to explain them in terms of electrical atmospheres—emphasizing as he did so the extent to which Canton had been committed at that stage to the notion of such atmospheres, as well as the advantages of denying their existence as Aepinus had done. He then discussed the subsequent investigations of Franklin, Wilcke,[39] and Aepinus, leading to the enunciation by the latter two of the basic induction principle underlying Canton's experiments. Only then did Priestley turn in detail to Aepinus' denial of the existence of atmospheres (an idea he carefully refrained from attributing also to Wilcke, even though he had until this point made no attempt to discriminate between the contributions of Wilcke and those of Aepinus). Immediately thereafter he remarked how Canton "now" shared this opinion: "It is now also Mr. Canton's opinion, that electric atmospheres are not made of effluvia from excited or electrified bodies, but that they are only an alteration of the state of the electric fluid contained in, or belonging to the air surrounding them, to a certain distance."[40]

The order of Priestley's presentation is highly suggestive here, especially when we remember how well placed he was to report Canton's attitudes correctly. Straddling the discussion of Aepinus' ideas, the account of Canton's changing views about atmospheres is presented very much in a "before-and-after" mode. In particular, the juxtaposition of Canton's change of heart with Aepinus' denial that atmospheres exist hints at a direct causal relationship between the two; it suggests that it was from reading Aepinus that Canton was led to change his mind.

[37] *Ibid.*, 1759, *29*:424–425.

[38] Priestley, *History*, pp. 322–324.

[39] Johan Carl Wilcke was Aepinus' student for some years and was associated with him in many of his electrical investigations. His most important contributions to electricity were set out in his thesis *De electricitatibus contrariis* (Rostock, 1757), which is the work Priestley cited. For an account of Wilcke's relationship with Aepinus, see C. W. Oseen, *Johan Carl Wilcke, Experimental-Fysiker* (Uppsala: Almqvist & Wiksell, 1939).

[40] Priestley, *History*, p. 263. For a further discussion of Canton's ideas about atmospheres, see my article "Franklin's Electrical Atmospheres," forthcoming in the *British Journal for the History of Science*.

It thus appears almost certain that Canton was thoroughly familiar with Aepinus' *Tentamen* by about the end of 1766 at the latest. Even if he was not, however, he must certainly have become familiar with it soon afterward, for in April 1772, shortly after his death, the book was listed in a catalogue of his library drawn up by the executors of his estate.[41]

The case of Benjamin Wilson is perhaps the most interesting of all. By profession one of eighteenth-century London's more popular portrait painters, Wilson had long displayed a penchant for natural philosophy and for the study of electricity in particular. In a series of publications during the 1740s and 1750s he had propounded at length a complicated and not very successful theory of electricity based ultimately on Bryan Robinson's version of an all-pervading Newtonian aether.[42]

With the advent of the tourmaline at the center of the scientific stage, Wilson seized his opportunity. Fortunately possessed of what he confessed to be "the only one known of here at that time,"[43] and subsequently furnished with ever larger stones by his friend the physician William Heberden, he quickly established himself as *the* British authority on the tourmaline. His pre-eminence was confirmed soon afterward when the Royal Society bestowed on him its Copley Medal for 1760, for (in the president's words) "the very many curious Experiments, relating to Electricity, that he has made upon the Tourmaline and other Bodies; and which he has laid before this Society since our last Anniversary Meeting."[44]

Inevitably their interest in the tourmaline drew Aepinus and Wilson together, and despite their inability to reconcile their seemingly contradictory experimental results, they appear to have carried on a cordial correspondence for a number of years. The most accessible item from this correspondence is an open letter to Aepinus which Wilson read to the Royal Society in two parts, on December 23, 1763, and in March 1764. This was subsequently printed in the *Philosophical Transactions* and was also published separately as a pamphlet.[45] From it we learn of other, earlier exchanges. Wilson, we read, had sent Aepinus a copy of his earlier paper announcing the discovery of pyroelectricity in stones other than tourmaline, while from his discussion of Aepinus' *Recueil de differents memoires sur la tourmaline* we learn that the traffic was not all in one direction: "Your treatise upon this stone, published in 1762, seems to be the same you formerly mentioned in a letter to me."[46]

From other sources, indeed, we learn that Aepinus not only wrote to Wilson, he also sent him copies of his books, including the *Tentamen*. First, in a letter to the Swedish scientist Torbern Bergman dated June 29, 1763, Wilson remarked at one point, "I have not seen Mr. Æpinus's book upon the Tourmalin but I suppose he will send it

[41] Royal Society, Canton Papers, Vol. II, p. 97.

[42] *An Essay towards an Explication of the Phaenomena of Electricity, deduced from the Aether of Sir Isaac Newton* (London, 1746); *A Treatise on Electricity* (London, 1750); and, in collaboration with Benjamin Hoadly, *Observations on a Series of Electrical Experiments* (London, 1756). Cohen discusses Wilson's debt to Robinson in his *Franklin and Newton*, pp. 418–423.

[43] Benjamin Wilson, *Phil. Trans.*, 1759, *51*:315.

[44] Royal Society, *Journal Book*, entry for Dec. 1, 1760.

[45] Benjamin Wilson, *Phil. Trans.*, 1763, *53*: 436–466. *A Letter from Mr. Benjamin Wilson . . . to Mr. Æpinus . . . Read at the Royal Society December 23, 1763, and March 1764* (London, 1764).

[46] Wilson, *Letter . . . to Mr. Æpinus*, p. 5. The earlier paper which Wilson sent was published in the *Phil. Trans.*, 1762, *52*:443–447.

me, as he sent me his other works."[47] Beyond this, from the entry in the *Journal Book* of the Royal Society recording the presentation noted above of three of Aepinus' works, the *Tentamen* included, to the Society, we learn that Aepinus had sent the books *via* Wilson.[48] In short, Aepinus not only communicated on a personal basis with Wilson, he also used him as his contact with the Royal Society and thus with British science generally. From the point of view we have adopted in this paper, it follows that Wilson must certainly have had ample opportunity to acquaint himself with Aepinus' contributions to electrical theory. Unfortunately, there is nothing to show that he profited from so doing.

V

From our discussion so far, it follows that contrary to the impression one gains from what Cavendish wrote in 1771, all the other important electrical investigators in Britain during the 1760s—Franklin, Canton, Wilson, and of course Priestley later on—were all well acquainted with Aepinus' work. The question which remains, therefore, is to what extent they were influenced by it. Unfortunately this is a question which is difficult to answer unequivocally.

To begin with, there are a few points at which we can pin down the influence of the *Tentamen* more or less definitely, and one of these has been noted already—Canton's decision to abandon the theory of electrical atmospheres. Another emerges from Franklin's correspondence, for in a letter dated July 10, 1764, and addressed to John Winthrop, Professor of Mathematics and Natural Philosophy at Harvard, Franklin confessed himself persuaded that he should abandon his earlier notion that glass must be given a special place in electrical theory because of its impermeability to the electrical fluid, in favor of Aepinus' view that impermeability is a property common to a greater or lesser extent to all electrics *per se:* "I think with him that Impermeability to the El. Fluid, is the Property of all El[ectric]s per se; or that, if they permit it to pass at all, it is with Difficulty, greater or less in different El[ectric]s per se."[49]

More generally, Aepinus' development of a theory of magnetism analogous to the theory of electricity seems to have won general favor, even though it was also generally recognized that he had not said the last word on the subject. Once again, Franklin's views are worth citing, since they seem to have been fairly typical. In the same letter to Winthrop from which the above quotation is drawn, Franklin expressed his approval of what Aepinus had done, but at the same time he put his finger on the most serious weakness in his account:

> I must confess I am pleas'd with his Theory of Magnetism. Perhaps I receive it the more readily on Account of the Relation he has given it to mine of Electricity. But there is one difficulty I cannot solve by it quite to my Satisfaction, which is that if a Steel Ring be

[47] *Torbern Bergman's Foreign Correspondence*, Vol. I, ed. Göte Carlid and Johan Nordström (Stockholm: Almqvist & Wiksell, 1965), p. 423. Evidently the situation of which Wilson here complains was soon rectified, for the paper he read to the Royal Society six months later (his *Letter . . . to Mr. Æpinus*) amounted to a detailed response to the *Recueil.*

[48] *Loc. cit.*: "These three presented by the hands of Mr. Wilson, F.R.S."

[49] Franklin, *Papers*, Vol. XI, p. 254. Cohen too has drawn attention to this remark (*Franklin and Newton*, p. 540). Franklin's belief in the uniqueness of glass had already been weakened some years earlier, for in a letter to Kinnersley dated July 28, 1759, he had suggested that a wooden board, if properly dried and heated, might take the place of the glass in the Leyden experiment (Franklin, *Papers*, Vol. VIII, p. 417).

> made magnetical by passing Magnets properly round it, and afterwards broken into two Semicircles, each of them will have strong N. and S. Poles, in whatever Part the Ring is broken.[50]

Beyond these few instances, however, Aepinus' influence in Britain during the 1760s is much harder to assess. What is certain is that we do not find (as we do in the period immediately following the publication of Franklin's *Experiments*, for example) a sudden flurry of publications inspired by what Aepinus had done. His name would not occur near the top of a survey of citation frequencies in the electrical writings of the period. It is possible, of course, that his most important ideas had simply become generally known and accepted without any fanfare, but in view of Robison's later remarks this seems unlikely. Apparently we must reconcile ourselves to the fact that although, as we have now established, Aepinus' work was both known and admired at an early date, it still had singularly little impact in Britain outside the few instances noted above. In particular, Aepinus' postulation of an additional electrical force between the particles of matter seems to have been totally ignored, despite his demonstration that such a postulate was necessary if the foundations of the theory were to be rendered consistent.

VI

It has been remarked already that the similar fate accorded Cavendish's paper a few years later suggests that the seemingly paradoxical situation we have thus uncovered in fact reflects some general features of British science at the time. To enter into a detailed analysis of these would clearly be beyond the scope of the present paper, though it is perhaps worth reiterating that the widespread inability of the experimental scientists of the time to cope with the intricacies of mathematical analysis—or even to understand the need for a rigorous mathematics-like consistency in the foundations of their theories—must have been a powerful contributory factor here. In Aepinus' case, however, besides such very general reasons, we can pinpoint certain specific features of his book which would have contributed to its apparent lack of impact. I shall therefore conclude this article by outlining these briefly.

To begin with, nowhere in his book (or indeed elsewhere that I am aware of) did Aepinus lay any stress whatsoever on the improvements he had wrought in electrical theory. Instead, he made it clear over and over again that his primary intention in writing the *Tentamen* was to develop a theory of magnetism and that any importance the book might have would come from its contribution to magnetic science. Thus he stressed right at the outset that he did not intend to give an exhaustive account of all the phenomena of electricity. Rather, he was only going to discuss those for which there was an analogue in the magnetic realm: "It would be completely foreign to the scope of this essay to bring in here the other phenomena of electricity, since my intention is to consolidate the theory of magnetism rather than that of electricity."[51] Similar statements recur throughout Aepinus' writings. As a result, it was all too easy for his readers to concentrate on what he was saying about magnetism and to overlook the novelty and importance of his contributions to electrical theory.

[50] Franklin, *Papers*, Vol. XI, p. 254.

[51] Aepinus, *Tentamen*, p. 14: "A scopo enim scriptionis omnino alienum foret, etiam caetera electricitatis phaenomena huc trahere, cum magnetismi potius, quam electricitatis condere theoriam, animus mihi sit."

Aepinus' deliberate policy of confining his attention to those electrical phenomena which had a direct magnetic analogue is relevant to our question from another point of view as well. Because he adopted the policy he did, on the electrical side his account was confined largely to an analysis of the attractions and repulsions between electrified and other bodies: he devoted much less attention to the phenomena associated with the charging and discharging of bodies. And yet it was just these latter phenomena—most notably in relation to Leyden jars and storm clouds—which were of primary concern to almost all the other leading electricians of the day! Not perceiving that Aepinus was exposing grave weaknesses in the foundations of their theory as a whole, many of his readers must have seen his emphasis on such "minor" problems as the attractions and repulsions as excessive, and his whole account as of only marginal relevance to their own major interests.

The way in which Aepinus chose to present that account only added to the problem. Neither his postulate invoking a repulsive force between the particles of ordinary matter nor his denial of the existence of atmospheres of electrical matter around electrified bodies was mentioned in his initial statement of the principles of the theory. The former only emerged, surreptitiously as it were, as the mathematical development got under way—after it had been shown that without it the theory could not be reconciled with the most elementary observation of all, namely that two electrically neutral bodies had no effect on each other.[52] The latter appeared much later still,[53] at a point where many a reader must have long since tired of following Aepinus through his exposition of all the minutiae of a vast multitude of possible situations.

Overall, therefore, the novelty and importance of what Aepinus was saying with respect to electricity could all too easily have been overlooked in a cursory reading of his book; and unfortunately Aepinus' style, with its extraordinarily thorough but equally tedious attention to detail, was highly likely to encourage cursoriness in his readers. All in all, John Winthrop's reaction to the book, while grossly unfair to Aepinus, may have been all too typical:

> Aepinus is a man of clear ideas & a comprehensive as well as inquisitive mind. His book contains many curious things. He has set the theory of Magnetism in a new light; & his hypothesis agrees so well with the various, puzzling phenomena of that mysterious Power, as renders it in a considerable degree probable; tho many of 'em might be accounted for in a different way. But he has the right German prolixity, even to a degree of tediousness; & I freely confess, if I had had the leisure, I should not have had the patience, to follow him minutely thro all his formulas, which fill a 4to of almost 400 pages;—a Vol. big enough to contain a System of Philosophy. As to Electricity, I don't find that he has advanced much, if at all, beyond our excellent Countryman, to whom he renders great & just honors; & I scarcely recollect any thing he has said on this subject, but what I had learn'd before, & with greater pleasure, from Dr. Franklin.[54]

[52] *Ibid.*, pp. 37–38. It is perhaps worth emphasizing that contrary to what one reads so often in the secondary literature, Aepinus did *not* introduce his new postulate simply or even primarily to enable the theory to cope with the observed mutual repulsion of two negatively charged objects.

[53] *Ibid.*, p. 257.

[54] Winthrop to Stiles, Feb. 21, 1764 (Stiles Papers, Beinecke Rare Book and Manuscript Library, Yale University). This passage was quoted (in not altogether successful Russian translation) by G. M. Lester in his article "Znakomstvo uchenykh severnoĭ Ameriki kolonial'nogo perioda s rabotami M. V. Lomonosova i Peterburgskoĭ Akademii Nauk," *Voprosy istorii estestvoznaniya i tekhniki*, 1962, *12*:142–147, but he did not cite the Yale MS as his source. Presumably the document he saw was a copy of this.

The *Tentamen* was clearly no easier a book to read in the early 1760s than it is now; very few Britishers seem to have profited very much from it. But in 1767, of course, Priestley brought the essential novelty of many of Aepinus' ideas to the attention of a much wider audience, albeit in the company of a rather sweeping derogatory comment. Despite this feature, others should thereafter have been inspired by Priestley's account to read what Aepinus had written. Even if (as seems most unlikely) Nourse's bookshop had only ever had in stock the single copy bought by Cavendish, and even if no other British bookseller ever stocked Aepinus' book, we have now established that there were nevertheless several copies scattered around London, so that anyone interested should have been able to borrow one. Most notably, a copy was available for loan at the Royal Society. That so few people show any evidence of having profited from this opportunity is a sad reflection on the state of British science at the time.

XVII

SCIENTIFIC LINKS BETWEEN BRITAIN AND RUSSIA IN THE SECOND HALF OF THE EIGHTEENTH CENTURY

Some years ago, the Soviet scholar M.I. Radovskii wrote an inte esting history of Anglo-Russian scientific relations. It is very striking, however, that he found almost nothing to say in this work about scientific connections between the two countries in the period of almost sixty years between the retirement of Sir Hans Sloane as President of the Royal Society of London in 1741 and the end of the eighteenth century. Indeed, apart from a short account of some contacts established by the St Petersburg academician Johann Taubert du ing a visit to England in 1748, Radovskii confined his discussion of this period to mentioning a few instances in which the work of Sir Isaac Newton influenced a Russian author.(1)

Yet it would be astonishing if contacts between the leading scientists of the two countries were in fact as sparse during these years as Radovskii's treatment suggests. The regular exchange of information has been a vital part of the scientific enterprise at least since the time of Mersenne, and has, indeed, been one of the chief factors in the rapid advances that science has made in the years since then. In England, it was and still is the principal *raison d'être* of the Royal Society of London, whose meetings have always been largely given over to the reading of research reports submitted by Fellows, and of scientific communications sent in by the Society's extensive circle of correspondents at home and abroad. So far as the tiny community of practising scientists in eighteenth-century Russia was concerned, clustered as it was in Peter the Great's Academy of Sciences in St Petersburg, regular links with scientists elsewhere were even more essential.

In contrast to the situation in England, where the Royal Socie was a voluntary and largely self-financing association of people see ing to advance the cause of science, in Russia the Imperial Academy Sciences was wholly dependent on the government for its existence. Not until the latter years of the eighteenth century were significan numbers of academicians of Russian birth. Not many Russians outside its walls sympathised with its aims, or with the new modes of thinking about the natural world which it espoused. In effect, for most of our period the St Petersburg Academy was so far as its membership was concerned a foreign, largely German, institution, maintained by the Russian government for reasons of national prestige and as a source of technical expertise which could be drawn on from time to time as the need arose. In its early years, the academicians' connections with everyday Russian life were minimal; and though their feeling of intellectual isolation gradually grew less as the upper levels of Russian society became more westernised, they still had every reason to look to the scientific centres of western Europe for inspiration, rather than to the nation which gave them a home.(2)

Given that most of the foreign-born St Petersburg academicians were of German origin, it was only natural that the Academy's closest links should be with the German scientific community of the day,rather than with scientists in the more distant parts of Europe. The published minutes of the Academy's meetings reveal a steady stream of correspondence throughout the eighteenth century with scholars from all parts of the German-speaking world.(3) Communication with regions further afield was much less regular. Nevertheless, as we learn both from the minutes and from other sources, it sometimes did occur, including -- this being my principal concern here -- with Britain. In what follows, I shall briefly describe some of these previously unremarked connections, which are of interest both in themselves and for what they tell us more generally about the eighteenth-century Academy.

One of my two examples concerns the important electrical and magnetic investigations of the St Petersburg academician F.U.T. Aepinus, the results of which were set out in Aepinus's great work, _Tentamen theoriae electricitatis et magnetismi_, published in St Petersburg in 1759. As is well known, Aepinus took as his starting point the ideas of the Anglo-American Benjamin Franklin, having become thoroughly acquainted with these some years earlier, before he ever left his native Germany or set foot in the Russian capital. In a very real sense, therefore, he himself served as one of the channels which brought the new English modes of thought concerning electricity to Russia; and though his book, despite its merit, seems not to have had a great influence on his new-found colleagues in the St Petersburg Academy, he himself did subsequently help to popularise Franklin's great practical invention, the lightning conductor, in circles outside the Academy itself.(4)

Most of the scientific contributions of the St Petersburg academicians were published in the Academy's journal, variously entitled during the course of the century its _Commentarii_, _Novi commentarii_, _Acta_ and _Nova acta_. In the main, therefore, the academicians relied upon the distribution of these annual volumes to make their work known in the wider scientific community of the day. This could be a slightly chancy business. The publication of the journal sometimes fell several years behind schedule (the nadir being reached with the volume for 1739, which did not appear until 1750), and, in addition, the channels by means of which copies were sent to the learned of western Europe, though usually effective, were not always as prompt or as reliable as they might have been. In Aepinus's case, since his work was a separate publication, the problems of distribution were compounded.

To begin with, Aepinus tried to take advantage of the journal's established distribution system by having his work declared a supplement to the Commentarii for that year; and, indeed, it is thus described on its title page.(5) But the Academy's administrative officers were not convinced, despite a further approach from the academic conference fifteen months afterwards, and in the end only twelve copies were released for distribution by this particular channel.(6) Aepinus was thus left with two options: either he coul leave the distribution of his work to the vagaries of the international book trade, or he could take a direct hand in the process himself (and presumably at his own expense). He chose the latter alternative, and began sending out presentation copies. What is of interest to us here is that he seems to have concentrated particularly on sending copies to Britain. The Royal Society received one, as did Franklin. So did Benjamin Wilson, F.R.S., with whom, moreover, Aepinus was corresponding at this time about the peculiar electrical properties of the tourmaline crystal. John Canton, the other leading English electrical investigator at this period, also acquired a copy, and this, too, almost certainly came directly as a gift from Aepinus.(7) A few years later, the Edinburgh mathematicia and natural philosopher John Robison, visiting St Petersburg as secretary to Admiral Sir Charles Knowles, met Aepinus and was given copy of the work.(8) In due course, the book trade also lent a hand Henry Cavendish found a copy at Nourse's, the well-known eighteenth-century London bookstore, and Lord Shelburne (almost certainly actin on the recommendation of his librarian, Joseph Priestley, who had already read Canton's copy) presumably likewise got his through the trade during the 1770s.(9)

Unfortunately, apart from Cavendish and Robison, Aepinus's British readers were ill-equipped to profit from his analyses, and i was many years before his contributions to electrical and magnetic science came to be valued by the British scientific community at the full worth. Nevertheless, his example shows how news of what was being done in St Petersburg could be disseminated even as far as Brita if only a sufficiently determined effort was made. Few of his colleagues seem to have understood this, however, or else they were simply not prepared to make the same effort, and as a result, for lo periods during the century there are very few references in, for example, the Journal Book of the Royal Society, to anything that was being done in the Russian Academy.

Beginning in the mid-1770s, however, scientific exchanges between the two countries suddenly became much more frequent; and it is this period which constitutes my second example. Here, too, we find, the closer links were brought about not by deliberate planning within either of the institutions involved, but by the enterprise of a single individual. The arrangements for dealing with foreign correspondence that had evolved within both the Royal Society and the St Petersburg Academy were designed entirely, we discover, to handle incoming mail, and not at all to promote the outward dissemination of scientific news. Both institutions were generally content to rely on their official published journals to keep others informed about their activities, and this notwithstanding the well known uncertainties to which the distribution of these journals was subject.

The central figure in my second example is a remarkable Portuguese by the name of João Jacinto de Magalhães, better known under the Anglicised version of his name as John Hyacinth de Magellan (or Magalhaens).(10) Born in the small Portuguese town of Aveiro in 1722, Magellan entered the Augustinian monastery at Coimbra at the age of 11, where he remained for over twenty years. Released by the Pope from his monastic vows, he left Portugal some time after 1755 and by 1760 he was in Paris, eking out a small income from some land-holdings in Aveiro by doing some translating work.(11) Here he made the acquaintance of the noted Portuguese doctor, Ribeiro Sanches, once personal physician to the Russian Empress Elizabeth, and now living quietly in Paris following his hasty departure from St Petersburg some years earlier. Sanches encouraged Magellan to settle in London rather than Paris, and, indeed, assisted him financially to do so. Subsequently he paid him an annual allowance for many years.(12)

Magellan was established in London before the end of 1763, and he soon began attending the weekly meetings of the Royal Society with great regularity as the guest of one or other of the Fellows.(13) During the next few years he quietly established a modest place for himself in London's scientific community, undertaking, for example, to help Emanuel Mendes da Costa build a collection of Portuguese 'natural productions',(14) attending enthusiastically at John Canton's striking experiments on the compressibility of water,(15) and submitting a number of minor papers to be read at Royal Society meetings.(16) He is generally reported, too, to have added to his income by acting from time to time as travelling companion to young gentlemen touring on the Continent.(17) The two leading features of his subsequent scientific endeavours, his interest in instrument design and his wide-ranging scientific correspondence, are already evident in some of his early communications to the Royal Society. In June 1765, for example, he wrote to the Earl of Morton, the Society's President, describing a new 'baro-thermometer' he had invented. Eleven months later, an extract from a letter he had received from the French astronomer Charles-Joseph Messier was read.(18)

In the course of the following two decades, Magellan's correspondence grew to immense and extraordinarily wide-ranging proportions, while from 1775 on, he also published a series of tracts on various scientific subjects, mostly to do with improvements in instrument design and manufacture. Some of these were quite valuable contributions in their own right.(19)

First appearances notwithstanding, Magellan's twin activities were closely connected. London was at this period the world's leading centre for the manufacture of scientific instruments, and many of Magellan's letters to his correspondents elsewhere were given over to news about the latest developments in this field. Magellan, it is clear, had struck upon a novel means of making a living, one which fitted perfectly with his own interest and enthusiasm for science. J.P. Brissot, who sought to do something similar, met him in London, and reported in his memoirs that 'sans fortune et voulant vivre indépendant, Magellan se livrait à une science qui pouvait difficilement remplir ses vues ... Mais ... en se faisant le correspondant de tous les savants, il était parvenu a son indépendance chérie, à une indépendance honorable'.(20)

Through the excellent contacts he established with the leading scientists and scientific institutions of the day, Magellan helped to spread news of the latest discoveries throughout the learned worl By itself, however, this would not have brought in an income; yet hi life style reveals that somehow or other he must eventually have fou a means of supplementing his Portuguese rents and the allowance from Sanches upon a more regular basis than by an occasional excursion to the Continent as gentleman's companion. This must have been a means moreover, which did not infringe upon his own status as a 'gentleman', since that is how he was always described. It seems likely that the key to Magellan's success lay in the service he provided hi numerous correspondents in keeping them supplied not merely with new but also with the latest scientific books, instruments and laborator apparatus. Probably he was paid (most discreetly, of course, by means of discounts) a commission on sales by the booksellers and instrument makers whom he patronised. If this is correct, we can at once understand why, though Magellan's various tracts on instrument design were published in London, they were usually written in French the international scientific language of the day. Serious and sometimes very useful contributions to the scientific literature though they undoubtedly were, Magellan's writings may also have had an ulterior motive, namely, to whet the appetites of his readers -- to drum up business, in fact.

This new career, however, advanced only slowly. Despite Magellan's extremely regular attendance at meetings of the Royal Society, and notwithstanding that he was evidently on excellent term with many of the Fellows, it was not until 1774 that he became a Fellow himself, and not until a year after this that he published hi first scientific work. In the meantime, the circle of his correspor dents had been growing steadily wider. His connections with the Fre scientific community were particularly close. Henry Guerlac has told us how, during the early 1770s, Magellan was one of those chiefly responsible for bringing the important new British discoveries about gases to the attention of the French chemists, including Lavoisier.(21) Guerlac has gone so far, indeed, as to accuse Magellan of industrial espionage on behalf of the French, on the basis of an unsuccessful attempt to smuggle a fully assembled loom of a new design across the Channel to France.(22) But Magellan was also busily writing and sending shipments of materials to various other parts of Europe: to Torbern Bergman in Uppsala, Achard and Bode in Berlin, Ingenhousz in Vienna, Volta and Landriani in Italy, Christian Mayer in Mannheim, Van Swinden and later also Van Marum in Holland, the Abbé T.A. Mann in Brussels, Professor Spielmann in Strasbourg, Prince Massalski in Poland, Dr Nathaniel Wolf in Danzig, Arthur Young, Joseph Priestley, Willian Herschel, James Watt and others in England. The list could be expanded almost indefinitely. Magellan was appointed a Corresponding Member of the Paris Academy of Sciences as early as 1771, and afterwards he was also clearly acting in an at least semi-official capacity on behalf of the Academies of Sciences in Berlin, Brussels and Lisbon, as well as the governments of both Spain and Portugal. What is of primary interest to us here is that he also became very active in supplying the Academy of Sciences in St Petersburg with instruments, books, and the latest scientific gossip; and so pleased were the Russian authorities with his efforts on their behalf that they first elected him a foreign member of their Academy, and then in 1783 awarded him a pension of 200 rubles per annum.

The evidence concerning Magellan's contacts with the St Petersburg Academy is set out in the published minutes of the meetings of the Academic Conference.(23) Some of his letters are preserved in Leningrad,(24) but, to judge from the published minutes, others which once existed have been lost. A few of the letters Magellan received from St Petersburg have also survived, and are now preserved in the Bodleian Library in Oxford;(25) here, too, however, it is clear that the surviving record is far from complete.

From the minutes we learn that the initial contact came from Magellan, in what looks like a carefully orchestrated series of moves during the latter part of 1776. First, he despatched a packet of books to St Petersburg, including a couple of volumes sent on behalf of the Royal Society, a copy of Lord Stanhope's edition of the *Opera* of the mathematician Robert Simson, sent as a gift from Stanhope himself, and a copy of his own tract on navigational instruments, published in the preceding year. By the next mail he wrote announcing that the books had been sent, and suggesting that the Academy might establish a London outlet for its publications by contacting the bookseller John Nourse. Finally, he offered to both the Academy and individual members of it, 'tous les services qui dépendront de lui'.(26) Three days later, he wrote to the senior academician, Leonhard Euler, repeating the information about the package of books, and including news about some very interesting experiments on the electric fish, the *gymnotus*, which had recently been reported to the Royal Society.(27)

The St Petersburg academicians were clearly delighted by Magellan's approach, and even more by the opportunity it gave them to establish closer links with English science. Euler's son, Johann Albrecht Euler, the Secretary of the Academic Conference, was instructed to reply, thanking Magellan for the books, and taking up his offer of help by asking him to assist the Academy to complete its run of the Philosophical Transactions. (As a noteworthy indication of how irregular the Academy's contacts with England had become, it had not received any of the volumes for the preceding nine years.) At the same time, Euler offered to send the Royal Society any volumes of the St Petersburg Commentarii which it lacked. As to Nourse, the Academy was rather more cautious; Magellan was asked to suggest to him that he contact one of the English merchants in St Petersburg, who could purchase as many of the Academy's publications for him as he should desire.(28)

Magellan's next letter, dated 14 March 1777, successfully cemented his relationship with the St Petersburg Academy. He announced that the missing volumes of the Philosophical Transactions were shortly to be despatched and offered to procure for the Academy any other 'ouvrages Académiques' it needed. He let the Academy know which volumes of its Commentarii were needed to complete the Royal Society's run, and ordered several volumes of the series on his own account. He proposed arrangements for this kind of transaction which the academicians found very satisfactory, and concluded by passing on 'diverses nouvelles littéraires fort intéressantes'. The academicians accepted all his offers 'avec reconnoîssance', instructed their Secretary to ensure that the administrative hierarchy of the Academy was informed of what was going on, and resolved to send Magellan an order for a bulky packet of books including, for example, a complete run of the Nautical Almanac.(29)

Thereafter, letters were exchanged regularly between London and St Petersburg. Magellan proved assiduous in keeping the academicians informed of the latest scientific news, and they for their part enthusiastically availed themselves of his services, both individuall and as a body. Packets of books arrived in the Russian capital with great regularity, and soon the trade expanded to include clocks, scientific instruments and collections of seeds. As early as Decembe 1776, only a couple of months after Magellan's first contact with St Petersburg, his name was brought forward -- though unsuccessfully on this occasion -- when the Academy's complement of foreign members was being reconsidered. Less than two years later, his name by now being much better known to the academicians, he was duly elected.(30) Subsequently, he directed his correspondence for some years through the Academy's astronomer, A.J. Lexell, whom he had got to know during the latter's extended tour through the scientific centres of western Europe in 1780-81, and who was probably better qualified than any of the other academicians to judge the worth of the flood of new instruments being brought to their attention by the energetic Portuguese. Following Lexell's death in 1784, Magellan addressed his letters to the younger Euler once more. On at least one occasion the Academy's senior astronomer, Stepan Rumovskii, used Magellan as a channel by which to make an observation known to his fellow astronomers in western Europe.(31) In the later 1780s, however, Magellan's letters became less frequent, ceasing almost entirely in the last eighteen months of his life, in which, following a fall, be became a helpless invalid.(32) He died in February 1790, aged 67.

There were other contacts between Russian and British scientists during this period, besides those described so far. The St Petersburg academician P.S. Pallas, for example, maintained an active correspondence for a time with the English naturalist Thomas Pennant.(33) Likewise, a number of letters to the St Petersburg Academy from the English electrical investigator Benjamin Wilson survive in Leningrad.(34) For a brief period in the late 1760s, the Academy and the Royal Society collaborated over arrangements for observing the long-awaited transit of the planet Venus across the face of the Sun.(35) Various entries in the Society's Journal Book reveal that whenever news arrived from Russia, it was promptly passed on to the Fellows. For example, the arrival of an occasional letter from Euler (who was himself a Fellow of the Society) is recorded, as are the gift of a book and geological specimens from another of the senior academicians, Jakob von Stählin, and letters from one J.P. Wiel, writing from Moscow, and the academician G.F. Müller.(36)

All of these, however, except perhaps the collaboration over the transit observations, share one striking feature with the two leading examples discussed in this paper; they were wholly the result of private initiatives of one kind or another. Neither the Royal Society nor the St Petersburg Academy, we discover, made any particular effort, as institutions, to foster scientific connections between the two countries. (Here we might note the sharply contrasting case of the Swedish Academy of Sciences, which seems to have been exceedingly diligent throughout this period in maintaining links with scientific institutions elsewhere.) And with the responsibility left to them, individual scientists reacted in different ways. The isolation of the St Petersburg Academy makes Aepinus's strenuous efforts to bring his work to the attention of leading scientists elsewhere easy to understand. Few of his colleagues, however, seem to have been so anxious to publicise their work abroad.

So far as scientific exchanges in the other direction are concerned, one gets the feeling that Magellan's motives were almost wholly commercial. He seems not to have been especially diligent in passing on to the Royal Society, or even to the less formally constituted Chapter Coffee House Philosophical Society whose fortnightly meetings he also attended with great regularity,(37) the scientific news he was receiving from St Petersburg. Whether or not the Russian academicians realised what Magellan was up to (and there are signs once or twice that they did),(38) they clearly welcomed his initiative, and profited greatly from the books, news and scientific instruments he sent them. Yet it is extraordinary that they themselves had done nothing to foster such links before Magellan appeared on the scene. Particularly striking is their having allowed the Academy's run of such an important journal as the *Philosophical Transactions* to get so far behind.

From the instances I have described, it is clear that communications between the two countries was not difficult, if only one had a mind for it. The institutional records upon which I have largely depended in this paper suggest, however, that for surprisingly long periods most of the St Petersburg academicians simply could not be bothered making the effort. True, these records may not reveal the full range of contacts between individual scientists of the two countries: a great deal of further archival research would be needed to obtain a definitive answer on this point. Yet there are good reasons for expecting that a more detailed investigation would merely confirm the conclusions reached here. It generally seems to have been the case that whenever a member of either the Royal Society or the St Petersburg Academy received scientific news of any value from a fellow scientist elsewhere, he took care to table it soon afterwards at a meeting of his institution. Both the Royal Society's Journal Book and the St Petersburg Academy's minutes reveal numberless instances of this being done. Hence such sources probably do provide a fairly reliable record not only of the formal scientific links, such as they were, which were maintained between the two countries, but also of the extent to which up-to-date scientific news was being exchanged between individual scientists on the two sides.

It is an unflattering picture of the St Petersburg academicians which emerges. Content with their very comfortable lot in the Russian capital, few seem to have had any ambition to strut more boldly on the world's scientific stage, and (as the example of the *Philosophical Transactions* shows) they seem to have felt no urgent need to keep up with what was being done elsewhere. In short, their intellectual horizons were more limited than we might have expected, given the standing of their institution as a whole within the European scientific community of the day.

Notes

(1) M.I. Radovskii, Anglo-russkikh nauchnykh sviazei (Moscow-Leningrad, 1961). Newton's influence is also the central theme of Valentin Boss' recent study, Newton and Russia: The Early Influence, 1698-1796 (Cambridge, Mass., 1972).

(2) For an analysis of the increasingly Russian character of the Academy towards the end of the century, see the M.A. thesis by my student Ludmilla Burgess, 'The Russification of the St Petersburg Academy of Sciences and Arts in the Eighteenth Century' (University of Melbourne, 1974).

(3) Protokoly zasedanii konferentsii Imperatorskoi Akademii Nauk s 1725 po 1803 goda (Spb., 1897-1911), passim.

(4) Aepinus, 'Razsuzhdenie o sredstvakh sluzhashchikh k otvrashcheniiu povrezhdeniia proizvodimago gromovym udarom i molnieiu domam i drugim stroeniiam', Trudy Vol'nago ekonomicheskago obshchestva, XVI (1770), 257-85. An edited version of this essay was printed twice more before the end of the century.

(5) Cf. Protokoly, II, 428.

(6) Ibid., II, 455: and G.F. Müller to the Chancellery of the Academy, 19 September 1760; Arkhiv Akademii nauk SSSR, Leningrad, fond 3, op. 1, no. 257, l. 9. (The Chancellery's decision is noted on the back of this letter).

(7) R.W. Home, 'Aepinus and the British Electricians: The Dissemination of a Scientific Theory', Isis, LXIII (1972), 190-204. Wilson's presentation copy of Aepinus's book is preserved in the Silvanus P. Thompson Collection at the Institution of Electrical Engineers, London.

(8) Robison's annotated copy, marked on the title page 'Joh Robison 1772' and inscribed 'Ex dono illustri. Auctoris' is preserved in the Ronalds Library, Institution of Electrical Engineers, London.

(9) Home, 'Aepinus and the British Electricians', p. 194. Lord Shelburne's copy is preserved at the Royal Institution, London.

(10) The principal sources concerning Magellan are Ricardo Jorge, 'Amigos de Riveiro Sanches: J.H. Magellan', in Medicina contemporânea, 1910 (which I have, unfortunately, been unable to consult); Maximiano Lemos, Estudos de história da medicina peninsular (Pôrto, 1916), pp. 236-95; A. de Sousa Pinto, A vida e a obra de João Jacinto de Magalhães (Pôrto, 1931);

Joaquim de Carvalho, 'Correspondência científica dirigida a João Jacinto de Magalhães', Revista da Faculdade de Ciências da Universidade de Coimbra, XX (1951), 93-283, and also published separately (Coimbra, 1952); Arthur Birembaut, 'Sur les lettres du physicien Magellan conservées aux Archives Nationales', Revue d'histoire des sciences, IX (1956), 150-61; and David Willemse, 'Suites d'un voyage aux Pays-Bas: João Jacinto de Magalhães (1722-1790) et ses rapports avec Jean Henri van Swinden (1746-1823)', Arquivos do Centro Cultural Portugês, VII (1974), 225-78.

(11) Magellan, Journal étranger, April 1760, pp. 196-234; Carvalho, op. cit., p. 96; Lemos, 'Portuguezes illustres em França: Soares de Barros, João Jacintho de Magalhães e Ribeiro Sanches', Academia das Sciencias de Lisboa: Boletim da Segunda Classe, III (1910), 447.

(12) Lemos, Estudos, pp. 293-4. Few details are known of Magellan's activities during the early 1760s, but it appears that he earned Sanches's undying gratitude for performing a service for him, at some risk to his own safety, probably during a visit to Lisbon in 1762 (ibid., pp. 240, 293-4; and cf. Willemse, 'Suites ...', p. 230, who quotes from a letter from Magellan to Sanches, written from Portugal in April 1762). On Sanches, see Lemos, Ribeiro Sanches: a sua vida e a sua obra (Pôrto, 1911), and Willemse, Antônio Nunes Ribeiro Sanches - élève de Boerhaave - et son importance pour la Russie (Leiden, 1966).

(13) It has generally been maintained that Magellan arrived in London in 1764, though Carvalho ('Correspondência', p. 97) indicated that it could have been earlier. The Royal Society's Journal Book shows that he attended a meeting of the Society for the first time on 10 November 1763, as the guest of Dr. Thomas Birch. He attended a meeting every few weeks for the next few months, and then from the time the Society reconvened after its summer recess in 1764, for a period of many years he scarcely missed a meeting.

(14) British Library, Add. MS. 28539, f. 218.

(15) Royal Society, Canton Papers, II, 54-5.

(16) Royal Society, Letters & Papers, IV, 279, 288, 292.

(17) Michaud, Biographie universelle ancienne et moderne, art. 'Magellan'.

(18) See fn. (16) above, the last two items.

(19) E.g. his *Description des octants et sextants anglois* ... (London-Paris, 1775).

(20) Brissot, *Mémoires (1754-1793)*, ed. Cl. Perroud (Paris, 1911), I. 363.

(21) Guerlac, *Lavoisier - The Crucial Year: The Background and Origin of His First Experiments on Combustion in 1772* (Ithaca, 1961), chap. 2.

(22) *Ibid.*, p. 41; cf. Birembaut, *op. cit.*, pp. 153-4, for the details of the incident with the loom.

(23) *Protokoly*, III and IV, *passim*.

(24) Cf. I.I. Liubimenko, *Uchenaia korrespondentsiia Akademii nauk XVIII veka* (Moscow-Leningrad, 1937), which lists some letters not mentioned in the minutes, but fails to list others which are.

(25) University of Oxford, Bodleian Library, MS. Rigaud 38; published by Carvalho, 'Correspondência', as his nos. XXI, XXXVI, XLIII.

(26) *Protokoly*, III, 262.

(27) *Ibid.*, p. 261.

(28) *Ibid.*, p. 262.

(29) *Ibid.*, p. 295.

(30) *Ibid.*, pp. 274, 375. Magellan's name was also brought forward unsuccessfully on one other occasion in the intervening period (*ibid.*, p. 328).

(31) Rumovskii to Magellan, 6/17 November 1786; Carvalho, 'Correspondência', pp. 231-2. Cf. *Phil Trans.*, LXXVII (1787), 48-9.

(32) Magellan to Van Marum, 30 December 1788, and Ingenhousz to Van Marum, 7 December 1789; R.J. Forbes *et. al.* (eds.), *Martinus Van Marum: Life and Work* (Haarlem-Leiden, 1969-76), VI, 175, 288.

(33) C. Urness (ed.), *A Naturalist in Russia: Letters from Peter Simon Pallas to Thomas Pennant* (Minneapolis, 1967).

(34) Arkhiv Akademii nauk SSSR, Leningrad, fond 1, op. 3, no. 62, ll. 282-284; no. 63, ll. 1-2, 74-77, 103-104, 125; no. 65, ll. 51-52.

(35) Royal Society, Journal Book Copy, entries for 14 January and 21 January 1768, and 12 January, 9 November and 7 December 1769. Cf. H. Woolf, _The Transits of Venus: A Study of Eighteenth-Century Science_ (Princeton, N.J., 1959), pp. 179-81.

(36) Royal Society, Journal Book Copy, entries for 12 January 1769 and 18 March 1774 (Euler); 9 June 1774 (Stählin); 16 March 1769 (Wiel); and 8 December 1768 (Müller).

(37) The minutes of the Chapter Coffee House Society's meetings are preserved at the Museum of the History of Science, Oxford.

(38) Cf. _Protokoly_, III, 675.

XVIII

"PHYSICAL PRINCIPLES AND THE POSSIBILITY OF A MATHEMATICAL SCIENCE OF ELECTRICITY AND MAGNETISM"

One of Poisson's most successful scientific achievements was his application of mathematics to the traditional sciences of electricity and magnetism. Why did he succeed where others before him had failed? Some earlier attempts at mathematizing these sciences are considered, in order to determine the extent to which the particular physical principles that different workers adopted facilitated or hindered their efforts at mathematization.

In Poisson's day - especially during the early part of his career - many physicists remained opposed to the introduction of mathematics into the study of electricity, magnetism, and other traditional branches of physical inquiry.

I have charted the attitude on this matter of the generation before Poisson's, in my study of the reception of Aepinus's pioneering efforts at mathematizing the sciences of electricity and magnetism. I have argued that the reason Aepinus's work had little influence for a quarter of a century and more after it first appeared in print was that few physicists in his day could see the point of what he was doing.[1] We today have no difficulty in

1. Aepinus's Essay on the Theory of Electricity and Magnetism, introd. by R.W. Home, trans. by P.J. Connor (Princeton, 1979), chap.5.

discerning a number of important advances which Aepinus introduced into electrical and magnetic theory by bringing a rigorous mathematical analysis to bear on the subject. John Winthrop, however, could not, despite his position at that time as professor of both mathematics and natural philosophy at Harvard College, and wrote as follows to his friend Ezra Stiles concerning Aepinus's contributions to electrical theory: "As to Electricity, I don't find that he has advanced much, if at all, beyond our excellent Countryman, to whom he renders great & just honors; & I scarcely recollect any thing he has said on this subject, but what I had learn'd before, & with greater pleasure, from Dr. Franklin".[2]

Others, such as the London instrument maker George Adams, were openly hostile to what Aepinus had done. "Æpinus", Adams wrote, "by a mathematical theory of electricity, has closed the door on all our rese[a]rches into the nature and operations of this fluid". Aepinus's work manifested, so far as Adams was concerned, all the evils attendant upon the failure of mathematicians to attend closely enough to the phenomena: "when neglecting these, calculations are made to serve an hypothesis; the more elegant and beautiful they are, the more detrimental they become to science".[3] There is every reason to believe that Adams's criticisms were based on hearsay rather than on direct acquaintance with Aepinus's book. They were thus of the "in principle" kind, directed against the very notion of a mathematical science of electricity, and not against any specific weaknesses that Adams had discovered in Aepinus's presentation.

Kenneth L. Caneva in an important recent study has discerned a similar hostility towards the use of mathematics in physics among German physicists of the early nineteenth century.[4] As eighteenth-century "electricians" had done before them, these nineteenth-century German professors drew a sharp

---

2. Winthrop to Stiles, 21 February 1764; Stiles Papers, Beinecke Rare Book and Manuscript Library, Yale University.

3. Adams, Lectures on Natural and Experimental Philosophy (2nd ed., London, 1799), IV, 304.

4. Caneva, "From Galvanism to Electrodynamics : The Transformation of German Physics and Its Social Context", Hist. Studs. Phys. Sci., 9 (1978), 63-159; see also Kathryn M. Olesko, "The Emergence of Theoretical Physics in Germany: Franz Neumann and the Königsberg School of Physics, 1830-1890" (Cornell University Ph.D. thesis, 1980), esp. chap.IV.

distinction in their theorizing between the "concretely physical" and the "abstractly mathematical". They "self-consciously defended an ideal of science that held mathematical description in no great esteem on the grounds that mathematics tended to mislead the physicist from his proper field of study".[5]

Only in France was this attitude under serious challenge during the first years of the nineteenth century. There, under the leadership, initially, of Laplace and (to a lesser extent) Coulomb, a new kind of physics was being forged in which, in opposition to the traditional view, mathematics was to play an integral role.[6] Poisson's work in the field of electricity and magnetism was a major contribution to this general mathematizing tendency, and was so seen at the time. (The impact of his work on British physics, for example, is clearly shown in a recent article by Crosland and Smith.[7]) Nevertheless, Poisson's was not the first attempt at mathematizing these subjects. I have mentioned Aepinus's efforts already, and there were others. In what follows, I shall discuss some of this earlier work and try to show

---

5. Caneva, op.cit., p.86.

6. There is a rapidly growing literature describing these developments. See especially T.S. Kuhn, "Mathematical vs. Experimental Traditions in the Development of Physical Science", J. Interdisciplinary Hist., 7 (1976), 1-31 (reprinted in Kuhn, The Essential Tension : Selected Studies in Scientific Tradition and Change [Chicago, 1977], chap.3); R. Fox, "The Rise and Fall of Laplacian Physics", Hist. Studs. Phys. Sci., 4 (1974), 89-136; M. Crosland, The Society of Arcueil : A View of French Science at the Time of Napoleon I (London, 1967); C.C. Gillispie, Lazare Carnot, Savant (Princeton, 1971); idem., art. "Laplace" in Dictionary of Scientific Biography, vol.XV; R.H. Silliman, "Fresnel and the Emergence of Physics as a Discipline", Hist. Studs. Phys. Sci., 4 (1974), 137-162; C.S. Gillmor, Coulomb and the Evolution of Physics and Engineering in Eighteenth-Century France (Princeton, 1971); E. Frankel, "J.B. Biot and the Mathematization of Experimental Physics in Napoleonic France", Hist. Studs. Phys. Sci., 8 (1977), 32-72; idem., "Career-making in Post-revolutionary France: The Case of Jean-Baptiste Biot", Brit. J. Hist. Sci., 11 (1978), 36-48; H. Guerlac, "Chemistry as a Branch of Physics: Laplace's Collaboration with Lavoisier", Hist. Studs. Phys. Sci., 7 (1976), 193-276; I. Grattan-Guinness with J.R. Ravetz, Joseph Fourier, 1768-1830 (Cambridge, Mass., 1972); Grattan-Guinness, "Mathematical Physics in France, 1800-1835", pp.349-370 in H.N. Jahnke and M. Otte, eds., Epistemological and Social Problems of the Sciences in the Early Nineteenth Century (Dordrecht, 1981); idem., "Mathematical Physics in France, 1800-1840 : Knowledge, Activity and Historiography", in J.W. Dauben, ed., Papers in Honor of K.R. Biermann (New York, 1981).

7. M. Crosland and C. Smith, "The Transmission of Physics from France to Britain, 1800-1840", Hist. Studs. Phys. Sci., 9 (1978), 1-61.

why Poisson was so much more successful - as he certainly was - than his predecessors.

A *sine qua non*, presumably, for the mathematization of any branch of physics, is the establishment of quantitative laws linking relevant physical variables. Accordingly, I shall begin here by looking briefly at some early attempts to arrive at quantitative laws relating to electrical or magnetic action.

Most such attempts were, in fact, directly inspired by the grand example of Newton's treatment of gravity in his famous *Mathematical Principles of Natural Philosophy* (1687). If one followed Newton's lead, it appeared all too obvious that in the cases of electricity and magnetism, too, one should seek above all to establish the laws governing the variation, with distance, of the forces exerted by electrified or magnetized bodies. Accordingly, we find a number of attempts at doing this recorded in the eighteenth-century literature, beginning with those of Newton himself and his colleagues in relation to magnetic forces.

Newton reported his result in a single sentence added to the second (1713) edition of the *Principia*, without giving any details as to how he had arrived at it. The power of a magnet, he proclaimed, "in receding from the magnet decreases not as the square but almost as the cube of the distance, as nearly as I could judge from some rude observations".[8] The results reported to the Royal Society at about the same time by Francis Hauksbee and Brook Taylor were much less definite; indeed, their papers, at least in their published versions, merely presented sets of raw data, without any attempt to extract a functional relationship from the numbers.[9] Elsewhere, however, Taylor did take up the question, both in his initial report to the Royal Society from which the paper as published was extracted, and subsequently in a letter to Sir Hans Sloane published in the *Philosophical Transactions* a few years afterwards.[10] His discussion well illustrates the conceptual difficulties that long plagued those pursuing investigations in this area.

---

8. Newton, *Mathematical Principles of Natural Philosophy*, trans. Andrew Motte, rev. Florian Cajori (Berkeley, 1934), p.414.

9. Hauksbee, *Phil. Trans.*, *27*, (1710-12), 506-511; Taylor, ibid., *29* (1714-16), 294-295.

10. Royal Society of London, *Letter Book original*, *14*, 394; Taylor, *Phil. Trans.*, *31* (1720-21), 204-208.

Like both Newton and Hauksbee, and also most other early investigators of the magnetic force, Taylor was concerned with the "power" of a magnet and the way it varied with distance. "Power" here meant simply the general attractive force exerted by a lodestone on a piece of iron or a compass needle, and is not to be confused with the later and much more sophisticated idea of the force exerted by one isolated magnetic pole on another. Indeed, in Taylor's day and for many years afterwards, the poles of a magnet were generally seen, not as centres of magnetic force, but merely as the ends of the axis according to which a magnet would align itself with respect to the Earth. Taylor complained that "if it were known what point within the Stone, and what point in the Needle [it attracts] are the Centers of the Magnetical power, it would be easy to find the true powers of the Magnet at all the distances observed", but for want of such knowledge he had to make arbitrary assumptions about the locations of these points. Unable by that means to find a regular law, he then tried to work backwards and find points that he might use as the centres of power, such that "the Law, by which the Magnetism alters, could be reduced at all distances to any one certain power of those distances". But he found that for this to be the case, the centre of the lodestone's attractive power had to be assumed to lie outside the stone! Defeated, he was driven to the dissatisfying conclusion that "the power of Magnetism does not alter according to any particular power of the distances, but decreases much faster in the greater distances, than it does in the near ones".

Similar difficulties plagued the efforts of Pieter van Musschenbroek to determine the law of magnetic action, and for similar reasons.[11] Yet Musschenbroek's results were long regarded as definitive. His method consisted of suspending a small magnet from one arm of a balance above a larger fixed magnet, and then measuring the weight required to be added to the pan on the other arm of the balance (or subtracted if the force were repulsive) in order to restore equilibrium. To permit him to measure the force at different distances, the entire arrangement of balance arm and load could be raised or lowered by means of a pulley. His apparatus is shown on the accompanying illustration, taken from his Dissertatio physica experimentalis de magnete (Leyden, 1729). (The rather complicated mechanism

---

11. Musschenbroek, Phil. Trans., 33 (1724-25), 370-378; his Dissertatio physica experimentalis de magnete (Leyden, 1729), pp.13-23; and his Elements of Natural Philosophy, trans. J. Colson (London, 1744), I, 205-208.

XVIII

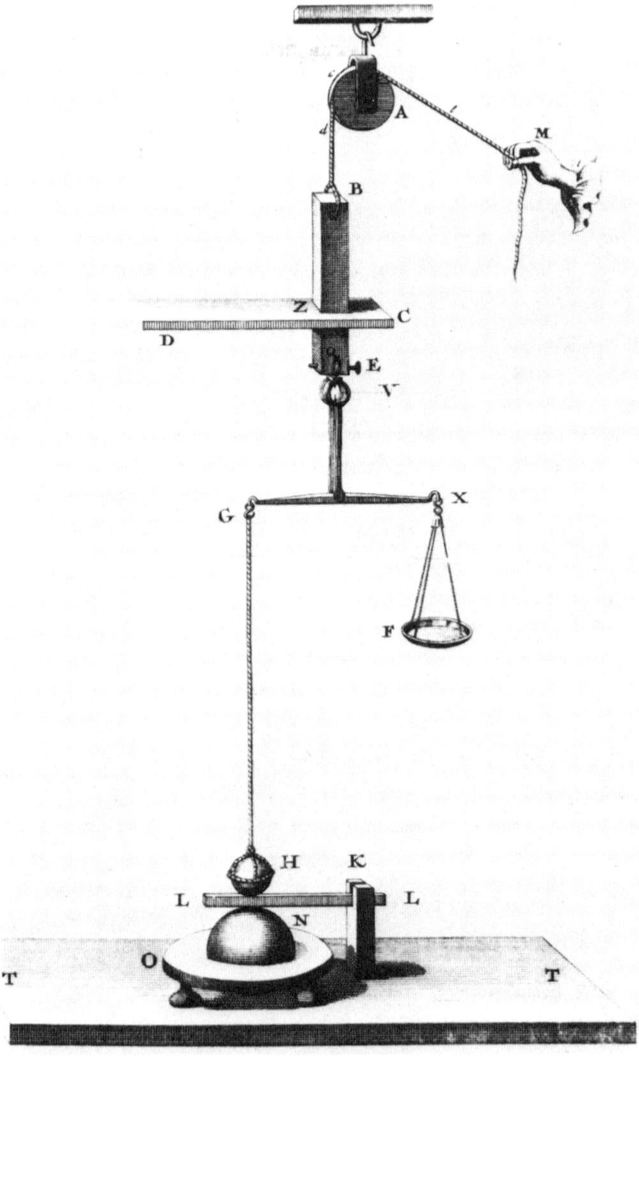

shown linking the balance arm to the rope over the pulley was designed to prevent the apparatus twisting to and fro above the large magnet NO.) It is clear that Musschenbroek, like Taylor, was concerned to measure simply the total force exerted by one magnet on another. If unlike poles were facing each other, this force was an attractive one, if like poles were adjacent, it was repulsive, and the apparatus served equally well to measure either. Musschenbroek's initial results gave no more hint of a regular relationship between force and distance than Taylor's had done: "I can only conclude", he said, "that there is no proportion between forces and distances". Later, he claimed to have discovered definite laws of magnetic action, but a different (and generally quite complex) law in each different set of circumstances, depending on the shapes of the magnets used.

It would be most unjust to criticize Musschenbroek here. He was an accurate and careful experimenter, and his apparatus was well designed to measure what he wanted to measure. In fact, the results he obtained were quite correct; in the circumstances of his experiment, there is no general law to discover, and such incidental regularities as might exist do depend upon the shapes of the magnets involved. In other words, the problem confronting Musschenbroek and the others was a conceptual one. _We_ know that the force exerted by a magnetic dipole on a single magnetic pole varies approximately as $1/r^3$ at distances large with respect to the length of the dipole, and this presumably accounts for the result which Newton obtained. But in reality it was not even this that he and his successors were measuring. In many cases (as may easily be seen, for example, in Musschenbroek's depiction of his apparatus), the distances involved were _not_ large with respect to the length of the primary magnet being used. Over and above this, however, these men were dealing with the resultant force exerted by the dipole not on a single pole but on a _pair_ of poles induced in the body attracted, in addition (at least in the case of Musschenbroek's experiment) to a pre-existing pair. Furthermore, the strength of the induced magnetism would itself have been varying with distance, and would have depended as well on the hardness and geometry of the iron used, and on the positioning of the iron with respect to the axis of the dipole.

None of these complications was apparent to Musschenbroek or to most other eighteenth-century workers, and their theoretical commitments (or, in Musschenbroek's case, their ostentatious lack thereof) gave them no guidance as to how the difficulties that had emerged in their experimental results might be resolved. Indeed, so far as magnetism was concerned there was a strong tendency during the middle years of the century, even among those such as Leonhard Euler who in other cases were if anything over-hasty in applying their mathematical skills to physical problems, to accept Musschenbroek's experimental data at face value and conclude that they were dealing with an almost irreducibly errant phenomenon.[12] If so, the subject was obviously not readily amenable to mathematical analysis.

Eighteenth-century attempts to determine a force law in the case of electricity have recently been discussed by Heilbron.[13] Perhaps the earliest was that of Kratzenstein in 1746, who used a method very similar to that adopted by Musschenbroek in the case of magnetism. Kratzenstein suspended a metal disc above the globe of his electrical machine by a metal wire attached to the arm of a steelyard, and measured the weight needed to balance the force between disc and globe.[14] Though his tabulated data revealed wide divergencies, he felt able to conclude from them that the force varied inversely as the distance from the globe. Some years later, Daniel Bernoulli reached a different conclusion from an essentially similar experiment. He measured the force acting between a disc attached to his electrical machine and a similar parallel disc directly beneath it by fixing the latter to the top of a hydrometer floating in water, and found that the force varied inversely as the _square_ of the distance between the discs.[15]

Let us not delude ourselves into thinking, however, that Bernoulli had therefore obtained the right result whereas Kratzenstein had somehow made an unfortunate blunder. In retrospect, we can see that, as in the experiments on magnetism discussed earlier, neither Kratzenstein's experiment nor Bernoulli's could in fact yield information of the kind later summarised in Coulomb's law. Since the source of electricity in both experiments was a

---

12. Cf. Euler, "Dissertatio de magnete", _Recueil des pièces qui ont remporté les prix de l'Académie Royale des Sciences_, V (Paris, 1752), 3-47, esp. pp.34-35.

13. J.L. Heilbron, _Electricity in the 17th and 18th Centuries: A Study of Early Modern Physics_ (Berkeley/Los Angeles, 1979), chap.19.

14. C.G. Kratzenstein, _Theoria electricitatis more geometrico explicata_ (Halle, 1746), pp.34-35.

15. A. Socin, _Acta Helvetica_, _4_, (1760), 214-230.

working electrostatic generator, while the disc attached to the measuring device is in each case effectively grounded, it follows that in both cases we are dealing with a capacitor (with parallel plates in Bernoulli's experiment but of a rather more complicated geometry in Kratzenstein's), the plates of which are held at a constant potential difference while the distance between them, and hence the charges on them, are systematically varied. Coulomb's law, on the other hand, refers as everybody knows to the variation with distance of the force between two given charges. The fact that Bernoulli nevertheless obtained an inverse-square relationship is an accidental by-product of the experimental geometry he had chanced to use![16]

It thus becomes apparent that in the cases of both magnetism and electricity, these early attempts at quantification contributed little or nothing to the mathematization of these sciences. Their failure was due, I would suggest, to their not being sufficiently controlled by theory, indeed, to their being in a sense unconnected to the theories of magnetism and electricity respectively that were accepted by the investigators concerned.

So far as magnetism was concerned, prior to the work of Aepinus most eighteenth-century investigators, including Newton himself, accepted one version or another of the theory of circulating subtle matter invented by Descartes.[17] (Musschenbroek was a lonely exception in this regard, but since he deliberately refrained on methodological grounds from formulating an alternative, his work lacked all theoretical perspective. His helpless floundering is striking proof, if such were needed, of the pointlessness of mere quantification for its own sake.) According to the circulation theory, magnetic poles were merely the regions at which the circulating streams of subtle matter envisaged by the theory entered or left a magnet. The theory offered no grounds for seeing them as centres of magnetic force: indeed, it offered good grounds for *not* so seeing them. More generally - and this surely explains the dilemma in which Brook Taylor found himself - the notion of "centre of magnetic force" itself, imported as it was from the very different conceptual framework of Newtonian gravitational theory, was simply not a well-founded notion within the context of the circulation theory. Even if some regular quantitative relationship linking force and distance could have been satisfactorily established experimentally, this would not have

---

16. Heilbron, op.cit., p.461.
17. *Aepinus's Essay*..., chap.4.

facilitated in any way the quantification of the theory, because the relevant variables of the theory had to do not with forces between identifiable centres of action but with fluxes of the subtle matter supposedly in motion within and in the vicinity of magnetized bodies. Conversely, the theoretical dynamics of fluids in motion were at this period very incompletely worked out. There was thus no established mathematical formalism that one could apply, once one assumed the broad outlines of the circulation theory, to the magnetic case. The theory therefore remained entirely qualitative and unable to offer guidance even as to which variable one should in one's experiments seek quantitative laws for.

A similar situation existed in Kratzenstein's day and for some time afterwards with regard to electrical theory. Prior to the development of Franklin's theory of electricity in the middle years of the eighteenth century, there was no generally understood notion of electrical "charge", but merely a rather loosely formed idea that bodies could be electrified more or less strongly. Most theorists tried to account for electrification in terms of an agitation of some kind set up in the subtle matter supposedly lying within the pores of the body in question: Kratzenstein himself developed a somewhat idiosyncratic version, which he even attempted to express in mathematical terms, according to which this subtle matter swirled in Cartesian-style vortices around electrified bodies. In any theory of this kind, it was natural to associate degrees of electrification with differing quantities of motion imparted to the subtle fluid. (Such an idea is explicit, for example, in Richmann's discussion of the usefulness of the measuring instrument or "indicator" of electricity that he invented at about this time.[18]) As in the magnetic case, there were no obvious way of fitting a law of force of the Newtonian kind into such a theory; Kratzenstein, for example, despite his having formulated his theory in mathematical terms, was quite unable to connect the "law" that he announced to any of the variables of the theory.

Even a firm conviction, however, that electrical and magnetic phenomena were to be understood in terms of forces acting between the particles involved, was not by itself sufficient to render these sciences amenable to satisfactory mathematical treatment. In the magnetic case, this is best

---

18. G.W. Richmann, Novi commentarii Acad. Sci. Petrop., 4 (1752-53), 301-340.

illustrated by the work of Tobias Mayer and Johann Heinrich Lambert, each of whom sought with only limited success during the 1760s to develop a mathematical description of the interaction between two magnets.[19] Mayer in particular was an outspoken critic of the theory of magnetic circulations, and tried single-mindedly to reduce interactions between magnets to forces between the particles of which they were composed, as the foundation for a wider theory of terrestrial magnetism intended as a contribution to scientific geography. A knowledge of the law according to which these forces varied with distance was essential, and Mayer quickly concluded that this was inverse-square in character. However - and this is where the underlying weakness of his account immediately becomes apparent - his experimental data allowed this conclusion only if he simultaneously made the assumption, highly implausible on physical grounds, that "Each part [of a magnet] is thought of as attracted not...by all the unlike parts of the other magnet, but by one only, and similarly repelled by one only of the like parts; and the parts which act thus on one another are those which in similar magnets are the same distance from their centres or whose distance from their centres is proportional to the distance between their poles". If, on the contrary, Mayer made the physically much more reasonable assumption that any part of one magnet was attracted by all the unlike parts of the other, and repelled by all the like parts, theory and experiment were wildly divergent. Furthermore, Mayer also had to assume, as did Lambert, that the "power" of the different parts of a magnet was proportional to their distance from the geometrical equator of the magnet, which he further assumed always to coincide with its magnetic equator. Aepinus, who had a much better grasp than Mayer of the complexities involved, rejected both of these assumptions and hence Mayer's theory as a whole[20]; and even though the latter's subsequent calculations concerning terrestrial magnetism achieved some successes, we must, I think, agree with Aepinus that his basic assumptions provided a totally inadequate foundation for a theory of magnetism as such.

In the case of electricity, problems likewise remained even after Franklin's theory became widely accepted. To begin with, Franklin himself and most of his immediate followers associated with the notion of electrical charge a much less well defined notion of electrical "atomospheres" surrounding charged bodies. While opinions concerning such atmospheres

---

19. The Unpublished Writings of Tobias Mayer, ed. E.G. Forbes (Göttingen, 1972), vol.3; Lambert, Mém. Acad. Berlin, 22 (1766), 22-48; 49-77.

20. Aepinus, Novi commentarii Acad. Sci. Petrop., 12 (1766-67), 325-350.

tended to vary somewhat in detail from one worker to the next, it was assumed by virtually everybody that they played a mediating role in the attractions and repulsions exercised by electrified bodies.[21] Had any of those involved attempted seriously to construct a mathematical theory of electricity rather than a merely qualitative one, they would have had above all to subject these atmospheres to mathematical analysis. In fact, the process of mathematization took a different route, as first Aepinus and then Henry Cavendish abolished the atmospheres entirely from their theories, and constructed instead theories which relied unequivocally on forces acting at a distance between charged bodies.

A prerequisite for the quantitative determination of electrical phenomena was, of course, agreement as to what constituted equality of electrification. Unfortunately, there were two possibilities here, corresponding to what we call today equality of charge and equality of potential. We have seen already the importance of the distinction between charge and potential, in discussing Bernoulli's attempt to fix the force law. In fact, most early "indicators" of electricity showed potential rather than charge, but, failing to recognize the complications involved, those using them believed that they were measuring something quite unproblematic which they referred to by terms such as the "strength" or "force" or "degree" of electricity.

It was one of the great merits of Cavendish's treatment of electrical theory that he did perceive the difference, and distinguished carefully between the two concepts. In his terms, on the one hand bodies might be "undercharged" or "overcharged" in relation to the quantities of electric fluid natural to them but, on the other, when a number of conducting bodies were joined together, it was the degree of "compression" of the fluid that counted rather than the absolute quantities of it present. "It is plain", he wrote, "that the electric fluid must be equally compressed in all these bodies, for if it was not, the electric fluid would move from those bodies in which it was more compressed to those in which it was less compressed till the compression became equal in all. But yet it is possible that some of these bodies may be made to contain more than their natural quantity of electricity, and others less". As Maxwell long ago pointed out, such a notion "is precisely that of *potential*".[22]

---

21. *Aepinus's Essay*..., pp.97-106.

22. *The Electrical Researches of the Honourable Henry Cavendish, F.R.S.*, ed. J. Clerk Maxwell (Cambridge, 1879), p.95.

In contrast to Cavendish, Aepinus, and after him Coulomb, did not have a well established concept of this kind. Aepinus discussed the case of two electrical conductors in contact only very briefly, but it was one of the least successful sections of his work.[23] Elsewhere, his concern was always with induction (or "communication", as he called it), not conduction, and his technique consisted simply of summing the forces assumed to act at a distance between the pieces of ordinary matter and the quantities of electric fluid involved in each situation. Likewise, Coulomb's concern throughout his electrical investigations was with charges (or surface charge densities) rather than potentials, because it was those that he could measure with his torsion electrometer.

The examples of Aepinus and Coulomb show that it was possible to advance a considerable distance towards a mathematical theory of electricity, even without a clear idea of electrostatic potential. Some such notion would seem to have been essential, however, to any theory that aspired to encompass the sharing of charge between conductors in contact: in other words, the concept of charge alone, no matter how clearly formulated, is not sufficient to reduce all electrical phenomena to order. Yet from Poisson's papers on electricity we learn that all the difficulties are not necessarily resolved even when one has a concept of the potential that from a mathematical point of view is perfectly formed.

In fact, Poisson in his discussion of electricity simply adapts certain techniques already worked out by Lagrange and Laplace for the gravitational case.[24] The starting point and indeed the core of his analysis is a certain function V which we can recognize immediately as the potential function. Its value at a particular point due to a system of attracting masses is given by the sum of the masses divided by their respective distances from the given point. The components of the force acting at the point are then given by the partial derivatives of V. The system of attracting masses which Poisson considers is a thin layer of electric fluid spread within the surface of a conducting body or bodies.

The feature of Poisson's discussion to which I wish to draw attention here is that he remains unable to state that the surface of the conductor is an

---

23. Aepinus's Essay..., pp.350-351 and (for editorial comment) 125-126.

24. Poisson, Mém. de l'Institut, 1811, pt.1, pp.1-92; pt.2, pp.163-274.

equipotential surface. He cannot do so because, though he insists that the layer of fluid at the surface is "infinitely thin", for him it yet remains a real physical layer, the thickness of which varies from point to point as the surface charge varies. In Poisson's analysis, the resultant forces are everywhere normal to the inner surface of the fluid layer, so that that surface is a surface of uniform potential. But with physical layers of electric fluid of varying thickness, and hence non-confocal surfaces, it then follows that on the outer surface of the layer - and according to Poisson it is always the outer surface that coincides with the actual surface of the conducting body - the resultant forces are in general not normal to the surface. The difficulty arises, therefore, precisely because Poisson accepts quite literally the existence, as real fluids and not mere theoretical constructs, of the two electric fluids postulated by Coulomb.

The issue involved here was significant for the further development of electrostatic theory during the nineteenth century.[25] From the eighteenth-century perspective adopted in this paper, however, the problem, had it been perceived, would have been regarded as but a very minor blemish in an otherwise extraordinarily impressive achievement. Once Poisson's starting point, the potential function V, was established, his procedure was straightforward. In the case of the two situations for which Coulomb had published data on surface charge distributions, namely two spherical conductors of different radii either in contact or separated after contact, Poisson was able by means of his unsurpassed skill with series expansion methods to integrate his expression for V and then determine the distribution of electric fluid in each case by imposing the equilibrium condition that the resultant action on any point inside either conductor should be zero. The values he obtained were generally in good agreement with those which had been determined experimentally by Coulomb, and thus constituted a triumphant vindication of the importing of mathematical techniques into the traditionally disorderly realm of physics.[26]

---

25. Jed Z. Buchwald, "William Thomson and the Mathematization of Faraday's Electrostatics", Hist. Studs. Phys. Sci., 8 (1977), 101-136.

26. Worthy of remark here is Poisson's peculiar and entirely unsatisfactory method of "averaging" the differences between the values given by experiment and theory, in order to show that these are insignificant. His procedure is striking proof that in his day, adequate methods of assessing the validity of quantitative data had still to be established.

A decade later, Poisson developed a mathematical theory of magnetism along very similar lines. The analysis in this case was made more complicated by the molecular conception of magnetism that Poisson inherited from Coulomb, according to which the two magnetic fluids, the supposed existence of which formed the basis of the theory, were supposed confined to individual molecules or "magnetic elements" of iron and other magnetizable materials. This postulate, introduced by Coulomb in order to account for a variety of experimental data, implied that the magnetism of any magnetized body was spread throughout the body. Poisson succeeded in showing, however, that despite this, for homogeneous bodies at constant temperature, "the magnetic actions of a body of any form whatever are equivalent to those of a layer of fluid of very small thickness covering the entire surface". By applying his general formulae to the case of spherical bodies, he was able to compare his results with experimental data that had been published shortly before by the Woolwich professor, Peter Barlow. The agreement, he acknowledged, was not perfect. Nevertheless, he claimed, "Mr. Barlow's numerous observations are a remarkable confirmation of the theory of magnetism with which this Memoir is concerned". In a second paper, Poisson extended his analysis to spheroidal bodies in general, the dimensions of which could even be stretched so as to encompass both flat iron discs and magnetized needles.[27]

Despite Poisson's curious failure to see the implications for the underlying assumptions of his theory of the remarkable discoveries made shortly before by Oersted and Ampère - he seems to have felt that, at most, these might lead one to conclude that magnetism, rather than being "a particular fluid" (un fluide particulier) is "only the electric fluid modified by some special properties of [magnetizable] bodies and distributed in a particular way within them"[28] - his papers on magnetism also constitute a remarkable achievement. Like his electrostatics papers, they build upon theoretical foundations laid by Coulomb, and upon his demonstration (which Poisson acknowledges to be less secure in the magnetic than in the electrical case[29]) that the forces acting are inverse-square in character. In the electrostatics case in particular, I would suggest, Poisson's success depended in part upon his redefining the objective of electrical theory. His account in fact nowhere addresses any of the traditional problems of electrical science

---

27. Poisson, Mém. de l'Acad., 5 (1821-22; publ. 1826), 247-338; 488-533. The quotations come from pp.261 and 336 respectively.

28. Ibid., p.249.

29. Ibid., p.248.

concerning the interactions of electrified material bodies. It cannot, because nowhere does ordinary matter enter into Poisson's account. Instead, he is concerned throughout solely with the resultant forces exerted by different distributions of the two electric fluids on other masses of such fluid, and hence solely with electrostatic induction and the various distributions of charge to which this can give rise in bodies which offer no resistance to the motion of the fluids.

It can be seen, therefore, that the route to Poisson's quantified sciences of electrostatics and magnetostatics was by no means as straightforward as it is usually depicted. To begin with, a general resistance to the importation of mathematical techniques into these fields had to be overcome. But even those who sought in these rapidly developing sciences to emulate Newton's achievement with respect to gravity more often than not came to grief. I have argued that the fundamental problem here was an incompatibility between the quantifying ambitions of these eighteenth-century physicists and their basic theoretical presuppositions. Only in the theories of Aepinus and, after him, Cavendish, Coulomb and Poisson, was it unequivocally the case that the techniques of mathematical gravitational theory - including the notion of an empirically determined law of force - could be adapted to these sciences. I have suggested in addition that no other route to a full mathematization of these sciences was open at this period - as it was, by contrast, in Maxwell's day - since the mathematical theories of stressed media and of vortical and other fluid motions were as yet wholly inadequate to such a task. (The question also arises here, which I have not addressed at all in my paper, of whether during the eighteenth century even gravitational theory was sufficiently developed to provide a satisfactory route to mathematization.) In effect, the choice for the would-be mathematizing physicist was, at best, an action-at-a-distance theory after the gravitational model, or nothing. A major problem then still remained which only began to find its answer in the work of Aepinus, namely to show how the complex behaviour that real magnets and electrified bodies exhibited could be reconciled with such a theory. Poisson's genius was to see how at least some aspects of their behaviour could be accounted for on this basis, quantitatively and in detail.

XIX

POISSON'S MEMOIRS ON ELECTRICITY: ACADEMIC POLITICS AND A NEW STYLE IN PHYSICS*

Siméon Denis Poisson (1781–1840) was a major figure in French science throughout the first forty years of the nineteenth century. Though his papers lack the brilliant mathematical creativity of some of those published by even more gifted contemporaries such as Joseph Fourier (1768–1830) and Augustin-Louis Cauchy (1789–1857), they nevertheless display a formidable talent for mathematical analysis, applied with great industry and success in a large number of investigations ranging over the whole domain of mathematical physics. Several were of such importance that even on their own they would have sufficed to win him lasting fame.

Poisson's working life coincided with the most exciting period French physics has known. The last years of the eighteenth century and the first decades of the nineteenth were precisely the period in which physics, dominated and inspired by the French, became for the first time a genuinely mathematical science. Poisson played a major part in bringing about this transformation. His 1807 memoir on sound, though largely devoted to purely mathematical questions, has been seen by some as one of the early highlights of the new programme;[1] his development shortly afterwards of a mathematical theory of electrostatics was an unqualified triumph for the French school, a *tour de force* in which, at one stroke, a major and until then still largely qualitative field of inquiry was subjected to exact analysis.[2] Subsequent publications on elasticity, heat and various other topics consolidated his position as an outstanding exponent of the new approach.

Poisson's work on electrical theory also marked an important milestone in his own scientific career, since his undertaking it was intimately connected—though in a way that has hitherto been inadequately understood—with his election to membership of France's premier scientific institution, the Academy of Sciences (or, as it was called during

Research for this paper was supported by a grant from the Australian Research Grants Committee. I am grateful to Maurice Crosland, Ivor Grattan-Guinness and Jim Cross for comments on an earlier draft.

[1] Poisson, 'Mémoire sur la théorie du son', *Journal de L'École Polytechnique*, *7* (1808), 319–392. Cf. the judgement of Horace Lamb, *The Evolution of Mathematical Physics* (Cambridge, 1924), p. 9.

[2] Poisson, 'Mémoire sur la distribution de l'électricité à la surface des corps conducteurs', *Mém. de l'Institut*, *12* (pt. 1), 1–92; 'Second mémoire sur la distribution de l'électricité à la surface des corps conducteurs', ibid., *12* (pt. 2), 163–274.

the period 1795–1815, the First Class of the Institut de France). The prestige associated with membership ensured that elections to positions in the Institut were always hotly contested, and even for the most outstanding candidates, success usually depended upon attracting the patronage of influential members of the Class. As we shall see, Poisson's case was no exception in this regard.

The role of patronage in shaping the careers of French scientists at this period has been the subject of a number of recent studies. In addition, the suggestion that the patronage system might have helped to determine the fate, in France, of a number of specific scientific doctrines has been pursued at some length. Crosland has documented the power wielded within the Academy by Berthollet and Laplace in particular.[3] Fox has described within this context the rise and subsequent fall from favour of a wide-ranging research programme based on Laplace's vision of a science in which all physical effects were reduced to determinate forces acting between material corpuscles according to known laws.[4] Grattan-Guinness has argued that there were in fact two different approaches to mathematical physics, of which the Laplacian vision was one, struggling for supremacy within French science at this time, and he has connected these with two different approaches to mathematics itself that he discerns in the writings of the period. He has stressed, too, the bitterness that often marked the personal relationships between those involved.[5] Frankel has pursued, in connection with a single individual, Jean-Baptiste Biot, the question of the extent to which the exigencies of building a career in science at this period might have determined the research choices of a young scientist.[6]

Poisson's career has so far received less attention from historians than it deserves.[7] So far as its early stages are concerned, Frankel's question is as apposite as it is in Biot's case. Indeed, because it took Poisson longer than it did Biot to become a member of the Academy, the influence of his career ambitions on his initial programme of research is even more strikingly apparent. His case is of interest from other points of view as well. Though

[3] Maurice Crosland, *The Society of Arcueil: A View of French Science at the Time of Napoleon I* (London, 1967).

[4] R. Fox, 'The Rise and Fall of Laplacian Physics', *Historical Studies in the Physical Sciences, 4* (1974), 81–136.

[5] I. Grattan-Guinness with J.R. Ravetz, *Joseph Fourier 1768–1830* (Cambridge, Mass., 1972); Grattan-Guinness, 'Mathematical Physics in France, 1800–1835', pp. 349–370 in H.N. Jahnke and M. Otte, eds., *Epistemological and Social Problems of the Sciences in the Early Nineteenth Century* (Dordrecht, 1981); idem., 'Mathematical Physics in France 1800–1840: Knowledge, Activity and Historiography', pp. 95–138 in J.W. Dauben, ed., *Mathematical Perspectives: Essays on Mathematics and Its Historical Development* (New York, 1981); idem., 'Recent Researches in French Mathematical Physics of the Early 19th Century', *Annals of Science, 38* (1981), 663–690.

[6] Eugene Frankel, 'Career-Making in Post-Revolutionary France: The Case of Jean-Baptiste Biot', *British Journal for the History of Science, 11* (1978), 36–48. See also idem., 'J.B. Biot and the Mathematization of Experimental Physics in Napoleonic France', *Historical Studies in the Physical Sciences, 8* (1977), 33–72.

[7] See, however, the recently-published volume marking the bicentenary of his birth, *Siméon-Denis Poisson et la science de son temps*, ed. M. Métivier et al. (Paris, 1981).

his exceptional mathematical talent gained him powerful patronage from a very early stage, even this was not at first sufficient to gain him the ultimate accolade of election to the Institut. Thus the history of his candidature for a position there not only brings out yet again the importance of the patronage system in French science at this period, it also helps to define the limits within which that patronage system operated. In addition, the circumstances surrounding Poisson's eventually becoming a member of the Institut are of considerable interest in themselves, for they bring into particularly sharp focus the enormous gap that was developing at this time between French physics and the old-style qualitative physics still being done elsewhere.

Poisson was born on 21 June 1781 in Pithiviers, fifty miles south of Paris, of humble parentage. At an early age he was sent to live with an uncle, a surgeon, at Fontainebleau, in order to learn the uncle's trade. However, the lack of manual dexterity for which he later became notorious, combined with a tendency, so it is said, to faint at the sight of the simplest operation,[8] soon put a stop to all hopes of that kind. Fortunately, in Fontainebleau, Poisson was also able to attend classes at the local École Centrale. Here he came into contact with mathematics, and at once his true vocation became apparent, one for which the Revolution had recently created new opportunities for advancement. After two years of study, in 1798 Poisson took the entrance examination for the École Polytechnique—an examination based largely on mathematics—and performed brilliantly, being placed first in the *promotion* for that year.[9]

At the École Polytechnique, Poisson's teachers quickly learned that they had a prodigy in their midst. According to the best-known story from this period, only six weeks after Lagrange began his introductory course of lectures, Poisson presented him with an improvement on one of his own theorems. The master was duly impressed. Others among Poisson's professors were evidently impressed by his talents, too, and were prepared to bend the rules for such a pupil. When he proved totally incompetent at geometrical drawing he was allowed to forgo this normally compulsory subject; and at the end of the two-year course, he was exempted from sitting the usual final examinations. Instead, he was named immediately as *répétiteur-adjoint* to assist with the teaching of the school's course in analysis.

Such privileges and such rapid advancement bespeak powerful friends, the most important of whom can be readily identified. In particular, Lagrange, we are told, had opened his house to the young scholar.[10] Another strong supporter was Hachette, one-time deputy

[8] G. Libri, 'Lettres à un américain sur l'état des sciences en France. III: M. Poisson', *Revue des deux mondes*, ser. 2, *23* (1840), 410–437, p. 412. Arago, drawing ostensibly on Poisson's own testimony, does not mention fainting, but in a similar vein stresses Poisson's distaste for the career that had been chosen for him (Arago, 'Poisson', *Oeuvres complètes*, II [Paris, 1854], 593–698.

[9] Arago reports that Poisson was actually ready to take the examination a year earlier, but was prevented by ill health from doing so (Arago, op. cit., p. 598).

[10] Libri, op. cit., p. 416.

lecturer for Monge at the École and subsequently the professor there of descriptive geometry and mechanics. Hachette it was who formally proposed exempting Poisson from the final examination and appointing him instead to the staff of the school. At about this time, he and Poisson collaborated in preparing an addition to a memoir by Monge on analytical geometry.[11] Soon afterwards, on 21 frimaire an IX (i.e. 12 December 1800), Poisson made his initial appearance before the First Class of the Institut. At the tender age of nineteen, he presented a memoir on the calculus of finite differences that was subsequently reported on favourably by a sub-committee consisting of Legendre and Lacroix.[12] If he had not attracted the great man's attention already, this was sufficient to gain him his most powerful supporter of all, Laplace, who engaged him soon afterwards, we are told, to correct the proofs of his famous treatise, his *Mécanique céleste*, as these emerged from the press.

At first, Poisson's position on the staff of the École Polytechnique was a very modest one. As *répétiteur-adjoint*, he was paid a salary no greater than the allowance paid to a senior student, a *chef de brigade*.[13] Nevertheless, he had a toe in the door, and in the circles that counted in French science he had been noted as a rising star. Furthermore, the base he had acquired was in Paris: unlike such talented contemporaries as Biot, Malus and Fresnel, Poisson was never required to serve a scientific 'apprenticeship' in the provinces. Yet the patronage he now attracted, which saw him quickly promoted to the rank of acting professor at the École when Fourier went to Grenoble as *Préfet* of Isère, and then appointed on 20 November 1805 as *professeur titulaire* when Fourier eventually formally resigned his chair, was no mere dispensing of largesse by the aristocrats of French science to their favourite of the moment. Not even the most powerful of patrons could have procured promotions like these for someone lacking in talent, and nor would these particular patrons have tried. Poisson's advancement came about primarily because he displayed such prodigious promise as a mathematician, a point that is nicely captured in Libri's report of Laplace's attitude on the matter. To Poisson's repeated thanks for favours

[11] Ibid., pp. 415, 417. Their paper was published in *J. École Polytechnique, 4* (11th cahier), an X, 170–172.

[12] Académie des Sciences, *Procès-verbaux des séances de l'Académie tenues depuis la fondation jusqu'au mois d'août, 1835* (10 vols., Hendaye, 1910-22), II, 277, 283; *J. École Polytechnique, 4* (11th cahier), an X, 173–181.

[13] A Fourcy, *Histoire de L'École Polytechnique* (Paris, 1828), p. 272; Libri, op. cit., p. 418. Poisson was living at this time with Jean Guillaume Garnier, who ran a preparatory school for students intending to sit the entrance examination for the École Polytechnique in addition to serving as *suppléant* for Fourier, absent in Egypt with Bonaparte's famous expedition, at the École itself. Poisson did some of the teaching at Garnier's school in return for board and lodging (cf. Fourcy, op. cit., p. 154, and Garnier's autobiographical notes, published in A. Quetelet, 'Notice sur Jean Guillaume Garnier (1766-1840)', *Annuaire de L'Acad. Roy. de Bruxelles*, 1841, pp. 161–207; p. 171. Garnier claims to have been acting also as *professeur adjoint* to Lagrange during the period 1798–1802, but Fourcy does not mention this, and the story seems impossible anyway since Lagrange retired in 1799 and was replaced by Lacroix). Poisson is reported also to have been receiving financial support during these years from the well known social philosopher Henri de Saint-Simon (Crosland, *The Society of Arcueil*, p. 91).

granted him, Libri tells us, Laplace was always content merely to reply: 'Truly, that was your due'.[14] In addition—something that ought not to be forgotten—by the time Poisson's position at the École Polytechnique was made secure, he had already proved himself to be a first-rate and conscientious teacher, someone to whom the students looked for inspiration and guidance. In other words, patronage or no, in appointing him the school made an excellent choice.

For Poisson, additional appointments quickly accumulated in the traditional French manner: acting professor at the Collège de France deputizing for Biot during the latter's temporary absence from Paris; *géomètre-adjoint* at the Bureau des Longitudes; professor at the Paris Faculté des Sciences following Napoleon's reform of the university system in 1808. He became an active member of the Société Philomathique, and from 1807 one of the editors of the Society's journal. At the same time, he threw himself into his research, publishing a number of papers in the École Polytechnique's two series, its *Journal* and *Correspondance*, and in the *Bulletin* of the Société Philomathique; somewhat later, he presented to the Academy major papers on problems in higher analysis and on the theory of sound. These were as well received as his first papers had been. Meanwhile he began dispensing patronage himself, being instrumental in Arago's appointment at the age of twenty as a secretary at the Observatory.[15]

Long before Poisson submitted his new papers to the Academy, however, that body had given a clear indication that he was destined for still higher honours. On 14 germinal an XI (i.e. 4 April 1801), an election was held to fill a vacancy that had arisen in the geometry section following the appointment of Delambre as one of the two *Secrétaries perpétuels* of the organization. This was, in fact, the election in which Biot, Poisson's senior by several years, became a full member of the Institut by defeating the statistician and economist Duvillard in a vote the closeness of which (28 votes to 23, plus one vote to a third candidate) itself indicates that those renowned wielders of patronage, Laplace and Lagrange—both of whom were strongly supporting Biot's candidature on this occasion—had much less than full control of elections to the Academy, even when the vacancy that was being filled was in their own mathematics section within the organization. What is of particular interest in the present context is that the Academy's nominating committee, headed on this occasion by Lagrange, felt it worthwhile including the 21-year-old Poisson's name on the list of candidates to be placed in the ballot. Doubtless no-one expected him to be elected ahead of Biot—indeed, when the votes were counted, Poisson had

[14] Ibid.: 'Véritablement, cela vous était dû'. Garnier reports with considerable bitterness how, though Fourier had promised him the succession to his chair, Laplace had intervened to have Poisson appointed to the position instead (Quetelet, op. cit., p. 172).

[15] *Procès-verbaux*, III, 211, 222, 303–305, 563, 594–597. Arago, *History of My Youth*, trans. Baden Powell (London, 1855), p. 19.

not received a single one—but by the mere fact of his nomination, notice was given that already he was being seen as potential Academy material.[16]

Poisson's career was thus off to a very promising start. Furthermore, the group of senior mathematicians which had been backing Biot was able, now that he was safely elected, to swing its support behind Poisson instead. Biot himself, secure in his new position of power, was another supporter. Yet despite this array of influential friends, a further eleven years passed before Poisson actually became a member of the Institut. According to Costabel, author of the entry on Poisson in the *Dictionary of Scientific Biography*, Poisson's election constituted, even then, an unjustifiably rapid rise, being supported by an inappropriately modest body of published work. Presumably, therefore, Costabel regards this eleven-year hiatus as no hiatus at all.[17] In this, however, he appears to be applying twentieth-century criteria for Academy membership to a very different early nineteenth-century situation. When we compare Poisson's case with those of others elected to the Academy in the Napoleonic period, quite the reverse of what Costabel alleges appears to be true. Several of those appointed were younger at the time of their election than Poisson was when he was eventually admitted, and they had much less substantial publication records behind them. Certainly this is how it appeared to Arago, who even felt a need to provide some explanation for the 'late date' of Poisson's admission. The delay, he said, was a mere accident of fate. The Academy held scrupulously to the different specializations represented by its various sections. A place had been marked out for Poisson in the geometry section, but no vacancies appeared here for many years. 'Finally', Arago says, 'made impatient by seeing a man as eminent as Poisson outside its bosom, the majority of the Academy bent the rules and named him to the physics section'.[18]

The surviving documents show that Arago's account is accurate so far as it goes, but he does not tell the whole story by any means—indeed, he probably did not know it all. Poisson himself tried, we discover, though with little success at first, to take a hand in events by directing his research into new channels that seemed more likely to bring about his election. There is every reason, moreover, to suppose that in this he was acting on the advice of powerful friends who were attempting at the same time to manipulate affairs within the Academy in ways that would strengthen his candidature. It does not follow, however, that the quality of Poisson's work was adversely affected thereby: on the contrary, some of the papers he wrote at this period stand among his most enduring scientific achievements.

The years immediately after 1800 saw Poisson consolidating his

[16] Ibid., II, 640, 642.
[17] Pierre Costabel, art. 'Poisson' in *Dictionary of Scientific Biography*, XV, 480–490.
[18] Arago, op. cit., pp. 603–604.

reputation as a *géomètre*, at the same time as he built himself a career in the undisputed centre of French science, in Paris. A number of short publications on a variety of mathematical topics was followed by a sequence of major papers presented to the First Class of the Institut. The first two of these were read on 16 floréal and 21 prairial an XIII (i.e. 6 May and 10 June 1805). They were both reported on by Lacroix in very favourable terms on 13 January 1806, his fellow commissioners being Laplace in one case and Lagrange in the other. Poisson's work, the Academy was told on this occasion, 'presenting a considerable number of new results on a very important subject, and rendering uniform the solutions of the questions it treats, appears very worthy of the Class's approval'.[19] Both papers were recommended for publication in the Academy's subsidiary journal, the so-called *Savans étrangers* or *Mémoires présentés à l'Institut . . . par divers savans et lus dans ses assemblées*. (In fact, despite the consistency with which such recommendations were made with regard to Poisson's work, none of the papers he presented to the Academy prior to his himself becoming a member ever appeared in this series, even though volumes appeared in 1806 and 1811 in which they might well have been included. Poisson arranged for his papers to be published instead, and much more promptly than the Academy could have achieved, in the *Journal* of the École Polytechnique. In some cases, summaries appeared even more promptly still in the *Bulletin* of the Société Philomathique.[20])

Eighteen months later, on 24 August 1807, Poisson read a long paper to the Academy on the theory of sound.[21] On 5 October of the same year, Biot reported on behalf of his fellow commissioners Lagrange, Monge and Lacroix that 'this work is entirely worthy of the great proofs of talent that M. Poisson has already given on several occasions'. Once again publication in the *Savans étrangers* was recommended, though again to no avail.[22]

Shortly before Poisson read this paper, a further vacancy had at last appeared in the Academy, as a result of the death of the astronomer Lalande. Unfortunately for Poisson, it was in the astronomy rather than the geometry section, and he had established no credentials at all as an astronomer. Yet some hope remained, because at its meeting on 25 May 1807 the Academy had decided, on the advice of the remaining members of the astronomy section (Burckhardt, Bouvard, Cassini IV, Lefrançois-Lalande and Messier), that the vacant position should not be filled immediately. The reasoning that lay behind this decision is not recorded, but it would appear that, with no obvious replacement for Lalande in

[19] *Procès-verbaux*, III, 305.

[20] See n.15 above. Lists of Poisson's publications are given in Royal Society of London, *Catalogue of Scientific Papers (1800–1863)* (London, 1867–72), IV, 964–969; in Arago, op. cit. (n. 8), pp. 672–689; and in Métivier et al., op. cit. (n. 7), pp. 212–225. The last of these supersedes the other two. Poisson's experience with the *Savans étrangers* was by no means unusual. He was more fortunate than some in having such ready access to alternative journals in which his work could be published.

[21] See n.1 above.

[22] *Procès-verbaux*, III, 563, 594–597.

sight, it was felt appropriate to give aspirants for the position an opportunity to stake their claims more convincingly. Initially it was resolved to reconsider the situation after six months; on 7 December 1807, however, it was decided to defer the election again, after which no further action appears to have been taken for almost two years.[23]

With no sign of any vacancy arising in the geometry section, it is possible that Poisson's backers engineered this delay in order to create an opportunity for him; later reports certainly suggest that Laplace in particular had hopes of obtaining the position for him. On the other hand, the sparseness of the list of candidates whose names were brought forward when the election was eventually held, and their generally very modest records of achievement even then, suggests that there were objective reasons for not filling the position at first. In any event, it is difficult not to see Poisson's suddenly taking up research on problems of celestial mechanics at precisely this time as a response to the situation in the Academy. Given Poisson's ambitions vis-à-vis the Academy, this was undoubtedly a logical move on his part. If he aspired to Lalande's place, he would have to earn it by means of his mathematical skills, since he had neither talent nor experience as an observational astronomer. At the same time, work in this area would stand him in equally good stead in any future election to the geometry section.

The new orientation of Poisson's work was announced modestly enough when, on 11 April 1808, he presented to the Academy a copy of the new edition he had prepared of Clairaut's classic text, *Théorie de la figure de la terre*.[24] Soon afterwards, on 20 June 1808, he read a very important memoir to the Academy on the secular inequalities in the planetary motions, a paper in which he established the long-term stability of the solar system on a much more satisfactory analytical footing than had previously been done. The Academy's commissioners on this occasion were Lagrange, Laplace and Biot. Biot presented their report on 16 August: 'The Class', he declared in fulsome style, 'will note with interest the constant progress of the talent of this young geometer; it will see with pleasure that he appears to be directing his research towards the theory of the system of the world, and it should engage him to continue to apply himself to this kind of analysis which is the lamp of astronomy and its chief instrument'.[25]

On 20 March 1809, Poisson read another memoir, on the rotation of

[23] Ibid., III, 532, 623, 626. Though it was certainly within the Academy's rules not to proceed at once to an election (cf. *Réglement intérieur de la Classe des Sciences Mathématiques et Physiques de l'Institut National* [Paris, an XI], p. 7), one nevertheless wonders at the legality of the academicians' refraining for so long from filling a vacancy in their midst. Perhaps the Academy's officers also had doubts on the matter, for on both occasions the unanimity of the decision was carefully recorded.

[24] A. Clairaut, *Théorie de la figure de la terre* (2nd ed., Paris, 1808). Cf. *Procès-verbaux*, IV, 45. Poisson's contribution to the new edition amounted to no more than a short preface and the elimination of typographical errors from the first edition.

[25] *Procès-verbaux*, IV, 78, 89–92; cf. Poisson, 'Sur les inégalités séculaires des moyens mouvements des planètes', *J. École Polytechnique*, *8* (1809), 1–56.

the Earth. Laplace and Lacroix presented a favourable report only three weeks later.[26] Yet another paper followed on 23 October 1809, on a very important topic in relation to problems of celestial mechanics, the method of variation of arbitrary constants. Laplace and Lacroix were again appointed commissioners, reporting two months afterwards that 'the very ingenious manner in which M. Poisson treats the important questions that are the object of this memoir ought to earn him the same praise as the researches he has presented previously'.[27]

Between them, these three papers were judged sufficiently important contributions to astronomy for Poisson to be awarded the Academy's Lalande Prize in 1810, in succession to Gauss.[28] They also included a contribution of lasting value to mathematical science, the so-called 'Poisson brackets', and furthermore they directly inspired some important new work by Lagrange on the same topic. They were not, however, sufficient to win Poisson the coveted membership of the Institut. Even before he had presented the third of them, the Academy had at last proceeded, on 18 September 1809, to fill the vacancy in the astronomy section. The successful candidate was Arago, Poisson's one-time student at the École Polytechnique and still at this point only twenty-three years old.[29]

Arago's election at such an early age was, in fact, an astonishing episode in the history of the Academy. At the time of Lalande's death two and a half years earlier, Arago was merely a bright young man with a promising future. He had collaborated with Biot in some useful research and had presented one fairly short memoir of his own to the Academy, on the speed of light. Had the Academy proceeded as it normally did and chosen Lalande's successor when the vacancy first appeared, he would surely have stood little chance of being elected. Shortly after reading his paper, however, Arago had travelled to Spain, at first in company once again with Biot, in order to extend the measurement of the meridian through Paris southwards to the Balearic Islands. A number of reports from Biot of the expedition's work kept the younger man's name before the Academy for a time, but when war broke out anew between France and Spain, he was stranded. At one point he was even though to have perished, but eventually, after some hair-raising adventures in Spain and North Africa, he managed to make his way back to France, still bearing with him the record of his geodetic observations.

[26] *Procès-verbaux*, IV, 181, 190–192; cf. Poisson, 'Sur le mouvement de rotation de la terre', *J. École Polytechnique*, *8* (1809), 198–218.

[27] *Procès-verbaux*, IV, 266, 290–291; cf. Poisson, 'Sur la variation des constantes arbitraires dans les questions de mécanique', *J. École Polytechnique*, *8* (1809), 266–344.

[28] *Procès-verbaux*, IV, 405. This prize had been endowed by Lalande towards the end of his life, to be awarded each year for 'the most curious observation or the most useful memoir for the progress of astronomy' during the previous twelve months (resident members of the Institut being ineligible for the award).

[29] Ibid., IV, 253.

The Academy seems to have been swept collectively off its feet by the euphoria surrounding Arago's return, being encouraged in this by his supporters, especially, on this occasion, Biot.[30] On 17 July 1809, Biot announced to the Academy Arago's safe return to France. At the meeting on 28 August, Arago was welcomed back in person, and a special invitation was extended to him to take part in the proceedings. A week later, he read a report on his work in Spain, while Delambre presented a much-delayed report prepared in conjunction with Laplace on the by now three-year-old memoir on the speed of light. At the same meeting, the Academy resolved to proceed at last to fill Lalande's place. The next week, the astronomy section brought forward three nominations for the position, their stated order of preference being Arago, Nouet and then Mathieu. On 18 September, the election was held.[31].

If Poisson and his backers in the geometry section had held hopes of having him elected as an astronomer, Arago's reappearance in such romantic circumstances put paid to these. Yet Poisson's chances must always have been slight, anyway. The astronomy section had always had a strong orientation towards observational work, and the failure of its remaining members even to include Poisson's name on the list of candidates for the election, despite the two excellent astronomically-oriented memoirs he had recently presented, shows how jealous they were of the boundaries of their discipline, and how effectively they could resist the blandishments even of a Laplace.

In the last few days before the election, urgent negotiations took place behind the scenes. According to one report, Biot won over Lagrange to Arago's cause by convincing him that a renowned observational astronomer such as Lalande ought to be replaced by someone also skilled as an observer.[32] Laplace, however, remained strongly committed to Poisson's cause and continued to work for his election even after all seemed lost.

[30] Gratten-Guinness had laid considerable stress on the disputes that later erupted between Biot and Arago (op. cit., n. 5). He sees the first signs of these in a misunderstanding that Arago described many years later over their 1806 joint paper on atmospheric refraction (Arago, *Oeuvres complètes*, XI (Paris, 1859), 702–703). However, either any tension aroused at that stage must soon have dissipated, or else Arago succeeded very well in bottling up his resentment, because during and after the Spanish expedition the two men seem to have worked together in close and friendly co-operation.

[31] Crosland has suggested that Lalande's place was not filled for so long because it had been earmarked for the absent Arago from the start (Crosland, *Society of Arcueil*, pp. 163–165). An indication that Arago's claim to the position was not as clear-cut as that is provided by the terms in which the Academy decided to award the 1808 Lalande Prize to his rival, Mathieu. To be sure, the report adopted by the Academy leaves no doubt that Arago's merits were regarded as superior to Mathieu's; but it also reveals that his work was not seen as sufficiently outstanding to prevent Mathieu's nevertheless being given the prize. 'In the absence of new observations and important works', we read, it had been decided to award the prize 'as encouragement to M. Mathieu for the work to which he has devoted himself in conjunction with M Biot, to determine the length of the pendulum. The Commission points out that it would have proposed giving the award to M. Arago, had it not judged him sufficiently encouraged by his nomination to the place of *adjoint* at the Bureau des Longitudes' (*Procès-verbaux*, IV, 149).

[32] F. Lefort, 'Un savant chrétien: J.-B. Biot', *Le Correspondant*, *72* (1867), 955–995; p. 965. Though Lagrange may have paid lip service to Biot's argument, it does not automatically follow that he cast his vote for Arago in what was, after all, a secret ballot.

This is, at least, the implication of the story later told by Arago that before the election took place, Laplace tried to have it postponed yet again, until such time as a vacancy should also appear in the geometry section that would allow Poisson and Arago to be admitted to the Academy together.[33] According to Arago, Laplace even suggested to him at one point that he might himself write to the Academy declining to stand for election until there should also be a place available for Poisson. One can well imagine how the Academy would have responded to such a piece of presumption! In the event, Arago had an overwhelming victory, gaining no fewer than 47 votes against one for Nouet and four cast for Poisson in a last-minute write-in campaign. Both Arago and Biot's nineteenth-century biographer, Lefort, state that, in the end, even Laplace cast his vote for Arago. It was a secret ballot, however, and one cannot help but suspect that whatever Laplace might have said afterwards, his was actually one of the votes that went to Poisson. Almost certainly the others would also have come from members of the geometry section. If so, Arago's election would stand as a resounding defeat of an attempt by the Academy's mathematicians to instal one of their own in what had been until then a section of the organization devoted primarily to empirical work.[34]

With the astronomy section thus apparently closed to him, Poisson would have had to pin his hopes on election to one of the Academy's other sections instead. Along with its sections for *géométrie* and astronomy, the Academy included in its mathematical division sections for mechanics, geography and navigation, and general physics. A vacancy had appeared in the mechanics section in 1807, but Poisson had not been nominated for it. A glance at the pattern of membership of this group tells us why: the section had a strong engineering orientation, and in the 1807 election a leading naval engineer, Berthoud, was replaced by another, Sané. The three-member section for geography and navigation was likewise closed to Poisson, being reserved, it seems, for the navy's hydrographers and explorers. Elections were held in this section in 1810 and 1811, but Poisson was not considered on either occasion. Two of the navy's most senior hydrographers were appointed instead, Beautemps-Beaupré in 1810 and Rossel in 1811.[35]

There remained the section for *physique générale*. Yet Poisson could not have felt optimistic about being elected to that, either. Though by no

[33] Arago, *History of My Youth*, p. 106.

[34] Louis L. Bucciarelli and Nancy Dworsky (*Sophie Germain: An Essay in the History of the Theory of Elasticity* [Dordrecht, 1980], p. 130) offer as their guess, though without giving any reasons for their choice, that Poisson's four votes might have come from Laplace, Berthollet, Biot and Gay-Lussac. This, however, is to neglect both Lefort's claim that Biot was an active campaigner on Arago's behalf and the manifold evidence available of Poisson's strong links with other members of the geometry section besides Laplace, most notably with Lagrange and Lacroix. If we accept Lefort's statement that, in addition to Biot, two other mathematician members of the Academy, Delambre and Legendre, were Arago supporters, we are left with Laplace, Lagrange, Lacroix and the 79-year-old Bossut as the four members of the First Class most likely to have cast their votes for Poisson on this occasion.

[35] *Index biographique de l'Académie des Sciences du 22 décembre 1666 au 1er octobre 1978* (Paris, 1979).

means as close-knit a group as the astronomers, its members shared a strong commitment to experimental methods of inquiry for which, as we have seen, Poisson had no talent at all. Elections to the section had been held in 1806, when Gay-Lussac replaced Brisson, and 1807, when the aging balloonist Montgolfier was appointed to Coulomb's place, and Poisson had not been considered on either occasion.[36] Nor was his name put forward for the election which followed Montgolfier's death in 1810, a year after the debacle over the election to the astronomy section. On this occasion Poisson's supporters had another candidate in the field who was Poisson's senior by several years, Malus. Though very competent at mathematics, Malus did not share Poisson's genius for the subject. However, he also had an outstanding record as an experimenter, especially in optics where his work had provided strong support for Laplace's theoretical position,[37] and this proved more than sufficient to win him a place in the section ahead of the engineer Girard.

As for Poisson, he seemed to have no option but to wait for a vacancy to occur in either the geometry section or, with much less certain hopes of success, in the one for astronomy.[38] His position is reflected in his published output during the couple of years following his defeat by Arago, which included the third of his celestial mechanics papers, a two-volume *Traité de mécanique* based upon his lectures at the École Polytechnique,[39] and some

[36] The other members of the section at this time were then well known balloonist and public lecturer in experimental physics J.-A.-C. Charles, the navigator turned expert on optical instruments A.-M. Rochon, the professor of experimental physics at the Collège de France, Louis Lefévre-Gineau, and the hydrographer and one-time balloonist Pierre Levêque.

[37] Eugene Frankel, 'Corpuscular Optics and the Wave Theory of Light: The Science and Politics of a Revolution in Physics', *Social Studies of Science*, *6* (1976), 141–184; André Chappert, *Étienne-Louis Malus (1775–1812) et la théorie corpusculaire de la lumière* (Paris, 1977).

[38] Bucciarelli and Dworsky (op. cit. [n. 34], pp. 35–38) suggest that Laplace was scheming to have Poisson elected to the 'general physics' section as early as the first part of 1809, that is, several months before Arago captured the vacant astronomy position. This is, at least, their suggested explanation for the Academy's announcement in April of that year of a special prize competition on the mathematical theory of the vibrations of elastic solids which they interpret as 'part of Laplace's efforts to be helpful to Poisson and to turn his attention to terrestrial problems of physics that were of interest to Laplace himself'. Three points, however, tell against this suggestion. First, Poisson was hard at work at precisely this time on the final two parts of his trilogy of papers on problems in celestial mechanics, and we have Libri's testimony concerning Poisson's work habits that he always confined his attention exclusively to one question at a time (Libri, op. cit., [n. 8), p. 432). Since his chances of election to the Academy must at that point have seemed much brighter in astronomy than in physics, it appears unlikely that he would have diverted his attention at this stage to the physics-oriented topic of the prize competition. Secondly, Laplace already had a favoured candidate, Malus, for the next opening in the physics section. Any plans he might have had for Poisson in this connection must, therefore, have been very long-range indeed. Finally, as Bucciarelli and Dworsky themselves admit, there is no evidence that Poisson took up the question posed by the prize competition until several years after this. If he did take it up once his celestial mechanics memoirs were completed, the work appears to have come to nothing at that time.

[39] Among Ampère's papers at the Archives de l'Académie des Sciences, Paris (box 26, *chemise* 392) is a voluminous MS headed 'Cours de Mécanique fait par M^{r}. Poisson en 1812. 1er Division. École Royale Polytechnique' and signed at the end 'Le 20 fevrier 1816 pour copie conforme a mes régistres. Le Brun', which includes frequent cross-references to Poisson's *Traité*. Grattan-Guinness has recently pointed out that the *Traité* is in fact a re-working of a 500-page set of lecture notes published in 1810 under the title *Cours de mécanique* (Grattan-Guinness, *Ann. Sci.*, *38* [1981], 663-690; p. 690).

minor papers on the development of series and the integration of partial differential equations.

Suddenly, however, towards the end of 1811, the situation changed once more. Malus, so recently elected to the physics section, was found to have consumption. His condition quickly deteriorated, and it came to be recognized that he would not recover. In December, only a few weeks after Malus attended what was, in fact, to be his last meeting at the Academy, the mathematics section was called upon to propose a topic for the next prize competition on a mathematical subject. In the light of what happened subsequently, we can see that it made its recommendation with Poisson specifically in mind. At the Academy's annual public meeting, held on 6 January 1812, the competition was officially announced in the following terms:

> To determine by calculation and to confirm by experiment the manner in which electricity is distributed at the surface of electrical bodies considered either in isolation or in the presence of each other—for example, at the surface of two electrified spheres in the presence of each other. In order to simplify the problem, the Class asks only for an examination of cases where the electricity spread on each surface remains always of the same kind.

In line with the usual practice with such competitions, entries would close on 1 October of the following year (i.e. 1813 in this case) to enable the judges to announce the result at the Academy's public meeting two years after the prize was first announced (i.e. in January 1814).[40]

The terms in which the competition was set bear such a remarkable resemblance to the problems treated soon afterwards by Poisson in his published papers on electricity that it cannot be doubted, not only that the competition was set for his benefit by people who were well aware of what he was doing, but also that he had already completed much of the analysis for it at the time it was set.[41] It is possible, therefore, that he had taken up the investigation some time before. Yet it seems more than coincidental that Poisson should have turned his attention to questions of physics at precisely this time. With the place that had once been occupied by the great Coulomb appearing likely to fall vacant again in the fairly near

[40] *Procès-verbaux*, IV, 560, 562–3. The announcement of this competition has seldom been noted by historians, presumably chiefly because the prize was never awarded. The full text of the competition topic is given in Archives de l'Académie des Sciences, Paris, packet for the *séance* of 6 January 1812, as follows:

> Déterminer par le calcul, et confirmer par l'expérience, la manière dont l'électricité se distribue à la surface des corps électriques, et considérés soit isolément, soit en présence les uns des autres, par exemple, à la surface de deux sphères électrisées, et en présence l'une de l'autre. Pour simplifier le problème, la Classe ne demande que l'examen des cas où l'électricité répandue sur chaque surface reste toujours de la même nature.

The renewal of the competition on elastic plate vibrations was announced at the same meeting.

[41] The published version of the competition topic was, in the detail it provided, even more closely aligned with the direction that Poisson's work was taking than that approved initially, which had merely specified a prize for research on 'the laws of electrical equilibrium of two spheres in the presence of each other'. (*Procès-verbaux*, IV, 562–3).

future, what more suitable task could he have found for his next major piece of research than to establish a firm theoretical basis for Coulomb's own data? It is surely significant, too, that the prize competition was set not by the physics section but by the mathematics section of the Academy, that is, by Poisson's leading supporters—Laplace, Lagrange, Lacroix, Legendre and Biot. (Bossut, the sixth member of the mathematics section, was an old man and no longer active in these day-to-day academic affairs.) What better means of demonstrating the suitability of their candidate for Malus' position, once the inevitable occurred and the position became vacant, than the award of an Academy prize on a subject so manifestly appropriate to the physics section as electricity?

Malus' health deteriorated much more rapidly, however, than anyone appears to have envisaged, and his death was announced at a meeting of the Academy only seven weeks later, on 24 February 1812. Thereafter, events moved rapidly as various aspirants for his position sought to present their candidature in the best possible light. Poisson had, of course, not had time in the circumstances to complete the investigation he had planned, but nevertheless by 9 March he had a paper ready entitled 'Distribution de l'électricité' that he presented to the Academy. Significantly, no member of the physics section was included in the group appointed to prepare a report on this; instead, the commissioners chosen were Lagrange, the mineralogist Haüy (who, on account of his work on pyroelectric crystals, was the Academy's leading authority on electricity at the time) and Biot. At the same meeting, Girard, defeated by Malus in the election eighteen months before, presented a paper on some questions on hydraulic engineering for which Monge, Lazare Carnot and Gay-Lussac were appointed commissioners. Biot and Haüy brought forward a very favourable report on a paper on 'Observations on radiant caloric' that another hopeful, De la Roche, had read to the Academy some time before. Also at this meeting the Academy received a letter from an ambitious but little known inventor named Trouville who sought unsuccessfully to have his name included on the list of candidates at the forthcoming election.[42]

Under normal circumstances, the reports commissioned on papers presented by non-members to the Academy took anything from a few weeks to several years to be presented. With an election in the offing, however, matters proceeded rather more rapidly. By the following meeting, on 16 March, the commissioners assigned to the papers of both Poisson and Girard had their reports ready. We do not know what Biot had to say on behalf of the group appointed to peruse Poisson's paper, because, contrary to the normal practice, he failed to lodge a copy of the report with the secretary. (Perhaps it was felt inappropriate that he should do so once

[42] On Trouville, cf. 'Extraite d'un memoire sur la manière d'employer le syphon pour élever l'eau dans le machine du C. Trouville, par le C. Prony', *Bull. Soc. Philomathique*, *2*, 92–94 (in No. 36, Ventose, an VIII). However, this gives no details at all about Trouville himself.

Poisson had been elected to the Academy; perhaps, indeed, a formal written version was never actually prepared.) If Poisson's work was so well advanced at this stage as it appears to have been, however, the commissioners' report would surely have been very favourable. Girard fared less well; the commissioners commenting on his paper were muted in their praise and suggested that he should be encouraged to extend his experiments over a wider range of temperatures than he had managed so far. Also at this meeting, the physics section brought forward its recommendations as to whose names should go into the ballot. Poisson headed a list of no fewer than seven candidates.

The election was held a further week later, on 23 March. Prior to the votes being cast, however, a very favourable report was read to the Academy by the chemist Vauquelin on a work that had been submitted some time previously by yet another of the listed candidates, Hassenfratz. A copy of the work in question, the author's *L'art de traiter les minerais de fer*, had been presented to the Academy over eighteen months before. Presumably, Hassenfratz, professor of physics at the École Polytechnique and an unsuccessful candidate at several previous elections for a place in the physics section, did not have a paper ready that he could submit at short notice, and pressing for a report on his book was the best that he could do instead. The effort, however, was in vain, for he did not receive a single vote. Girard, with 24 votes, again polled well—he was, in fact, to succeed at the next election to the section, in 1815—but Poisson, with 30 votes out of the 57 cast, obtained the necessary absolute majority on the first ballot and was declared elected.[43]

When the dust settled, the Academy found itself with some unfinished business on its hands in the form of the prize competition it had announced shortly before on the distribution of electricity on bodies. As a member of the First Class, Poisson was no longer eligible to enter this contest that had been tailor-made for him. Yet he was close to having a complete solution to the problem that had been set, and so he continued to work on it. A few months after his election, on 3 August 1812, he read a second memoir on the subject. A third followed another year later, on 6 September 1813.[44]

[43] For the events surrounding the election, see *Procès-verbaux*, V. 20–40. Bertrand provides some additional colourful details about the lobbying that took place beforehand (*Revue des deux mondes*, *137*, 1896, 287–288). (I owe this and several other references to Jim Hofmann). Costabel (op. cit. [n. 17]) notes that Poisson's election was confirmed by Napoleon within three weeks. He sees this promptness as a sign that 'the authorities had not forgotten the acquiescent attitude that [Poisson] had adopted in 1804' in heading off a rebellion among the students at the École Polytechnique against the proclamation of the Empire. However, a delay of only three weeks was by no means exceptional in this regard; on the contrary, this was about the average time it took for approval to be granted.

[44] In the published version, the first two of Poisson's presentations were combined into a single paper, and in the absence of the original manuscripts it is impossible to determine how much of the work Poisson had actually completed at the time of the election. Further to confuse the issue, the date on which the first part was read is wrongly given as 9 May 1812 rather than 9 March, not only in the 'official' publication in the Academy's *Mémoires* but also in the various summaries that were published earlier. This was no typographical error, therefore, but a deliberate attempt to mislead, probably

Had Poisson been permitted to submit these papers as an entry to the prize competition, they would have constituted about as complete an answer to the problem set as could reasonably have been expected. In effect, therefore, they pre-empted the contest, so that those in charge must have been hoping that no entries would be forthcoming.[45] In this, however, they were to be disappointed, for a long Latin essay arrived a few days before the closing date set for the competition, the author of which was later discovered to be one François Joseph Gardini, 'docteur en Médecine, professeur de physique, Mathematique et histoire naturelle, Membre de plusieurs Sociétés savantes' and at the time a teacher at Sandamiano in the Napoleonic *département* of Marengo, today a part of Italy. Indeed, they were threatened with a second entry, since another sealed note arrived just before the competition closed, but without the entry itself that should have accompanied it. The would-be entrant was another schoolteacher of Italian origin, Vincent-Frédéric Da Olmi, professor of natural science at the *collège* at Sorèze in the south of France. One can only presume that his essay went astray in the mails.[46]

One glance at Gardini's essay would have been sufficient to show that it did not meet the Academy's expectations for the competition. Nevertheless, there were formalities that had to be observed. A commission of judges was appointed comprising Poisson, Charles, Laplace, Legendre and Gay-Lussac. On 27 December 1813 it reported that the prize should not be awarded. This advice was accepted and, after what must surely have been a very brief discussion, it was decided to withdraw the prize rather than to renew the competition for a further period.[47]

Prima facie, therefore, the contest came to a disappointing end. Yet in fact it had served its purpose so effectively that, together with the events surrounding it, it marks a watershed in the history of physics. Poisson was the first person whose scientific record was exclusively mathematical in orientation to be elected to the physics section of the Academy. To be sure, men who were competent mathematicians—Coulomb and Malus in particular—had previously been elected to the section; but they had also done outstanding experimental work, and in every case this appears to have been the principal consideration in the minds of those who had voted for them. Poisson, however, had no record at all as an experimenter—worse, he was known to be hopelessly lacking in the kinds of skills required

perpetrated in order to lend an air of legitimacy to the publication in the *Mémoires* of a paper, part of which, having been communicated by someone who was not at the time a member of the Academy, should by rights have appeared in the *Savans étrangers* instead.

[45] Poisson was later criticized by some of his colleagues in the Academy for presenting a paper on vibrating elastic surfaces while the long-running Academy competition on that subject was still current (Bucciarelli and Dworsky, op. cit. [n. 34], p. 65). In the present case, however, no similar criticism was offered, and there can surely have been no question of impropriety about his actions, the contest having so manifestly been set for his benefit in the first place.

[46] The two once-sealed notes, together with Gardini's essay, are preserved in the packet of papers dealing with this competition at the Archives de l'Académie des Sciences, Paris.

[47] *Procès-verbaux*, V, 250, 292.

for such work. Furthermore, as we have noted already in relation to Poisson's earlier, unsuccessful bid for a place in the astronomy section, when the Academy came to electing new members, the disciplinary boundaries between its various sections were guarded jealously. The procedure whereby the list of candidates for each election was drawn up by the section in which the vacancy had appeared provided the means for this because, in effect, it gave to the remaining members of the section the right to decide of each potential candidate whether or not his work fell within the ambit of the discipline in question. In the election itself, the section's stated order of preference among the candidates listed might be overturned by the Academy at large, but, as Poisson and his backers were made to recognize anew in the astronomy election in 1809, someone whose name did not appear on the list had little chance.[48] As we have seen, Poisson had never previously been listed as a candidate for a place in the physics section. It is, therefore, particularly striking that for the election to replace Malus, the members of the physics section itself ranked him first among the candidates whose names they brought forward. Their doing so surely testifies to the powerful impression that Poisson's success in subjecting electricity to mathematical analysis had made upon them, because this was the only work he had done since the previous election that could possibly have led to such a dramatic re-evaluation of his merits as a physicist. Their action amounted to a public acknowledgement, forced upon them by Poisson's achievements in electrical theory, that henceforth, mathematics would have a central role to play in general physics, a discipline which had until then been based almost wholly on experiment.

Just as Poisson had never previously been considered for a place in the physics section of the Academy, so, too, he had never previously been invited to join the experimentally orientated Société d'Arcueil, even though his patron Laplace was one of the prime movers behind this. The fact that he now also became a member of this latter group tells against Arago's allegation that his election to the physics section of the Academy was merely a device to enable the Academy to recognize his skills as a mathematician. It confirms that Poisson really had, quite suddenly, achieved standing as a physicist and not merely a mathematician at this time; and for this to have happened, physics itself must have come to be conceived of in a rather different way, at least among the leaders of the French scientific community of the period.

The nature of the change involved can be brought out in striking if perhaps exaggerated form by comparing Poisson's two memoirs on electricity with the entry submitted by Gardini for the Academy's prize competition, because Gardini's essay is an archetypical example of the

[48] The Academy's electoral procedures are described and their significance in maintaining disciplinary boundaries noted by Crosland, 'The French Academy of Sciences in the Nineteenth Century', *Minerva, 16*, 1978, 73–102.

approach of the previous generation to questions of physics. Gardini had learned his physics over forty years before from Giambatista Beccaria, one of the leading Continental contributors to the development of Franklinist electrical theory.[49] He had subsequently published a number of substantial works on electricity, including prize-winning essays submitted to competitions set by learned Academies at Lyon (1779, 1782), Mantua (1788) and Brussels (1809). His speciality was the role of electricity in the animal and vegetable kingdoms, but he had also attempted a more general study of his paper for the Mantua contest, subsequently published as *De electrici ignis natura dissertatio* (Mantua, 1792). In other words, Gardini had been well regarded as an electrical investigator for many years prior to his submitting an entry to the Paris Academy's prize competition.

Gardini's 195 pages of discussion are almost wholly devoid of quantitative or mathematical considerations. He takes up, *inter alia*, the analogies and differences between electric fire and various other supposed subtle matters such as common fire, light, the magnetic fluid and so on, and the 'anomalies' and 'perturbations' that can arise in the distribution of natural and artificial electricity; he discusses different kinds of 'electromotors', that is, methods of disturbing the usual equilibrium in the distribution of the subtle electric 'fire' which, following Beccaria and ultimately Franklin, he believes to be responsible for the various phenomena of electricity; and he describes at length his own experiments and observations on natural electricity arising from bodies belonging to all three kingdoms of nature. Near the beginning of the essay (pp. 13–15), he gives a bibliography of the works he has consulted that reveals as much by its omissions as by what is included in it. Joseph Priestley's *History of Electricity* (in the French edition of 1771) enjoys pride of place, along with all of Beccaria's major publications on electricity and a number of letters, apparently now lost, that he had written to Gardini on the subject. Various of Volta's papers and letters are mentioned, as are several works of two lesser Italian investigators, Antonio Maria Vassalli-Eandi and Carlo Amoretti, and also Sue's *Histoire du galvanisme* (1802). And, of course, Gardini cites his own previous publications. Strikingly, however, a quarter-century after Coulomb had announced that the law of electrical action was, like gravity, inverse-square in character,[50] Gardini fails altogether to mention his work. Likewise, he does not mention Aepinus's *Tentamen theoriae electricitatis et magnetismi* (1759), which had provided Coulomb with his starting point, nor even Haüy's much more recently published *Exposition raisonnée* of Aepinus's work (1787).[51] Nor, finally, does

[49] On Beccaria, see my paper, 'Franklin's Electrical Atmospheres', *British Journal for the History of Science*, 6, 1972, 131–151, and J.L. Heilbron, *Electricity in the 17th and 18th Centuries: A Study of Early Modern Physics* (Berkeley, 1979), pp. 362–372.

[50] Coulomb, *Hist. de l'Acad.*, 1785 (publ. 1788), Mém. pp. 578–611.

[51] On Aepinus, see *Aepinus's Essay on the Theory of Electricity and Magnetism*, introductory monograph and notes by R.W. Home, translation by P.J. Connor (Princeton, 1979).

he refer to a brief report by Biot of Laplace's application of some formulae from his *Mécanique céleste* to the problem of the distribution of electricity in an ellipsoidal conductor.[52] In short, Gardini appears to have been completely unacquainted with precisely those works that in fact had provided the background to the Academy's setting of the prize for which he was competing, the works that we now recognize as the major contributions prior to Poisson's to the mathematization of electrical theory.

Not even Gardini, however, could entirely overlook the fact that the Academy in announcing the competition had indicated that calculations were required. Hence, perforce, he attempts towards the end of his essay (pp. 170–178) to add a quantitative dimension to his work. Unfortunately, it quickly becomes apparent that he has little command of mathematical procedures, let alone any understanding of what the Academy was referring to when it set the problem. Unaware, apparently, that for a few special cases Coulomb had provided detailed sets of data on the static distribution of electric charge on the surface of insulated conductors, and thus not recognizing that this was what the Academy was alluding to in its announcement of the prize, Gardini decided that what was required were calculations concerning the redistribution of electric fluid between electrified bodies when these were brought into contact with each other. These he based upon an intuitive notion that different bodies have different 'capacities' for electric fluid, and the following 'law' that was also presumably founded upon intuitive notions about the flow of fluids in general, since he offered no other justification for it: 'Whatever quantity of absolute electricities two bodies have, when these come in contact and communicate with each other by any means, that quantity is distributed between them in the ratio of their capacities'.[53] As the discussion develops, it becomes clear that for Gardini this is equivalent to saying that when the two bodies come into contact, the electric fluid in them redistributes itself in such a way that its density becomes uniform throughout; that is, he understands the word 'capacity' in a quite literal sense as the volume of the body in question. Given his 'law', Gardini proceeds on the basis of simple proportionalities to draw up a table showing in general terms how, given various initial states of electrification, the electric fluid will redistribute itself when the two bodies are brought in contact. He then illustrates this with some numerical examples, e.g. he calculates how a charge of 10 units on a body of capacity 8 units will be redistributed when the body is brought into contact with another of capacity 5 units. (According to his principles, in this case 8/13 of the initial charge should remain on the first body;

[52] Biot, 'Sur un probléme de physique, relatif à l'électricité', *Bull. Soc. Philomathique*, *3*, 21–23 (in No. 51, Prairial, an IX).

[53] 'Quod summa electricitatum absolutarum quaecumque quam habeant duo corpora, ista dum in contactum veniunt, et quoquomodo inter se communicant, distribuitur ipsa summa in ipsis juxta rationem capacitatis eorum'.

regrettably, he complicates the calculation unnecessarily along the way, and commits a simple arithmetical blunder in the process!) Finally, in the next chapter of his essay, he describes some experiments, many of them due to Beccaria, which are said to confirm the proportions he has established. Inevitably, the evidence is far from convincing.

This section of Gardini's essay is, it must be emphasized, but a very minor part of the whole. By far the largest portion of the work is devoted to an entirely qualitative and experimentally orientated discussion of Gardini's favourite topic, the generation of electricity in various natural processes. Furthermore, the calculations that have been described are entirely vitiated by the fact that his 'law' is inconsistent with the inverse-square law of electrical action. Much more satisfactory is the alternative equilibrium condition first stated forty years earlier by Henry Cavendish, that a uniform *pressure* should be established throughout the electric fluid in conducting bodies in contact.[54] Unfortunately, it appears that Gardini was not familiar with Cavendish's work, either. He simply takes it for granted that his 'law' is valid, because he draws uncritically and too literally the analogy between ordinary fluids such as water and the supposed subtle fluid that he holds to be responsible for the various electrical appearances.

Poisson's memoirs on electricity stand in striking contrast. On the one hand, he rests his calculations on a secure experimental foundation provided by Coulomb, namely the inverse-square law. On the other, in place of Gardini's elementary algebraic manipulations, he adapts and brings to bear on the science of electricity the powerful analytical procedures developed by Lagrange, Laplace and others for the case of gravity. By this means he is able to generate precise quantitative values for surface charge distributions on conductors in certain cases. These he then compares with Coulomb's experimentally determined values for the same cases and shows that they agree closely with them. Furthermore, though founded upon and guided throughout by well established quantitative experimental results, Poisson's work is also, as a result of his exploration of the potential function, a significant piece of mathematical research as such.

At least as significant, however, is the fact that Poisson confines his attention to a single, very narrowly restricted question which is, moreover, not one that electrical theorists had previously concerned themselves with. In effect he and likewise the Academy itself in setting the topic that it did for its prize were redefining the objective of electrical theory. Because Gardini failed to perceive this, he submitted an irrelevant essay. Yet the questions he considered were in fact ones that had always previously been held to be of crucial importance to any adequate theory of electricity, while his manner of considering them was altogether typical of 'electricians' of his generation!

[54] Cavendish, *Phil. Trans.*, *61* (1771), 584–677.

The new approach to the subject was proclaimed by the Academy within the context of a prize competition in mathematics, set by the *géométrie* section. Initially, its endorsement by the Academy as a whole was probably no more than a formality, since for the most part the administration of the regular prize contests, including the selection of topics for them, was left to the section concerned. More particularly, it seems that none of the members of the section for *physique générale* was involved in choosing the topic on this occasion, even though it dealt with a subject that should in principle have been of great interest to them. On the contrary, the surviving documents suggest that they and their fellows in the other sections merely acquiesced as usual in the *géométrie* section's choice of a topic for the biennial competition it administered, without giving the matter any detailed consideration at all.[55] In other words, the new objective for electrical theory implied by the prize question seems to have been imposed on the subject by the Academy's mathematicians alone, with no more than formal consultation with their more experimentally orientated colleagues.

It is, therefore, easy to imagine what a dramatic impact Poisson's work must have on the *physicien* members of the Academy when he presented his first results to the Academy shortly after the prize was announced. Perhaps even more strikingly than in Fourier's treatment of heat flow, crowned in the *géométrie* section's previous, just concluded contest, it showed that the oft-proclaimed unification of mathematics with physics was no longer merely a dream that might one day be realized in some ideal future science, but was actually beginning to be demonstrated in the here-and-now. Poisson's election to the physics section a fortnight later amounted, as we have seen, to public recognition of this, and thus marks an important stage in the history of the subject.

[55] *Procès-verbaux*, IV, 560, 562–3.

INDEX

BOWLING GREEN STATE UNIVERSITY LIBRARY DISCARD

A113 0855676 7

SCI QC 507 .H66 1992

Home, Roderick Weir.

Electricity and experimental
physics in eighteenth-